FERMENTED FOODS

ECONOMIC MICROBIOLOGY
Series Editor
A. H. ROSE

ECONOMIC MICROBIOLOGY

Volume 7

FERMENTED FOODS

edited by

A. H. ROSE

School of Biological Sciences
University of Bath,
Bath, England

1982

ACADEMIC PRESS

A Subsidiary of Harcourt Brace Jovanovich, Publishers

LONDON   NEW YORK
PARIS   SAN DIEGO   SAN FRANCISCO   SÃO PAULO
SYDNEY   TOKYO   TORONTO

ACADEMIC PRESS INC. (LONDON) LTD.
24/28 Oval Road
London NW1

United States edition published by
ACADEMIC PRESS INC.
111 Fifth Avenue
New York, New York 10003

710 034 351 −0

Copyright © 1982 by
ACADEMIC PRESS INC. (LONDON) LTD.

No part of this book may be reproduced in any form by photostat, microfilm, or any
other means, without written permission from the publishers

British Library Cataloguing in Publication Data

Fermented foods.—(Economic microbiology; v. 7)
1. Food industry and trade   2. Food—Microbiology
I. Rose, A.H.   II. Series
630'.2'76   TP370.5
ISBN 0-12-596557-5
LCCCN 77-77361

Printed in Great Britain at the Alden Press, Oxford.

CONTRIBUTORS

ROBERT O. ARUNGA, Kenya Industrial Research and Development Institute, P.O. Box 30650, Nairobi, Kenya.

J. G. CARR, Long Ashton Research Station, University of Bristol, Long Ashton, Bristol BS18 9AF, England.

KO SWAN DJIEN, Department of Food Science, Agricultural University, De Dreijen 12, 6703 BC Wageningen, The Netherlands.

H. P. FLEMING, Food Fermentation Laboratory, U.S. Department of Agriculture, Agricultural Research Service, Southern Region and North Carolina Agricultural Research Service, Department of Food Science, North Carolina State University, Raleigh, North Carolina 27650, U.S.A.

B. A. LAW, National Institute for Research in Dairying, Shinfield, Reading RG2 9AT, England.

ERKKI OURA, Research Laboratories, State Alcohol Monopoly (Alko) P.O. Box 350, SF-00101 Helsinki 10, Finland.

H. J. PEPPLER, 5157 North Shoreland Avenue, Whitefish Bay, Wisconsin 53217, U.S.A.

A. H. ROSE, School of Biological Sciences, University of Bath, Claverton Down, Bath BA2 7AY, England.

H. SUOMALAINEN, Research Laboratories, State Alcohol Monopoly (Alko), P.O. Box 350, SF-00101 Helsinki 10, Finland.

E. R. VEDAMUTHU, Microlife Technics, Box 3917, 1833 57th Street, Sarasota, Florida 33478, U.S.A.

RISTO VISKARI, Oy Karl Fazer Ab, Bakery Group, P.O. Box 40, SF-15101 Lahti 10, Finland.

B. J. B. WOOD, Department of Applied Microbiology, University of Strathclyde, George Street, Glasgow G1 1XW, Scotland.

PREFACE

There are two volumes in this series dealing with the activities of micro-organisms in food. Volume 8 describes ways in which unwanted microbial activity in foods can be minimized, while it also contains chapters describing certain groups of food-spoilage micro-organisms. The present volume gives an account of processes in which microbial activity in food materials is purposefully encouraged in order to produce biologically safer and more flavourful foods.

A very important stage in the development of the human species came when early Man acquired the ability to process raw food materials so as to make these materials more tasty and also allow them to be preserved. These were the first steps in the development of the art of cooking. Early Man, for quite understandable reasons, greatly revered food, its acquisition and processing, and most of the early civilizations worshipped food gods in some form or other. Food raw materials, since they are nutritious for Man, can also support growth of micro-organisms. Many of the early processes for dealing with food raw materials were carried out with the aim, albeit unwittingly, of preventing growth of pathogenic microbes and of organisms that elaborate toxic compounds. This process of trial and error must have taken many years, and must have been attended with numerous tragic failures. However, during this time, it came to be appreciated that certain types of deterioration, now known to be caused by micro-organisms, were to be encouraged, since they produced foods which were not prone to further contamination by pathogenic and toxic microbes, while at the same time the foods had improved taste and flavour. These were the first of what we today refer to as fermented foods.

The present volume describes the microbiology involved in the production of the major fermented foods that are today manufactured World-wide. They represent in many respects a very contrasting group of products. Many, especially the indigenous fermented foods, breads and cheeses, have origins lost in antiquity, whereas others, including some cultured milk products and yeast extracts, are hardly a century old. The processes involved also differ considerably with regard to the extent to which they are understood in microbiological and biochemical terms. The micro-organisms involved in the fermentation process are very poorly understood when one considers indigenous fermented foods, whilst on the other hand a vast body of knowledge exists on properties required in baker's yeasts and on the changes that take place during dough fermentations.

Finally, it must be pointed out that not all of the major food fermentations are described in this volume. Production of alcoholic beverages, which represent a very large sector of the World's fermented foods, is described in Volume 1 of the series.

January 1982   ANTHONY H. ROSE

CONTENTS

3. Soy Sauce and Miso
BRIAN J. B. WOOD

4. Breadmaking
ERKKI OURA, HEIKKI SUOMALAINEN and RISTO VISKARI

5. Cheeses
B. A. LAW

6. Fermented Milks
EBENEZER R. VEDAMUTHU

7. Fermented Vegetables
H. P. FLEMING

8. Coffee
ROBERT O. ARUNGA

9. Cocoa
J. G. CARR

10. Yeast Extracts
H. J. PEPPLER

1. History and Scientific Basis of Microbial Activity in Fermented Foods

ANTHONY H. ROSE

School of Biological Sciences, Bath University, Bath, Avon, England

I. INTRODUCTION

Microbes have been, from time immemorial, involved in the preparation and processing of items in Man's daily diet. Although there are some exceptions, such as cereal grains which are too dry to support extensive growth of microbes, the majority of plant and animal materials which Man traditionally uses as food are sufficiently moist to be liable to extensive contamination with microbial growth. Lacking any knowledge of micro-organisms, or of ways in which contamination by them could be avoided, Man had, as a vital part of his evolution, to learn to live with microbially infected food material. Quite often, the microbes that infected food material ultimately made the food unacceptable, either by altering the appearance of the food to a point at which it ceased to be appetizing or by infiltrating the food with poisonous toxins, some of which were lethal. Occasionally, however, microbial infection of food material made it appear more attractive and taste more appetizing. Microbial infections of these food materials were, as a consequence, encouraged and ultimately exploited by Man, so that fermented foods and beverages now form a large and important sector of the food industry.

This volume in the series deals with the role of micro-organisms in the manufacture of fermented foods. Clearly, very closely related are beverages, whose production involves microbial activity. The vast majority of these beverages involve an alcoholic fermentation of sugars to give alcohol and carbon dioxide, although this fermentation is often accompanied by activity of other microbes which contributes significantly to the flavour of the final beverage. The role of micro-organisms in production of alcoholic beverages is described in Volume 1 of the series (Rose, 1977), and the reader is referred to this work for further details.

II. HISTORY AND TYPES OF FERMENTED FOODS

The origins of the first two classes of fermented food described in this volume, namely indigenous foods and sauces, are almost certainly lost in antiquity. The indigenous fermented foods described by Djien in Chapter 2 are mainly the products of Oriental agriculture that, millenia ago, were acknowledged to be more appetizing and digestable when they had undergone a limited type of fermentation. However, these foods are not confined to the Orient. A fermented rice, for example, known variously as arroz fermentado, arroz amarillo and arroz requemado, has long been prepared and consumed in the Andes mountains in Ecuador (van Veen *et al.*, 1968). Moreover, many of these foods are ceasing to be indigenous, as they are gradually introduced into countries in the western World often as fad foods. As van Veen and Steinkraus (1970) observed, growth of micro-organisms in these materials does not usually lead to a large increase in the protein content. It should be emphasized that the number of known fermented foods produced World-wide greatly exceeds those referred to in Chapter 2. A publication by Appleton *et al.* (1979) lists almost 80 foods, together with the nature of the raw material used in their production, the country of origin and the principal micro-organisms involved in the fermentation. Other reviews which have described these fermented foods have come from Djien and Hesseltine (1979), Hesseltine (1965) and Hesseltine *et al.* (1976).

The most important of the fermented sauces, which are dealt with by Wood in Chapter 3, is undoubtedly soy sauce. The origins of soy sauce, like other fermented foods, go back millenia. The first recorded reference to the soy bean, from which soy sauce is made, is thought by Lo (1964) to be Shen Nan's Materia Medica, produced in China in 2838 B.C. Soy sauce is now well known in the western World, particularly in North America where it has been popularized by communities with Chinese origins. Only very recently, however, have the changes brought about by micro-organisms in soy sauce manufacture begun to be described, as indicated in Wood's chapter.

Bread is a baked product made from a dough, most frequently prepared from wheat, that has been raised or leavened by carbon dioxide formed by yeast fermentation or by some other gas-forming agent. Breadmaking, which is carried out in some form or another in virtually every country of the World, has a vast historical literature. We shall certainly never know when Man first prepared flour and bread, although some authorities believe that it occurred immediately after the development of cannibalism and meat-eating and probably as far back as 7,000 B.C. (Jensen, 1953). The likelihood is that early Man discovered that cereal grains and similar plant products such as acorns could be dried in the sun and stored for many months and even years without spoiling. Precisely how he upgraded his use of these foods from simply chewing raw grains to breaking, winnowing and sieving them to produce a flour, and then to prepare a dough or gruel from the flour which could be baked, has occasioned a long trail of surmise interjected by just a few morsels of evidence (Storcke and Teague, 1952). It is generally agreed, however, that the forerunners of the operations that today constitute modern baking probably took place in Ancient Egypt. The Egyptians made a giant step forward when they noted that, if the bread dough was allowed to stand for a few hours after mixing, the dough was expanded after baking, and had a light spongy texture which was preferred to breads prepared from doughs that had not been allowed to stand. From a microbiological standpoint, subsequent development of the art of breadmaking was concerned largely with acquiring techniques for retaining portions of fermented dough for inclusion in later doughs, for it was not until the middle of the Nineteenth Century that methods were developed for mass-producing baker's yeast. Concurrently, more refined techniques were developed for cultivating high-quality grain and for producing flour from it. Because it is a staple item in the diet of Man, bakers of bread often became prominent individuals in a community—there were around 250 bakeries in Rome as early as 100 B.C. (Pederson, 1980)—and soon formed themselves into guilds in order to protect their art. Suomalainen and Oura describe the role of micro-organisms in modern-day bread baking in Chapter 4.

Mammalian milk is an oil-in-water emulsion, the stability of which is maintained by adsorption of phospholipids and proteins on the surface of the fat globules. Milk from domesticated animals has been used by Man for millenia. Pederson (1980) refers to rock drawings discovered in the Libyan desert, thought to date back to 9,000 B.C., and which depict cow worship and the milking of a cow. Unlike dried cereal grains, milk is very prone to microbial contamination, especially since, when freshly obtained from an animal, it invariably contains a large flora of microbes. Moreover, milk that has become sour represents a much less attractive food than fresh milk. The two fermented milk products described in Chapters 5 and 6 are foods in which the major components of milk have been preserved in forms that are much more stable than raw milk. Growth of bacteria in milk causes separation of curds from the whey. Again, it is impossible to state when the first removal of curds from whey was made, but it was the first step in cheese making to be followed by draining and drying of the curds to produce a cheese. Cheeses are manufactured throughout the World from a variety of mammalian milks and under a wide range of conditions that lead to growth of different micro-organisms in the cheese. The huge range of cheeses available World-wide, each with its own individual taste and aroma, undoubtedly constitutes one of the most sophisticated items in Man's day-to-day diet (Simon, 1956; Eekhof-Stork, 1976). The microbiology of cheese manufacture is described in Chapter 5 by Law. Also worth referring to are the three classic volumes on cheeses by Davis (1976).

Stabilization by lowering the pH value rather than by decreasing the water content of milk forms the basis for manufacture of the fermented milks described in Chapter 6 by Vedamuthu. Fermented milks are probably as old as cheeses. One can imagine how the discovery that certain types of spontaneous souring of milk gave a product that was acceptable and stable, and that vessels in which this souring had taken place were used over and over again to obtain the desired product. The role of micro-organisms in the modern-day manufacture of fermented milks is recounted in Chapter 6.

In the days before refrigerating facilities became available, a number of techniques were devised for preserving seasonally produced vegetables. One of the most efficient of these involved packing vegetables tightly in a vessel with salt or brine. The technique is thought to have originated in the Orient where, even today, it continues to be used extensively for preserving vegetables. Only in the last 50 years has it been shown that this method of preserving vegetables involves a microbiological fermentation, an area of research in food microbiology which has been richly contributed to by Carl Pederson and his colleagues (Pederson, 1980). It is now clear that, in the salt-containing vegetable mix, lactic-acid bacteria proliferate, and produce acids which lower the pH value of the mix. The combined action of the salt and acid lowers the activity of those enzymes in the vegetables that are responsible for breakdown of vegetable tissue, while at the same time inhibiting certain noisome oxidative changes in the tissue and preventing growth of microbes that may cause undesirable changes in the vegetables. Fleming, in Chapter 7, describes in more detail the manner in which vegetables are fermented.

The foodstuffs dealt with in Chapters 8 and 9 differ in several respects from those described in earlier chapters. While the fermentation processes involved in their production probably affect the flavour of the final beans, the primary function of the fermentation is to liberate the beans from the plant tissue which envelopes them in the harvested product. In addition, both of these fermented foods originated in parts of the World which were discovered by the western World only in the Sixteenth Century, so that their antiquity is even more obscure than that of the fermented foods described in other chapters of this book. The coffee bean is obtained from cherries of the coffee plant, *Coffea arabica*. There are records of the use of coffee in Abyssinia in the Fifteenth Century traditionally associated with the legend of the goatherd Kaldi (Kolpas, 1979). It was developed as a major World beverage by the Arabs, who are claimed to have consumed it to keep themselves awake during the prolonged religious services attended by the Islamic faithful. Nowadays, the main coffee-producing areas are in South America, with Brasil being the main supplier. Each cherry from the coffee plant contains two beans, and the fermentation process serves to free these beans from the parchment-like endocarp and pulp which surround them. Details of this fermentation are described by Arrunga in Chapter 8. Carr, in Chapter 9, provides comparable information about production of cocoa beans. Beans from the cacao plant, *Theobroma cacao*, were first used as a foodstuff by the American Indians. The exact origin of the cocoa tree is obscure, with claims being made for the Amazonas region of Brasil and the Orinoco Basin in Venezuela. The Aztec cocoa beverage was made from a mixture of cocoa beans and corn. The beverage as we know it today was probably developed by the Spaniards who colonized Central and South America (Pederson, 1980). A mature pod from the cacao plant contains 30–40 beans. The beans, which are enveloped in a white pulp, are scooped out of the pods and the mass is allowed to ferment. The fermentation removes the pulp, but also adds to the flavour of the cocoa bean.

It was possible to describe only the major fermented foods in this volume, with emphasis on those fermentations that have been researched reasonably fully in microbiological terms. There are, however, several fermented foods which have an important place in Man's cuisine, but which are not dealt with in this text. Early Man learned to preserve fresh meat by mixing it with garlic and spices and storing it in a packed form. This led to the development of what we today know as sausages of the salami and cervelat type. After packing, lactic-acid bacteria develop in the meat, and by lowering the pH value their activity helps to preserve the meat while contributing also to its flavour. Pederson (1980) has presented a well written account of the microbiology of sausages. Other items of food whose production involves a fermentation step include vanilla, citron and ginger. What little is known of the microbiology of these fermentations has been well covered in Pederson's (1980) text.

III. MICROBIAL ACTIVITY IN PRODUCTION OF FERMENTED FOODS

A. Factors Affecting Microbial Activity in Foods

The raw materials used in the manufacture of fermented foods, such as plant and animal tissue, mammalian milk and cereal extracts, are rich sources of microbial nutrients, and are excellent substrates for growth of micro-organisms. Information on nutritional requirements of micro-organisms can be found in any of the basic texts on microbial physiology (Dawes and Sutherland, 1976; Rose, 1976; Moat, 1979). Because plant and animal tissues contain virtually all of the nutrients required for growth of micro-organisms, merely allowing tissues to support a microbial flora could result in a fermented food that contained toxic compounds, including bacterial toxins and mycotoxins. The methods devised over the centuries for manufacture of fermented foods in a pragmatic fashion developed to a point at which the product was attractive, nutritionally valuable and microbiologically safe. With present-day knowledge of microbial physiology, it is possible to explain why fermented foods are microbiologically safe. The two main explana-tions are that foods are fermented under conditions that lead to a decrease in the pH value to a point at which growth of very many other microbes is prevented, while at the same time the water content of the food is lowered to a value that permits growth of only a few types of micro-organism.

1. Effect of Acidity on Microbial Growth

Textbooks on microbial physiology (e.g. Rose, 1976) invariably state that individual groups of micro-organisms react in different ways to the pH value of the environment in which they are held. Although it is accepted that there are exceptions, the generalizations usually made are that, for growth, bacteria prefer pH values near neutrality, fungi and yeasts slightly acid pH values, and actinomycetes pH values a little above neutrality. For the vast majority of micro-organisms, pH values below about 2.0 inhibit or prevent growth, although there is evidence that these acidic environments do not kill microbes.

In physiological terms, there is a decided dearth of information on the effect of low and indeed high pH values on micro-organisms (Lang-worthy, 1978). Very few studies have been made on micro-organisms growing in cultures in which the pH value is kept constant by addition of acid or alkali. It is generally agreed however that, although the pH value of the environment may vary considerably during growth of a culture, the pH value inside cells remains reasonably constant, probably near or just below neutrality. Any mature reflection in physiological terms must lead the reader to question the concept of an internal pH value in micro-organisms, a concept best described by Siesjö and Pontén (1966) as being theoretically awkward. Nevertheless, it would appear likely that the inhibitory effects of low external pH values on growth of micro-organisms are most probably explained by the effect of high concentrations of hydrogen ions on the activity of proteins exposed on the outside of the plasma membrane. These proteins include those concerned with transport of solutes across the plasma membrane, and enzymes that catalyse reactions leading to synthesis of envelope components.

However, it is also possible that high concentrations of hydrogen ions affect the properties of lipids in the plasma membrane, which in turn could influence the activity of membrane-bound proteins. The evidence for a role for plasma-membrane lipids is, admittedly, indirect. At low pH values, there is the possibility of hydrolysis of phospholipids in membranes, which could explain why species of *Thermoplasma* and *Sulfolobus*, which are able to grow at low pH values, synthesize not the usual types of phospholipid but diether derivatives of glycerol phosphate and long-chain alcohols. These lipids are more resistant to acid hydrolysis (Langworthy, 1977). Acidophilic bacteria also synthesize cyclized fatty-acyl residues; for example, thiobacilli synthesize phospholipids with cyclopropane fatty-acyl residues. It is thought that these cyclized structures confer rigidity on the membrane and decrease membrane permeability (De Rosa *et al.*, 1974). It might be inferred, therefore, that low pH values, in microbes that are not able to grow in the presence of high concentrations of hydrogen ions, cause instability in the membrane and increase membrane permeability, both of which phenomena could prevent growth.

2. *Moisture Requirements of Micro-Organisms*

Water accounts for between 80 and 90% of the weight of a micro-organism. All chemical reactions that take place in living organisms require an aqueous environment, and water must therefore be in the environment if the organism is to grow and reproduce. It must be, moreover, in the liquid phase, and this confines biological activity to temperatures ranging from around $-2°C$ (or lower in solutions of high osmotic pressure) to approximately 100°C; this is known as the biokinetic zone.

The need for water to obtain microbial activity has been recognized for centuries, and forms the basis of one of the oldest methods for preventing decay of perishable materials, namely desiccation. The water requirements of micro-organisms can be expressed quantitatively in the form of the water activity (a_w) of the environment or substrate; this is equal to p/p_o, p being the vapour pressure of the solution and p_o the vapour pressure of water.

Values for a_w can be calculated using the equation:

$$\ln a_w = \frac{-vm\phi}{55.5}$$

where v is the number of ions formed by each solute molecule, m the molar concentration of solute and ϕ the molar osmotic coefficient, values for which are listed for various solutes in a number of textbooks. Water has an a_w value of 1.000; this value decreases when solutes are dissolved in water.

Micro-organisms can grow in media with a_w values between 0.99 and about 0.63. For any one organism, the important values within this range are the optimum and minimum a_w values. These have been determined for a number of micro-organisms, and they seem to be remarkably constant for a particular species and to be independent of the nature of the dissolved solutes. The general effect of lowering the a_w value of a medium below the optimum value is to increase the length of the lag phase of growth and to decrease the growth rate and the size of the final crop of organisms. On the whole, bacteria require media of higher a_w value (0.99–0.93) than either yeasts or moulds. Staphylococci and micrococci characteristically have lower optimum a_w values in this range. With *Salmonella oranienburg*, the a_w value of the medium has important effects on the physiology of the bacterium; only at a_w values below 0.97 is proline required for growth of the bacterium. Moreover, in media lacking amino acids, accumulation of potassium ions by this bacterium increases to a maximum as the a_w value of the medium is lowered to 0.975, and then decreases as the water value is lowered further to 0.96. Yeasts also vary in the optimum a_w values required for growth, but the minimum values for these organisms (0.91–0.88) are lower than those for the majority of bacteria. A few yeasts, e.g. *Saccharomyces rouxii*, can grow in media of a_w value as low as 0.73 and these are known as osmophilic yeasts. *Saccharomyces rouxii*, unlike non-osmophilic yeasts, contains glycerol and arabitol, and it is believed that these intracellular polyols in some way enable the yeast to grow in media with low water activities. Moulds are in general better able to withstand dry conditions than other micro-organisms, and for some strains, e.g. members of the *Aspergillus glaucus* group, the lower limit of a_w value may be near 0.60. A few moulds, e.g. *Xeromyces bisporus*, have an upper a_w limit of approximately 0.97, and are described as xerophilic.

The low, or relatively low, moisture content of many fermented foods is a major factor in ensuring their microbiological stability. The water content of foods in relation to their microbiological flora has generated an extensive literature. This ranges from Scott's (1957) review, in which he advocated the concept of a_w value in relation to food stability, to the more recent text by Troller and Christian (1978) which deals comprehensively with the subject.

B. Consequences of Microbial Activity in Foods

The principal outcome of microbial activity in a fermented food is to alter the flavour and aroma of the food such that, invariably, it is more attractive than the raw material used. In some fermented foods, microbial activity also alters the texture of the food. The two main examples are bread, where carbon dioxide evolved by yeast fermentation creates the honeycomb-like texture of the baked bread, and certain cheeses, such as Gruyère, in which holes arise as a result of carbon dioxide evolved by lactic-acid bacteria becoming trapped in the curds.

Microbial activity in a food alters the flavour and aroma of that food as a result of the micro-organisms excreting flavour compounds and also chemically altering constituents of the raw material to produce new or additional flavour compounds. The biochemistry of flavour is an exceedingly complex subject, and only in recent years, with the availability of mass spectrometers and sophisticated analytical equipment such as gas–liquid and high-performance liquid chromatographs, have significant advances been made in the subject. Recent years have witnessed an upsurge of interest in the production of flavour compounds by micro-organisms, caused largely by the quest for natural flavour compounds that might replace synthetic chemicals as food additives. The text edited by Schultz *et al.* (1967) and that written by Margalith (1981) contain a wealth of information on production of flavour compounds by micro-organisms, and on flavour compounds that are present in many of the raw materials used to make fermented foods.

The indigenous fermented foods, which form the subject of Djien's chapter, have received very little attention from the standpoint of the origins of their flavour and aroma. This is explained very largely by the fact that, with the exception of soy sauce which is described in greater detail by Wood in Chapter 3, none of these foods supports a large Western-style manufacturing industry. Of the raw materials used in the manufacture of these fermented foods, fish meat has been most intensively studied as regards flavour and aroma. Fish fats and their oxidation products make an important contribution to fish flavour and aroma (Olcott, 1962) as do amino acids, amines (particularly trimethylamine) and nucleotides (Jones, 1967). Fish meat forms an excellent substrate for microbial growth, and the changes brought about in flavour and aroma compounds by microbial activity must be very extensive.

Freshly baked bread has a subtle flavour and aroma that few people can resist. However, they are a flavour and aroma that are rapidly lost as the bread cools and is allowed to stand. This might be explained by loss of flavour and aroma compounds by volatilization, and also by some of these compounds becoming oxidized. Coffman (1967) reviewed earlier work on the flavour and aroma compounds of white bread, while Collyer (1964) has done likewise for darker breads. Margalith's (1981) book describes more recent research on both products. The main flavour compounds that have been extracted from breads are, not surprisingly, similar to those detected in beers, which have been researched much more extensively in this respect. The compounds include alcohols, acids, esters, aldehydes and ketones.

Although there are very many varieties of cheese made throughout the World, they have much in common as regards qualitative flavour composition. The flavour and aroma of a cheese are derived from milk components which are modified chemically by bacteria and moulds that grow in the cheese and possibly by heat treatment (Day, 1967). It is generally held that the most significant factor in the development of hard-pressed cheeses, such as Cheddar, is breakdown of protein, whereas in mould-ripened cheeses it is lipolysis (Chapman, 1977). Considerable research has been made on compounds responsible for the typical flavour of Cheddar cheese, with only limited success. Somewhat more success has come from similar investigations on mould-ripened cheeses. Here, methylketones, especially 2-heptanone and 2-nonanone, are considered to be key flavour compounds. Short-chain fatty acids, particularly those with between four and eight carbon atoms, are thought to be responsible for the 'peppery' taste of these cheeses. Law takes the matter further in Chapter 5. The flavour and aroma of fermented milks, which are described in Chapter 6 by Vedamuthu, have very similar origins to those of hard-pressed cheeses. Diacetyl is a key flavour component in these products (Lindsay, 1967).

Hardly anything is known of the role of micro-organisms in changing the flavour of fermented vegetables, described by Fleming in Chapter 7. In all likelihood, the flavour changes are minimal, particularly as the principal objective in fermenting vegetables is to preserve flavour rather than to modify it. Microbial growth similarly makes no contribution to the flavour and aroma of coffee, production of which is described by Arunga in Chapter 8. Indeed, the flavour of green coffee is not very appealing. The attractive and desirable flavour and aroma of coffee, as we know it, come from roasting of the beans. Biochemical changes caused by a mild pyrolysis of coffee beans have been reviewed by Gianturco (1967). Not so, however, with production of cocoa beans, where end products of microbial activity have long been recognized as making a contribution to the flavour of the final product (Pederson, 1980). Nevertheless, very little research has been carried out to identify flavour and aroma compounds.

REFERENCES

Appleton, J.M., McGowan, V.F. and Skerman, V.B.D. (1979). 'Micro-organisms and Man'. UNESCO/UNEP Publication under Contract No. 258117.

Chapman, G. (1977). *International Flavours and Food Additives* **8** (2), 61.

Coffman, J.R. (1967). *In* 'The Chemistry and Physiology of Flavors' (H.W. Schultz, E.A. Day and L.M. Libbey, eds.), p. 185. Avi Publishing Co., Westport, Connecticut.

Collyer, D.M. (1964). *Baker's Digest* **38,** no. 1, 43.

Davis, J.G. (1976). 'Cheese', vol. III, 'Manufacturing Methods'. Churchill and Livingstone, Edinburgh.

Dawes, I.A. and Sutherland, I.A. (1976). 'Microbial Physiology'. Blackwell Scientific Publications, Oxford.

Day, E.A. (1967). *In* 'The Chemistry and Physiology of Flavors' (H.W. Schultz, E.A. Day and L.M. Libbey, eds.), p. 351. Avi Publishing Co., Westport, Connecticut.

De Rosa, M., Gambacorta, A., Minale, L. and Bu'Lock, J.D. (1974). *Journal of the Chemical Society. Chemical Communications 1974*, 543.

Djien, K.S. and Hesseltine, C.W. (1979). *In* 'Microbial Biomass' (A.H. Rose, ed.), p. 115. Academic Press, London.

Eekhof-Stork, N. (1976). 'The World Atlas of Cheese'. Paddington Press Limited, London.
Gianturco, M.A. (1967). *In* 'The Chemistry and Physiology of Flavors' (H.W. Schultz, E.A. Day and L.M. Libbey, eds.), p. 431. Avi Publishing Co., Westport, Connecticut.
Hesseltine, C.W. (1965). *Mycologia* **57,** 149.
Hesseltine, C.W., Swain, E.W. and Wang, H.L. (1976). *Developments in Industrial Microbiology* **17,** 101.
Jensen, L.B. (1953). 'Man's Foods: Nutrition and Environments in Food Gathering Times and Food Producing Times'. Garrard Press, Champaign, Illinois.
Jones, N.R. (1967). *In* 'The Chemistry and Physiology of Flavors' (H.W. Schultz, E.A. Day and L.M. Libbey, eds.), p. 261. Avi Publishing Co., Westport, Connecticut.
Kolpas, N. (1979). 'Coffee'. John Murray, London.
Langworthy, T.A. (1977). *Journal of Bacteriology* **130,** 1326.
Langworthy, T.A. (1978). *In* 'Microbial Life in Extreme Environments' (D.J. Kushner, ed.), p. 279. Academic Press, New York.
Lindsay, R.C. (1967). *In* 'The Chemistry and Physiology of Flavors' (H.W. Schultz, E.A. Day and L.M. Libbey, eds.), p. 315. Avi Publishing Co., Westport, Connecticut.
Lo, K.S. (1964). *Soyabean Digest* **24,** no. 7, 18.
Margalith, P.Z. (1981). 'Flavor Microbiology'. Charles C. Thomas, Springfield, Illinois.
Moat, A.G. (1979). 'Microbial Physiology'. John Wiley and Sons, New York.
Olcott, H.S. (1962). *In* 'Symposium on Foods: Lipids and their Oxidation' (H.W. Schultz, E.A. Day and R.O. Sinnhuber, eds.), p. 183. Avi Publishing Co., Westport, Connecticut.
Pederson, C.S. (1980). 'Microbiology of Food Fermentations', 2nd edition. Avi Publishing Co., Westport, Connecticut.
Rose, A.H. (1976). 'Chemical Microbiology', 3rd. edition. Butterworths, London.
Rose, A.H. ed. (1977). 'Alcoholic Beverages'. Academic Press, London.
Schultz, H.W., Day, E.A. and Libbey, L.M. eds. (1967). 'The Chemistry and Physiology of Flavors'. Avi Publishing Co., Westport, Connecticut.
Scott, W.V. (1957). *Advances in Food Research* **7,** 83.
Siesjö, B.K. and Pontén, U. (1966). *Annals of the New York Academy of Sciences* **133,** 78.
Simon, A. (1956). 'Cheeses of the World'. Faber and Faber, London.
Storcke, J. and Teague, W.D. (1952). 'A History of Milling Flour for Man's Bread'. University of Minneapolis Press, Minneapolis.
Troller, J.A. and Christian, J.H.B. (1978). 'Water Activity and Food'. Academic Press, New York.
Van Veen, A.G. and Steinkraus, K.H. (1970). *Journal of Agricultural and Food Chemistry* **18,** no. 4, 576.
Van Veen, A.G., Graham, D.C.W. and Steinkraus, K.H. (1968). *Archivos Latinamericanos de Nutricion* **18,** no. 4, 363.

2. Indigenous Fermented Foods

KO SWAN DJIEN

Department of Food Science, Agricultural University, Wageningen,
The Netherlands

I. INTRODUCTION

Fermented foods which are reviewed in the following chapters of this volume, including bread, fermented dairy products, vegetables and sauces, coffee and cocoa, have been scientifically studied during many decades, and knowledge about these foods keeps pace with developments in modern science. This resulted in production methods that follow advances in modern technology. Fermented foods, which are popular in Japan and include *shoyu*, *miso*, *natto* and *saké*, have followed the same modern trend.

In contrast, some other fermented foods of Asian and African countries including *tempe*, *gari*, *kimchi* and some others have come to the attention of modern scientists only in the last two decades, and many others have not yet been studied at all. Consequently, most of these foods are still manufactured according to traditional, less technologically advanced methods, using simple equipments and produced on a small village industrial scale or just at home for family consumption.

With the exception of a few scientists, including Boorsma (1900), Vorderman (1893), Wehmer (1900), Went and Prinsen Geerligs (1896), since a decade after World War II more and more modern scientists became aware of the fact that many important and nutritious fermented foods, other than those which are known in Western countries, are important for the food supply in many parts of the World.

Basically, fermented foods are agricultural products which have been converted by enzymic activities of micro-organisms into desirable food products whose properties are considered more attractive than those of the original raw materials. In addition to its external attractive properties, its nutritional values and keeping qualities are in many cases better than the raw materials. Moreover, if the manufacturing procedures are properly followed, the foods are usually safe for consumption. All of these beneficial properties of the final product increase the economic value of the original agricultural commodity.

Generally, traditional methods of manufacturing fermented foods are not complicated, and expensive equipment is not required. Therefore, fermentation of indigenous foods is considered as an inexpensive and effective means of food production that could be utilized in alleviating World food problems (Pederson, 1979).

With this background in mind, in recent years, a growing number of food scientists and microbiologists have recognized the value of indigenous fermented foods as a potential source of food supply for many parts of the World. To exchange knowledge and expertise on indigenous fermented foods and to stimulate co-operation between scientists in this field, a number of symposia, workshops and special sessions of conferences were organized in recent years (Table 1).

Lists of indigenous fermented foods with short descriptions of the nature of the food products, raw materials and micro-organisms involved in the fermentation have been compiled (Batra and Millner, 1976; Beuchat, 1978a,b; Hesseltine, 1965, 1979a,b; Mackie *et al.*, 1971; Sundhagul *et al.*, 1975), while a 'Handbook of Indigenous Fermented Foods' (K.H. Steinkraus, editor, 1982, personal communication) and a 'Glossary of Fermented Foods' (H.L. Wang and C.W. Hesseltine, personal communication) are currently in preparation. Recognition of the importance of indigenous fermented foods is also demonstrated in the publication of review articles as special chapters in scientific series on microbiology or on foods (Beuchat, 1978b; Hesseltine and Wang, 1972; Ko and Hesseltine, 1979; Wang and Hesseltine, 1979), while chapters on fermented foods are expanded or added in newer editions of textbooks on food microbiology (Frazier and Westcoff, 1978; Jay, 1978). Moreover, a book on the 'Microbiology of Food Fermentations' was published by Pederson firstly in 1971 and a second edition appeared in 1979.

Table 1

Conferences discussing aspects of indigenous fermented foods

Symposium/Workshop on Indigenous Fermented Foods,
November 21–26, 1977, Bangkok, Thailand.

World Conference on Vegetable Food Proteins,
October 29–November 3, 1978, Amsterdam, The Netherlands.

Symposium on Fermented Foods,
November 22, 1978, London, England.

International Symposium on Oriental Fermented Foods,
December 10–14, 1979, Taipei, Taiwan.

United Nations University Workshop on Research and Development Needs in the Field of Fermented Foods,
December 14–15, 1979, Bogor, Indonesia.

VIth International Fermentation Symposium,
July 20–25, 1980, London, Canada.

Eighth Conference of Association for Science Cooperation in Asia (ASCA),
February 9–15, 1981, Medan, Indonesia.

Also this volume on food fermentations would not be complete without a chapter on indigenous fermented foods. In this short review, some aspects of indigenous fermented foods, which are almost unknown in the Occident, are very briefly discussed, just for orientation in this field of food microbiology but which contain many unexplored areas. Special attention is paid to the micro-organisms and their role in the fermentation process. Consequently, in this chapter, indigenous fermented foods are classified according to the micro-organisms involved in the process. Some foods are mainly fermented by moulds, others mainly by bacteria, another category by a mixture of moulds and yeasts, and a number are firstly fermented by moulds followed by a fermentation with a mixture of bacteria and yeasts. Details and examples are limited to keep this review surveyable.

II. FOODS FERMENTED BY MOULDS

The most well-known foods which are manufactured mainly by the activity of moulds are *tempe*, *oncom* and *angkak*. *Tempe*, made by fermentation of soybeans with *Rhizopus oligosporus*, was discussed in details in a previous chapter in this series (Ko and Hesseltine, 1979). *Oncom* from peanut presscake and fermented by *Neurospora* species was also briefly described in the same chapter. *Angkak* is made by fermentation of rice with *Monascus purpureus* to produce red-coloured rice kernels. The product itself is not a food, but it is used to give an attractive red colour to certain products made of fish and soybeans and to certain alcoholic beverages (Palo *et al.*, 1960; Wang and Hesseltine, 1979).

A. Roles of the Moulds

1. Synthesis of Enzymes

One of the most important functions of the mould in food fermentations is synthesis of enzymes. These enzymes generally decompose complex compounds, including proteins, carbohydrates and fats, into smaller molecules. At the same time, other compounds may be synthesized from the food substrate. These complex chemical changes are accompanied by changes in the original properties of the raw material. Taste, flavour, texture, colour, palatibility, and other properties of the raw material are usually modified in such a way that the final product becomes more attractive to the consumer. In addition to this general function of producing enzymes, in certain products the mould has a special role.

2. Mould Growth

Mould growth on certain products contributes to the appearance of the food which is desired by the consumer. *Neurospora* spp. provide *oncom* cakes with a coating of its pink–orange-coloured and powdery conidia. *Rhizopus oligosporus* covers *tempe* with a clean white mycelium surface-layer and additionally has the function of binding together the soybeans into a solid, compact cake.

3. Synthesis of Colouring Compounds

The function of *Monascus purpureus* during fermentation of *angkak* is to produce the red-coloured compound monascorubrin ($C_{22}H_{24}O_5$) and the yellow pigment monascoflavin ($C_{17}H_{22}O_4$) in soaked rice (Wang and Hesseltine, 1979).

4. Protection of the Product

In spite of the deep-seated prejudice against mouldy products in the Occidental culture (Whitaker, 1978), which seems to be justified by the discovery of aflatoxin and other mycotoxins in the last two decades, studies on certain mould species, which are traditionally used for fermentation of Oriental foods, showed that they do not produce toxins, but, on the other hand, they resist accumulation of certain toxins which otherwise will be produced by other micro-organisms in the food. This could be considered as a protection of the product against other harmful micro-organisms.

A good example of such a protective role is demonstrated by *R. oligosporus*, the mould species which is used for fermentation of *tempe*. This mould species does not produce aflatoxin. On the contrary, if aflatoxin is already present in the growth substrate, its content could be lowered by *R. oligosporus* to about 40% of its original content (Van Veen *et al.*, 1968). In addition, it was found that *R. oligosporus* inhibits growth, sporulation and aflatoxin production by *Aspergillus flavus* (Ko, 1974). Extended studies showed that other species of *Rhizopus*, including *R. arrhizus*, *R. oryzae and R. chinensis*, do not produce aflatoxin and, in addition, inhibit aflatoxin production by *A. flavus* and *A. parasiticus* if they grow in the same substrate. Another study showed that when, during fermentation of coconut presscake into *tempe bongkrek*, contamination takes place with *Pseudomonas cocovenenans*, toxin production by these bacteria under certain conditions is inhibited by *R. oligosporus* (Ko, 1979).

III. FOODS FERMENTED BY BACTERIA

This type of food is made of different raw materials. In each of these raw materials, the bacteria involved have a different role.

A. Fermented Vegetable Products

Aspects of these foods are discussed in details by Pederson (1979). It is mentioned that production of fermented vegetables on a large scale took place for the first time during construction of the Chinese Great Wall in the third Century B.C. The foods, which were made of cabbage, radish, turnip, cucumber and other vegetables, were supplied as a portion of the workers' rations.

In general, bacterial activity during fermentation of vegetables follows the same pattern. The majority of micro-organisms on vegetables consists of strains of soil and water species of genera such as *Pseudomonas*, *Flavobacterium*, *Achromobacter*, *Aerobacter*, *Escherichia* and *Bacillus*. Lactic-acid bacteria, which are required for the desired fermentation, are originally present in comparatively low numbers.

At the start of the fermentation, salt is added in such a concentration (2.5–6.0%) that undesirable putrefactive bacteria are kept under control, long enough to favour lactic acid-producing bacteria to proliferate. Under these conditions, *Leuconostoc mesenteroides* soon proliferates in the expressed juice of the vegetable and produces organic acids and carbon dioxide. These lower the pH value rapidly and inhibit development of spoilage micro-organisms. Moreover, carbon dioxide replaces air from the vegetable mash and provides an anaerobic condition, which suppresses aerobic flora.

At the same time, the anaerobic environment together with the lower pH value make the condition more favourable for growth of other lactic-acid bacteria, which is required to obtain a desired food product.

The complex of an anaerobic condition, low pH value, the combination of salt and acid which have a much higher inhibitive effect on spoilage bacteria than salt alone, contribute ultimately to a situation that suppresses any activity of undesirable micro-organisms and provides the opportunity for lactic-acid bacteria to become dominant. The growth sequence of lactic-acid bacteria includes *Leuconostoc mesenteroides*, *Lactobacillus brevis*, *Pediococcus cerevisiae* and *Lactobacillus plantarum*. Growth of each species depends on its initial presence on the vegetable, the sugar and salt concentrations and the temperature. The species vary in their characteristics, particularly in regard to their tolerance of salt and acid and to their temperature ranges for growth. They are responsible for the complex changes in vegetable fermentations.

Leuconostoc mesenteroides is considered to establish proper environmental conditions for proliferation of other species of lactic-acid bacteria rather than producing favourful compounds. *Lactobacillus brevis* is important in imparting character to the final product and is often characterized by its ability to ferment pentose sugars. *Lactobacillus plantarum* is the high acid-producing species, and together with *P. cerevisiae* plays a major role, particularly in fermentation of vegetables in brine. In brine, an acidity beyond 1% is rarely attained, while in a dry-salted vegetable an acidity of 2.0–2.5% can be reached if sufficient sugar is present (Pederson, 1979).

B. Fermented Fish Products

Large quantities of small fish are fermented to produce fish or shrimp sauces and pastes in Southeast Asia. The basic procedure is to mix the freshly netted small and trash fish with sea salt to such an amount that the extracted fish juices contain about 20% salt in the final product. The fish sauces are salty condiments to be added in small quantities to other foods (Table 2). Similar processes in which less hydrolysis occurs lead to fish or shrimp pastes (Table 3). The pastes may be mixed with cereals. Despite their high protein content, these products are of limited nutritional value because of the low level of consumption (Panel on Microbial Processes, 1979).

The role of micro-organisms in the fermentation process is clearly different from that in fermented vegetable products. The high salt content of these products leaves only salt-tolerant micro-organisms to

Table 2
Origins of various fish sauces

Name	Country
Kecap ikan	Indonesia
Nampla	Thailand
Nuoc-mam	Vietnam
Patis	Philippines
Shottsuru	Japan

survive. These originated from the natural microbial population of the fish or shrimp itself, and from the salt and micro-organisms introduced during the manufacturing process from fermentation tanks, equipments and workers. It was suggested that protein digestion results from a combination of the autolytic action of the natural fish enzymes and the effects of microbial fermentation (Burkholder *et al.*, 1968).

Only a few studies have been reported on the role of bacteria in development of particular properties of fermented fish products. It was reported that halophilic *Bacillus* spp. were responsible for the production of volatile acids in a type of 'fish sauce' produced in Thailand (Saisithi *et al.*, 1966). *Leuconostoc mesenteroides*, *P. cerevisiae* and *L. plantarum* played the major acid-producing role in *burong dalag*, a fish paste made in the Philippines (Orillo and Pederson, 1968).

A study of the microflora of four fermented fish sauces, including *nampla* from Thailand, *patis* from the Philippines and *kaomi* and *ounago* from Japan, showed that the microflora in these products were

Table 3
Origins of various fish pastes

Name	Country
Bagoong	Philippines
Belachan	Malaysia
Kapi	Thailand
Mam-ton	Vietnam
Ngapi	Burma
Padec	Laos
Prahoc	Kampuchea
Shiokara	Japan
Trassi	Indonesia

halotolerant rather than halophilic. *Bacillus* species predominated in three of the products. The presence of *B. pumilus* in the early stages of the *patis* fermentation and *B. licheniformis* during all stages of the *nampla* fermentation was considered to suggest that spore-forming bacilli may play an active role early in the fermentation process. The occurrence of *Micrococcus colpogenes* and *M. varians* in one-month-old *patis* indicated the possible involvement of non-spore-forming micro-organisms in the early stages of some fish-sauce fermentations, although they were not present in the one-month *nampla* sample. With the exception of *Candida clausenii* in the *patis* sample, yeasts were not found in the other fish sauces (Crisan and Sands, 1975).

C. Fermented Seeds

1. *Natto*

One of the few soybean products which is made by fermentation with bacteria is the Japanese *natto*. It is made of soaked and cooked soybeans which traditionally are wrapped in rice straw and left for one or two days in a warm place. The rice straw is considered to have various functions. It supplies the essential micro-organisms for the fermentation, provides the aroma of straw to the product which is considered pleasing, and absorbs partly the unpleasant odour of ammonia which is released during the fermentation. The essential micro-organism for this fermentation was found to be *Bacillus natto* which is classified as a related strain of *B. subtilis* (Hesseltine and Wang, 1972). During the fermentation, the beans become covered with a viscous, sticky polymer. The product is considered of good quality if long stringy threads are formed when two beans are pulled apart. Natto has a slimy appearance and has a grey colour; its flavour is strong and persistent. It is consumed with rice or used as a side dish.

Although *natto* is well known in Japan because of its exceptional properties, it is not as popular as other fermented soybean products, including *miso* and *shoyu*. Similar to other traditional Japanese fermented foods, *natto* has been scientifically studied, and production on a modern technological base using a pure-culture starter has been developed. One of the new developments is production of powdered *natto* which could be added to biscuits or soup (Hesseltine and Wang,

1972). More information on this product is supplied in Ko and Hesseltine (1979).

2. *Thua-nao*

Thua-nao is a food product of Thailand which closely resembles *natto*. Unlike *natto*, however, *thua-nao* is lightly mashed into a paste to which salt and other flavouring agents are added. A review of this product was presented in Ko and Hesseltine (1979).

3. *Dagé*

Dagé is an Indonesian food product made of different oil-rich seeds (*Mucuna pruriens, Aleurites moluccana, Arachis hypogaea* and others) or their residues. Like traditional *natto*, it is subjected to a natural fermentation by bacteria (Vorderman, 1902; Heyne, 1950; Saono *et al.*, 1974). On a very limited scale, *dagé* is made at home and used as a side dish or as an ingredient in preparing certain dishes. Not much is known about the microbiology and biochemistry of this type of products which is of very limited importance.

D. Fermented Starch-Rich Raw Materials

Fermentation of starch-rich substrates by bacteria into fermented foods is practised mainly in Africa. The most important raw materials are maize (*Zea mays*) and cassava (*Manihot* sp.).

1. *Fermented Maize Products*

Not less than 20 different fermented maize products are known in African countries. Manufacture of these foods follows basically the same outline. Corn is soaked for one to three days, ground and mixed with some water to make a thick dough.

During preparation of *banku, kenkey, akpler* and other foods, the mash is left one to three days for fermentation. However, if *ogi* or *agidi* are to be made, the mash of ground maize is diluted with water, sieved to remove the bran and the filtrate left for fermentation (Whitby, 1968). Presumably, the fermentation process in the thick maize dough as well as in the diluted maize suspension follows the same pattern. The micro-organisms involved may be the same as in *ogi*. It was found that *Corynebacterium* spp. hydrolysed starch and initiated acidification; later on it was replaced by *Aerobacter cloacae*. The highest acidity, mainly attributable to lactic acid, was produced by *Lactobacillus plantarum*. At the beginning of the souring period, *Saccharomyces cerevisiae* proliferates rapidly while, at the end, *Candida mycoderma* was predominant. These two micro-organisms were considered to contribute to the flavour of the product and to its enrichment with vitamins. *Aerobacter cloacae* on the other hand decreased the thiamin and panthothenic-acid contents, but increased the riboflavin and niacin contents (Akinrele, 1970). In most cases, the fermented mash is concentrated by cooking until it becomes a thick porridge or cake.

2. Fermented Rice Products

Idli is a soft and spongy steamed bread, which is popular in the south of India. It is made from a mixture of milled rice and dehulled black gram (*Phaseolus mungo*) which is left for natural fermentation overnight. Instead of yeasts, which are usually required for manufacturing bread, the predominant micro-organism producing acid and gas in the natural fermentation of *idli* is *Leuconostoc mesenteroides*. This bacterium is generally present on Indian black gram. In the later stages of the fermentation, *Streptococcus faecalis* and *P. cerevisiae* may also play a role (Mukkerjee *et al.*, 1965). It was reported that the fermentation does not improve the nutritional value, but that it does improve the palatability of the product in terms of flavour, taste and texture (Steinkraus, 1973; Hesseltine, 1979a,b; Desikachar, 1979). When the dough has been raised sufficiently, it is steamed and is served while it is hot. It is a common snack or breakfast food used almost daily in the homes and restaurants in India.

Similar foods in India are *dosa* and *appam* which are made by fermentation of a mixture of rice, other legumes and certain ingredients. These are essentially pancakes which are prepared on a hot pan after the fermentation is finished (Desikachar, 1979).

Puto from the Philippines is essentially steamed bread or cake made by natural fermentation of fine-ground flour of glutinous rice. The carbon dioxide for leavening the dough is presumably produced by the same species of lactic-acid bacteria as for manufacture of *idli* (Pederson, 1979; Sanchez, 1977).

3. *Fermented Cassava*

In several countries of West Africa (Nigeria, Ghana), *gari* is an important staple food. For its manufacture, fresh cassava (*Manihot utilissima*) is peeled, grated and the greater part of the juice squeezed out. The remaining pulp is left for three or four days for natural fermentation to occur. It was suggested that this is a two-stage fermentation. In the first stage, *Corynebacterium* spp. attack the starch with production of organic acids. The lower pH value causes hydrolysis of a cyanogenic glucoside with liberation of gaseous hydrocyanic acid. Organic acids also make conditions favourable for growth of *Geotrichum candidum*. This micro-organism produces a variety of aldehydes and esters, which are responsible for the characteristic taste and aroma of *gari* (Collard and Levi, 1959). Other reports also mentioned the probable role of *Leuconostoc* spp. (Okafor, 1977) and *Streptococcus faecium* (Abe and Lindsay, 1978).

The fermented pulp is finally fried in iron or earthenware pans. Sometimes the pan is greased with palm-oil, in which case the *gari* comes out yellow. The resulting product is dry and granular and has the property of swelling up in cold water. It can be consumed simply with water, without any cooking. Sugar may be added, or it may be mixed with spices (pepper) or other foods like fish or egg (Whitby, 1968).

E. Fermented Plant Juice

Throughout the tropics, the sugary sap of palm trees of many species is spontaneously fermented into alcoholic beverages. The microflora isolated during this fermentation are rather complex, but it is considered certain that *Zymomonas* spp. are largely responsible for the alcoholic content (4–5%) and carbon dioxide formation. The fruity odour and the taste of palm wine may be also due to production of some acetaldehydes by *Zymomonas* spp. Lactic-acid bacteria contribute to the acidity by producing small amounts of lactic and acetic acids (Swings and de Ley, 1977).

Similar bacterial alcoholic fermentation occur during fermentation of the juice of certain species of *Agave* in Mexico to produce the alcoholic beverage *pulque* (Sanchez-Marroquin, 1977). Only small numbers of yeast cells are usually found in these alcoholic beverages. Because formation of alcohol and carbon dioxide from sugar fermentation is a characteristic of yeasts, alcoholic fermentation by bacteria is not a common process (Borgstrom, 1968).

IV. FOODS FERMENTED BY A MIXTURE OF MOULDS AND YEASTS

A. Ragi

Ragi itself is not a food; it is an inoculum used to induce fermentation of certain food products which will be described in this chapter. It is known in many Asian countries under different names (Table 4). In this review, it will be designated as ragi. Most probably, its origin is China, where *ch'ü* was already described in the old Chinese classics as the most important ingredient in the manufacture of alcoholic beverages. Later records gave thorough descriptions of its preparation and applications (Yamazaki, 1932). It is made of rice flour, which is moulded into flattened round cakes of 2–3 cm diameter.

Table 4

Names given in various countries to an inoculum used to manufacture certain food products

Name	Country
Bubud	Philippines
Ch'ü	China
Luk-paeng	Thailand
Nurook	Korea
Ragi	Indonesia
Ragi	Malaysia

B. Micro-Organisms

Numerous fungi and yeasts have been isolated from ragi (Boedijn, 1958; Dwidjoseputro and Wolf, 1970; Ko, 1965, 1972; Saono and Basuki, 1978, 1979). Of the many mould species which have been isolated from ragi, *Mucor* and *Rhizopus* species are the most important. They are amylolytically, lipolytically as well as proteolytically active (Saono and Basuki, 1978). *Chlamydomucor oryzae*, recently re-identified as *Amylomyces rouxii* (Ellis *et al.*, 1976), plays a major role.

Yeast strains isolated from ragi included species of *Candida*, *Endomycopsis* and also *Saccharomyces*. They produce alcohol from the sugar which is produced from starch by the mould. It was reported that most yeasts isolated from ragi were amylolytically active, but only some were lipolytically active and none was proteolytically active (Saono and Basuki, 1978). It was found that, for a good fermentation of glutinous rice into good quality *tapé* (see Table 5), a combination of *Chlamydomucor oryzae* and *Endomycopsis chodati* was essential (Ko, 1972). This combination of micro-organisms was the starting point for later studies on *tapé* fermentation (Cronk, 1975; Cronk *et al.*, 1977, 1979).

C. Fermented Starch-Rich Raw Materials

When ragi is inoculated into a starch-containing substrate like steamed rice, cassava, maize or sorghum, a soft juicy product with a sweet, mild sour taste and mild alcoholic flavour is obtained after two to three days incubation at 25–30°C.

1. Fermented Rice

Such a product made of rice is known in different countries under different names (Table 5). It is consumed as such without any preparation and is considered as a delicacy. In China, *lao-chao* has a

Table 5

Names given in various countries to fermented glutinous rice
(*Oryza sativa glutinosa*)

Name	Country
Binuburan	Philippines
Lao-chao ⎱ Chiu-niang ⎰	China
Khao-mak	Thailand
Tapai pulut	Malaysia
Tapé ketan	Indonesia

unique place in the diets of new mothers. It is believed that *lao-chao* helps them regain their strength (Wang and Hesseltine, 1970).

2. Brem

In Indonesia, a confectionery is made by separating the juice which is produced during fermentation of *tapé ketan*. Traditionally, the liquid is dried in the sun until a solid cake is obtained. It is considered a delicacy, particularly for children, and is known under the name of *brem*.

3. Rice Wine

When fermentation of rice is extended to several weeks, more alcohol and more liquid are produced. To produce rice wine, the liquid is expressed and kept for clarification and ripening during several months. The beverage is well known under different names in different countries (Table 6). Depending on the fermentation time, the alcohol content differs and a concentration of 15% can be reached. By distillation, a beverage can be obtained with approximately 50% alcohol.

Table 6
Different names given in various countries to rice wine

Name	Country
Brem Bali	Indonesia
Mie-chiu	China
Saké	Japan
Sato	Thailand
Sonti	India
Tapoi	Philippines
Yakju	Korea

4. Tapé-ketella

In Indonesia, peeled, washed and steamed cassava tubers are fermented by inoculation with ragi to produce *tapé-ketella* or *peuyeum*. The final product is a food with a soft texture, a lightly sour and sweet taste and a mild alcoholic flavour. It is considered as a delicacy and is usually consumed without additional preparation. Sometimes it is shortly deep-fried in coconut-oil before consumption. It may also be mixed with other ingredients to make a kind of pie.

V. FOODS FIRSTLY FERMENTED BY MOULDS, FOLLOWED BY A FERMENTATION WITH A MIXTURE OF BACTERIA AND YEASTS

The most studied foods in this category are those developed in Japan from traditional foods produced on a scale ranging from a house industry to high-industrialized products. These include *shoyu*, the Japanese type of soy sauce, and *miso*, a pasty product made of a mixture of rice and soybeans. For its fermentation, the traditional inoculum tane koji is applied.

A. Tane koji

Presumably, this starter originated from ancient China and was introduced into Japan about 200 A.D. (Yong and Wood, 1974). Once in Japan, it followed a typical Japanese way of development. From a traditional inoculum, it has developed into an inoculum made of pure cultures of mould strains whose properties are studied and controlled with the newest methods of modern microbiology. It is a green–yellow powder consisting of mould spores of one or more selected strains of *Aspergillus oryzae* and *A. soyae* which produce protease, amylases and a great variety of other enzymes. There are variations between strains in their ability to produce each of the enzymes; one strain may produce more proteases and less amylases than other strains and *vice versa*. Modern-day tane koji for a particular fermentation is composed of spores of several strains of *A. oryzae* mixed together in a definite proportion in order that various enzymes can be produced in the proper amounts during the different stages of fermentation.

B. Soy Sauce

Soy sauce is a liquid, brown coloured condiment which has different names in different countries (Table 7). It is made by a two-stage batch fermentation which involves the biochemical activities of moulds, bacteria and yeasts (Yong and Wood, 1974).

The first stage consists of growing one or more mould species on cooked soybeans or on a mixture of soybeans and wheat, oats, rye or

Table 7

Names given to soy sauce in different countries

Name	Country
Chiang-yu	China
Kan jang	Korea
Kecap	Indonesia
Shoyu	Japan

tapioca. Japanese *shoyu* is made from equal amounts of soybeans and wheat. For the fermentation of Japanese *shoyu*, the raw materials are inoculated with tane koji which contains spores of selected strains of *Aspergillus oryzae* and *A. soyae*. In less sophisticated traditional soysauce factories throughout South East Asia, mould species grow spontaneously on the soybeans by natural contamination from the air and from the bamboo trays on which soybeans of former batches were incubated (Bhumiratana *et al.*, 1980). The moulds involved are species of *Aspergillus*, *Rhizopus* or *Mucor*. Some Indonesian *kecap* manufacturers inoculate the cooked soybeans with tempe inoculum which contains spores of *Rhizopus oligosporus*. In this first stage of mould fermentation, enzymic breakdown of insoluble protein in the raw materials into soluble polypeptides, peptides and amino acids takes place by extracellular proteases of the mould. Starches are hydrolysed to disaccharides and monosaccharides by α-amylase secreted by the mould, while extracellular invertase hydrolyses sucrose in the soybeans.

When, after two to three days, mould growth has reached a certain level, the moulded raw material is placed into brine containing 15–20% sodium chloride in which the second stage of fermentation takes place. The contribution of bacteria and yeasts in this second part of the fermentation to the development of flavour and taste of soysauce was reviewed by Yong and Wood (1974). The chemistry and composition of Japanese soysauce, with emphasis on the flavour and aroma constituents, was reviewed by Yokotsuka (1960). After a fermentation of at least six months duration and traditionally two to three years, the liquid is removed by decanting or siphoning. In modern factories, the liquid is expressed with a hydraulic press from the residue. After final treatments including pasteurization, clarification, filtration and eventually addition of chemical preservatives, the brown liquid is bottled for distribution to the consumers.

Basically, there are two major kinds of soy sauce. The Chinese type is made of soybeans alone or from a mixture of soybeans and wheat with a higher percentage of the soybeans. In some cases, wheat could be substituted by wheat bran, oats, kaoliang and rye. An Indonesian *kecap* manufacturer uses tapioca-flour for this purpose. The Chinese type soy sauce is of high specific gravity and viscosity and high nitrogen content; it is dark in colour and is sometimes sweetened with cane sugar. The Japanese type soy sauce or *shoyu* is made of equal amounts of soybeans and wheat. It is lower in viscosity and lighter in colour than the Chinese type (Yong and Wood, 1974).

C. Other Fermented Soybean Products

Several other soybean products (Table 8) are basically similar to soy sauce with respect to micro-organisms and fermentation principles (Wang and Hesseltine, 1979), although the composition of raw materials, methods of manufacture, consistency and appearance of the final foods are all different.

In the first stage of the fermentation, moulds are grown on cooked soybeans or on a mixture of soybeans and cassave flour for the manufacture of *taoco*. For production of *miso*, the moulds are grown on steamed rice, while a mixture of soybeans and wheat flour is used for the manufacture of *hamanatto*. For *miso*, the moulds are introduced as spores of *Aspergillus oryzae* and *A. soyae* in tane koji. The other less sophisticated products could be purposely inoculated or naturally contaminated with strains of *Aspergillus oryzae*, *Rhizopus oryzae*, *R. oligosporus* or a mixture of two or more species.

Table 8

Soybean foods produced by a two-stage fermentation

Product	Country
Hamanatto	Japan
Miso	Japan
Soy sauce	Orient
Taoco	Indonesia
Tao-si	Philippines
Tou-shih	China

When, after two to three days, the mould has grown to a certain extent, the moulded substrate is mixed with other ingredients. These include salt and, depending on the method and the product desired, soybeans, spices, sugar and some water. The mixture is left for a period ranging from a few weeks to one year for the second stage of fermentation. During this period, the activities of the mould enzymes are accompanied by the activities of salt-tolerant yeast species of *Saccharomyces* and *Torulopsis* and species of *Pediococcus* and *Streptococcus*. Ultimately, the complex biochemical conversions result in products that have an aroma resembling that of soy sauce.

The fermented mash is worked up in different ways and results in final products with a different appearance and consistency. In the manufacture of Chinese *tou-chih* and Japanese *hamanatto*, the fermented beans are dried. *Hamanatto* is soft, having a high content of moisture. Chinese *tou-chih* has a much lower moisture content and is therefore not so soft (Wang and Hesseltine, 1979).

For the manufacture of Indonesian *taoco*, a solution of palm-sugar is added to the fermented mash. Then the mixture is concentrated by boiling until it has the consistency of a porridge. To make a dry variety of *taoco*, the soybeans are separated from the fermentation liquid and boiled with a concentrated solution of palm-sugar. Ultimately, the concentrated mash is sun-dried until a rather solid paste is obtained. In both types of *taoco*, the fermented soybeans are present as large particles.

Japanese *miso* has the consistency of peanut butter because, after fermentation and aging, the mash is ground into a uniform paste. *Miso* and hot water are the basic ingredients for preparing *miso* soup; vegetables and other ingredients are usually added. In many Japanese families, breakfast consists of cooked rice and *miso* soup. *Miso* is also used as seasoning agent to prepare fish and vegetable dishes (Shibasaki and Hesseltine, 1962). The aforementioned fermented soybean products are also used as flavouring agents in cooking as well as table condiments or as a side dish.

VI. SPECIFIC ASPECTS OF FERMENTED FOODS

A. Mould Species

The mould species that are traditionally used for fermentation of foods in different parts of the World belong to different genera.

Species of *Rhizopus*, *Mucor* and *Aspergillus* are used for fermentation of foods throughout the Orient, with the exception of Japan. In Japan, it is restricted to species of *Aspergillus* including *A. oryzae* and *A. soyae*. The difference between Japan and other Asian countries in the application of mould species is thought to be due to the development of tane koji and its general use as a starter for food fermentations in Japan. Tane koji is made by growing moulds on steamed rice.

In other Asian countries, on the other hand, ragi-type starters (Table 4, p. 27) are in common use. These starters are made by cultivating moulds on cakes made of rice- or wheat-flour which has not been steamed or cooked. The difference in preparing growth substrates for manufacture of the two types of inocula is thought to be the cause of a natural selection of the mould species which were developing in each of the type starters over many centuries, when they were produced with non-aseptic, traditional methods (Sakaguchi, 1972). Hesseltine (1979b) suggested that the relative humidity of the areas may be an important selective factor, so that fungi in the order Mucorales are found in the tropics while *Aspergillus* species grow in semi-tropical countries such as China and Japan. Exceptions in Asian countries are *Monascus purpureus* for manufacture of *angkak* from rice, and *Neurospora* species for production of *oncom* from peanut presscake. In Occidental countries, only *Penicillium* species are used for fermentation of foods including Roquefort and Camembert types of cheeses and certain fermented meat products (Mintzlaff and Christ, 1973).

B. Lactic-Acid Bacteria

In vegetable, fish as well as some soybean fermentations, one or more species of lactic-acid bacteria play an important role. Their production of organic acids not only contributes to the desired taste and flavour of the final product, but it also makes the substrate unfavourable for proliferation of spoilage and other undesirable micro-organisms. At the same time, the acids make the substrate more suitable for growth of desirable micro-organisms which improve the properties of the food. The role of carbon dioxide produced by *Leuconostoc mesenteroides* is in different products not the same. In vegetable fermentations, it provides an anaerobic condition which inhibits proliferation of aerobic micro-organisms. In fermentation of *idli* and *puto*, carbon dioxide is essential for leavening the dough.

C. Yeasts

When yeasts are involved in the fermentation process, production of alcohols improves the aroma of the product. In addition, alcohol at a certain concentration makes the substrate unsuitable for micro-organisms which may create undesirable properties in the product. Combined with the organic acids that are produced by lactic-acid bacteria, the inhibitory effect of alcohol on undesirable micro-organisms is increased.

D. Salt

Application of salt in fermentation of vegetables, fish and soybeans inhibits proliferation of putrefactive micro-organisms. Salt also affects the development of pathogenic and toxin-producing species. Growth of *Salmonella* spp. is prevented by concentrations of about 6% sodium chloride (Ingram and Kitchell, 1967). *Clostridium botulinum* is the species to which is attached most interest because of the fatal toxin produced when it multiplies in food, but all types of *Cl. botulinum* are inhibited by 10–12% salt (Mackie *et al.*, 1971). *Staphylococcus aureus* is able to resist up to 15% salt or occasionally even 20%, but 5% is the highest salt concentration at which toxin formation is recorded (Ingram and Kitchell, 1967). *Pseudomonas cocovenenans* died in 1% soya peptone medium containing 3.5% sodium chloride. Although 2% salt in a coconut medium did not suppress its growth, toxin production is completely suppressed (Ko *et al.*, 1977). On the other hand, some useful and unharmful species, including lactic acid-producing bacteria and some yeast species, are less affected by salt. *Lactobacillus delbrueckii* can adjust to growing readily on media containing 18% sodium chloride (Wood *et al.*, 1973).

The salt concentrations in many fermented foods are too high for micro-organisms to cause undesirable effects. Fermented fish pastes (Table 3, p. 22) contain 15–25% salt (van Veen, 1965; Mackie *et al.*, 1971; Soedarmo, 1972).

For the manufacture of fish sauces (Table 2, p. 22) approximately 30% salt is applied (van Veen, 1965; Saisithi *et al.*, 1966).

In the fermentation of vegetables, moderate salt concentrations of between 4 and 6% are used in the brine (Hsio Hui Chao, 1949). Notwithstanding the marginal salt concentrations, if the required conditions are supplied, lactic acid-producing bacteria will soon proliferate and produce enough acids. The combination of salt and acids will protect the product against undesirable effects of harmful micro-organisms. Considering also the protective role of certain mould species in fermented foods, it may be concluded that fermentation of indigenous foods is surrounded by many safety factors, provided the procedure of preparation is properly followed.

VII. ACKNOWLEDGEMENT

This manuscript was planned to be written together with Dr. C.W. Hesseltine, Chief, Fermentation Laboratory, Northern Regional Research Center, Peoria, Illinois of the United States Department of Agriculture. By unfortunate and unexpected circumstances concerning Dr. Hesseltine's health, the author has ultimately written this chapter alone. However, many parts of this paper reflect our mutual interest in the subjects, and our numerous conversations and correspondence of many years. The author wishes to express his indebtedness to Dr. Hesseltine's stimulating ideas.

REFERENCES

Abe, M.G. and Lindsay, R.C. (1978). *Journal of Food Production* **41**, 781.
Akinrele, I.A. (1970). *Journal of the Science of Food and Agriculture* **21**, 619.
Batra, L.R. and Millner, P.D. (1976). *Developments in Industrial Microbiology* **17**, 117.
Beuchat, L.R. (1978a). *Food Technology* **32**(5), 193.
Beuchat, L.R. (1978b). *In* 'Food and Beverage Mycology' (L.R. Beuchat, ed.), p. 224. Avi Publishing Company, Westport, Connecticut.
Bhumiratana, A., Flegel, T.W., Glinsukon, T. and Somporan, W. (1980). *Applied and Environmental Microbiology* **39**, 430.
Boedijn, K.B. (1958). *Sydowia* **12**, 321.
Boorsma, P.A. (1900). *Geneeskundig Tijdschrift van Nederlands Indie* **40**, 247.
Borgstrom, G. (1968). "Principles of Food Science", vol. 2. The Macmillan Company, New York.
Burkholder, L., Burkholder, P.R., Chu, A., Kostijk, N. and Roels, O.A. (1968). *Food Technology* **22** (10), 76.

Collard, P. and Levi, S. (1959). *Nature, London* **183**, 620.
Crisan, E.V. and Sands, A. (1975). *Applied Microbiology* **29**, 106.
Cronk, T.C. (1975). Ph.D. thesis: Cornell University, Ithaca.
Cronk, T.C., Steinkraus, K.H., Hackler, L.R. and Mattick, L.R. (1977). *Applied and Environmental Microbiology* **33**, 1067.
Cronk, T.C., Mattick, L.R., Steinkraus, K.H. and Hackler, L.R. (1979). *Applied and Environmental Microbiology* **37**, 892.
Desikachar, H.S.R. (1979). *In* 'Proceedings of the International Symposium on Microbiological Aspects of Food Storage, Processing and Fermentation in Tropical Asia' p. IV/2-1. Food Technology Development Center, Bogor, Indonesia.
Dwidjoseputro, D. and Wolf, F.T. (1970). *Mycopathologia et Mycologia Applicata* **41**, 211.
Ellis, J.J., Rhodes, L.J. and Hesseltine, C.W. (1976). *Mycologia* **68**, 131.
Frazier, W.C. and Westcoff, D.C. (1978). 'Food Microbiology'. McGraw-Hill, New York.
Hesseltine, C.W. (1965). *Mycologia* **57**, 149.
Hesseltine, C.W. (1979a). *Journal of the American Oil Chemists' Society* **56**, 367.
Hesseltine, C.W. (1979b). *In* 'Proceedings of the International Symposium on Microbiological Aspects of Food Storage, Processing and Fermentation in Tropical Asia', p. V/1-1. Food Technology Development Centre, Bogor, Indonesia.
Hesseltine, C.W. and Wang, H.L. (1972). *In* 'Soybeans: Chemistry and Technology' (A.K. Smith and S.J. Circle, eds.), vol. 1, p. 389. Avi Publishing Company, Westport, Connecticut.
Heyne, K. (1950). 'De nuttige planten van Indonesia', Deel I. N.V. Uitgeverij van Hoeve, 's Gravenhage.
Hsio Hui Chao (1949). *Food Research* **14**, 405.
Ingram, M. and Kitchell, A.G. (1967). *Journal of Food Technology* **2**, 1.
Jay, J.M. (1978). 'Modern Food Microbiology', 2nd edn. D.Van Nostrand Co., New York.
Ko, Swan Djien (1965). *In* 'Research di Indonesia' (M. Makagiansar and R.M. Soemantri, eds.), Vol. 2, p. 209. P.N. Balai Pustaka, Djakarta, Indonesia.
Ko, Swan Djien (1972). *Applied Microbiology* **23**, 976.
Ko, Swan Djien (1974). *In* 'Proceedings of the IV International Congress of Food Science and Technology' (E. Portelo Maris, ed.), Vol. 3, p. 244. Instituto Nacional de Ciencia y Tecnologia de Alimentos, Madrid, Spain.
Ko, Swan Djien (1979). *In* 'Proceedings of the International Symposium on Microbiological Aspects of Food Storage, Processing and Fermentation in Tropical Asia' p. V/2-1. Food Technology Development Center, Bogor, Indonesia.
Ko, Swan Djien, Kelholt, A.J. and Kampelmacher, E.H. (1977). *In* 'GIAM-V Proceedings' (Pornchai Matangkasombut, ed.), p. 375. GIAM-V Secretariat, Bangkok, Thailand.
Ko, Swan Djien and Hesseltine, C.W. (1979). *In* 'Economic Microbiology' (A.H. Rose, ed.), vol. 4, p. 115. Academic Press, London.
Mackie, I.M., Hardy, R. and Hobbs, G. (1971). *FAO Fisheries Reports,* no. 100.
Mintzlaff, H.J. and Christ, W. (1973). *Die Fleischwirtschaft* **53**, 864.
Mukkerjee, S.K., Albury, M.N., Pederson, C.S., van Veen, A.G. and Steinkraus, K.H. (1965). *Applied Microbiology* **13**, 227.
Okafor, N. (1977). *Journal of Applied Bacteriology* **42**, 279.
Orillo, C.A. and Pederson, C.S. (1968). *Applied Microbiology* **16**, 1669.
Palo, M.A., Vidal-Adeva, L. and Maceda, L.M. (1960). *The Philippine Journal of Science* **89**, 1.

Panel on Microbial Processes (1979). National Academy of Sciences, Washington, D.C.
Pederson, C.S. (1979). 'Microbiology of Food Fermentations'. The Avi Publishing Co., Westport, Connecticut.
Saisithi, P., Kasemsarn, B., Liston, J. and Dollar, A.M. (1966). *Journal of Food Science* **31**, 106.
Sakaguchi, K. (1972). *In* 'Fermentation Technology Today' (G. Terui, ed.), p. 7. Society of Fermentation Technology, Japan.
Sanchez, P.C. (1977). Paper presented at the Symposium on Indigenous Fermented Foods, Bangkok, Thailand.
Sanchez-Marroquin, A. (1977). Paper presented at the Symposium on Indigenous Fermented Foods, Bangkok, Thailand.
Saono, S., Gandjar, I., Basuki, T. and Karsono, H. (1974). *Annales Bogoriensis* **5**, 187.
Saono, S. and Basuki, T. (1978). *Annales Bogoriensis* **6**, 207.
Saono, S. and Basuki, T. (1979). *Annales Bogoriensis* **7**, 11.
Shibasaki, K. and Hesseltine, C.W. (1962). *Economic Botany* **16**, 180.
Soedarmo Moeljohardjo, D. (1972). Ph.D. thesis; Agricultural University, Wageningen, The Netherlands.
Steinkraus, K.H. (1973). *In* 'Fermented Foods' (D.L. Downing, ed.), Special Report no. 16, April 1974. Cornell University, Ithaca, N. York.
Sundhagul, M., Daengsubha, W. and Suyanandana, P. (1975). *Thai Journal of Agricultural Science* **8**, 205.
Swings, J. and de Ley, J. (1977). *Bacteriological Reviews* **41**, 1.
Van Veen, A.G. (1965). *In* 'Fish as Food' (G. Børgstrom, ed.), vol. 3, p. 227.
Van Veen, A.G., Graham, D.C.W. and Steinkraus, K.H. (1968). *Cereal Science Today* **13**, 96.
Vorderman, A.G. (1893). *Geneeskundig Tijdschrift voor Nederlands Indie* **33**, 343.
Vorderman, A.G. (1902). *Geneeskundig Tijdschrift voor Nederlands Indie* **42**, 395.
Wang, H.L. and Hesseltine, C.W. (1970). *Journal of Agricultural and Food Chemistry* **18**, 572.
Wang, H.L. and Hesseltine, C.W. (1979). *In* 'Microbial Technology' (H.J. Peppler and D. Perlman, eds.), 2nd ed., vol. II, p. 95. Academic Press, New York.
Wehmer, C. (1900). *Centralblatt fur Bakteriologie, Abteilung II*, **6**, 610.
Went, F.A.F.C. and Prinsen Geerligs, H.C. (1896). *Verhandelingen der Koninklijke Akademie van Wetenschappen* **4**, 1.
Whitaker, J.R. (1978). *Food Technology* **32** (5), 175.
Whitby, P. (1968). 'Foods of Ghana'. Food Research Institute, Accra.
Wood, B.J., Cardenas, O.S., Yong, F.M. and McNulty, D.W. (1973). *In* 'Lactic Acid Bacteria in Beverages and Food' (J.G. Carr, C.V. Cutting and G.C. Whiting, eds.), p. 325. Academic Press, New York.
Yamazaki, M. (1932). *Utsunomiya Koto-Narin-Gakko* (Bulletin Utsunomiya Agricultural College) **1** (2), 1.
Yokotsuka, T. (1960). *Advances in Food Research* **10**, 75.
Yong, F.M. and Wood, B.J.B. (1974). *Advances in Applied Microbiology* **17**, 157.

3. Soy Sauce and Miso

BRIAN J. B. WOOD

Department of Applied Microbiology, University of Strathclyde, Glasgow, United Kingdom

I. INTRODUCTION

In this review, I propose to give an outline account of the operations involved in producing soy sauce (otherwise known as *shoyu*). This will be followed by a discussion of the complex microbiology of the process.

Finally, I shall attempt to give an impression of the scale and importance of the World trade in the product. To begin with, however, it may be necessary to introduce the reader to this valuable condiment, until fairly recently a rather exotic thing to most Westerners, although a basic component of the diet in the Far East for many thousands of years (Yong and Wood, 1974).

Probably to the average European the first direct experience of soy sauce is as a bottle of red-brown salty fluid on the table of Chinese restaurants. In fact, as the careful perusal of the labels of many prepared, canned and 'convenience' foods and sauces will show, soy sauce is used extensively in Western food manufacturing industries. It is used, moreover, for precisely the same reasons that it is prized in Oriental cuisine, namely its remarkable ability to enhance and boost the flavour of the food to which it is added. Now, too, it is becoming readily available in shops and supermarkets, and considerable amounts are sold in the burgeoning health-food, whole-food and vegetarian shops.

It comes as a considerable surprise to many people, even microbiologists, to learn that this preparation is made by fermentation, or, to be more precise, by a series of linked fermentations. Viewed in its Oriental context, however, it may be recognized as but one of a whole family of fermented foods, including *miso* (a salty paste rapidly gaining popularity in the West), *saké* (rice wine) and a host of local products differing in greater of lesser degree from one another. In turn, this soy sauce 'family' can be regarded as a member of a still larger grouping, embracing on the one side the Oriental fish sauces, and on the other, soy curd cheeses, *tempeh* and many other products (Hesseltine, 1965; Wood and Yong, 1974; Wood, 1977). There are also, although more distant, affinities with commodities more familiar to Occidental tradition, such as ginger beer, the traditional bread levens or barms (Wood *et al.*, 1975) and the Belgian lambic beers.

In essence, the soy sauce 'family' of fermentations consists of processes in which a vegetable substrate is prepared, inoculated with a strain of selected mould, incubated to permit optimum mould growth, and then subjected to a second phase of (anaerobic) fermentation involving yeasts and bacteria (especially lactic-acid bacteria). Finally, the product is prepared for marketing by filtration and clarification as appropriate to the material in question.

Soy sauce is, moreover, no more a single entity then is Scotch Whisky. To the Eastern palate there are varietal differences at least as great and important as these between the different malt whiskies, ranging from the lightness of a Glenmorangie to the weight of Tallisker. Much of the soy sauce offered to the Western consumer could be compared in its blending of 'semichemical' and 'fully brewed' soy sauces (*vide infra*) to blended whisky. In the soy-based sauces of, for example, Indonesia (*ketjap*), one can see analogies with the existence of American Bourbon and Irish Whiskey, alongside the Scottish product, very different and each good in its own right to the discerning palate.

II. PREPARATION OF SOY SAUCE

A. Introduction

The process may be summarized as follows:

(i) Soybeans are steeped in water for 16 hours or so.

(ii) The soaked beans are dehulled and cooked.

(iii) The beans, either alone or (more commonly) in admixture with wheat flour or grits, are inoculated with *Aspergillus oryzae* and incubated for three days at 30°C with occasional stirring; this is the *Koji* stage.

(iv) The resulting material is mixed with brine.

(v) A fermentation dominated by yeasts and lactic acid bacteria then develops, this being the *Moromi* (salt mash) stage.

(vi) After an incubation of one month to three years, a dark salty liquid with a pleasing savoury aroma is drained from the fermentation vessel, clarified, pasteurized and packaged for distribution and sale.

This product, which has a characteristic aroma, salty taste and dark-brown colour, is sometimes referred to as 'fully brewed' soy sauce. Its use as a condiment reflects not only the addition of its own flavour and salt to the food, but also its flavour-enhancing qualities, accounted for in part at least by the glutamic acid which it contains. This method of preparation is of great antiquity, and has been practised in China for thousands of years. There is no reason to believe that this, or equivalent, fermentation is of less antiquity in other parts of the Far East or South East Asia, but the high scholarship of ancient China has left us more detailed records than are available elsewhere (Yong and Wood, 1974). These authors also discuss speculations that the technology for soy sauce production arose out of yet older fermentations of fish in brine, a group of remarkable fermentations worthy of separate discussion.

These fermentations are typical of the extensive range of traditional food fermentations found throughout Africa and Asia which are now coming to the attention of the West. Soy sauce (*Shoyu* in Japanese) and *miso* (a soy paste whose preparation has similarities to, and interesting differences from, soy sauce) are, however, outstanding in their World economic importance, and their penetration of European and North American markets; this situation is, in no small part, due to the very positive marketing policies of the manufacturers in Japan, Singapore, Korea and Hong Kong, four countries which supply a major part of the soy sauce and *miso* sold in the West.

The demands of manpower, space, specialized equipment and technical skills needed to produce soy sauce and *miso* on a large scale are of a very different order from those used in domestic or village-scale production. It is impossible to quantify this latter production (indeed it is impossible to quantify accurately the total large commercial-scale production of these foods) but it seems certain that it is still a substantial, although diminishing, fraction of the total annual World output.

With space, time and manpower all becoming increasingly costly, it is not suprizing that attempts have been made to shorten the long traditional fermentation. Although good soy sauce, like good wine or whisky, undoubtedly improves in organoleptic quality with storage, excellent soy sauce can be produced in quite a short time by control of conditions and inoculum. However, it was inevitable that more radical economies should be tried, resulting in the production of 'chemical' and 'semi-chemical' soy sauces. The chemical product is the result of hydrolysing soybean meal and wheat flour by refluxing with hydrochloric acid for 12 to 16 hours, followed by adjustment to pH 4.5 with sodium hydroxide, filtration, ageing and bottling (Minor, 1945). As might be expected, such a product has only limited acceptability, and therefore a combination of acid hydrolysis, followed by fermentation with yeast (Yokotsuka, 1960), was developed. Yong and Wood (1974) cite various studies of this product, which apparently commands a reasonable market, even in Japan.

After this short outline, let us turn our attention to a more detailed consideration of the complex and interesting biochemistry, microbiology and technology of the fermentation of soy sauce. To keep this chapter and its bibliography within reasonable bounds, the discussion which follows is based on Yong and Wood (1974, 1976, 1977a,b), Wood and Yong (1974) and Wood (1977) which provide an extensive list of original sources.

B. Preparation of Raw Materials

1. The Beans

The importance of using good quality beans, free from mould or other deterioration in storage, will be clear. Less seems to be known about the degree of importance attached to the germinative capacity of the beans, and it will be interesting to study this problem, since observations made by Dr. Keith Steinkraus (personal communication) might suggest that young fresh beans are preferable; if so it will be important to arrange the storage conditions so as to maintain maximum viability.

Beans are soaked overnight for about 16 hours in clean water which should be changed at intervals to minimize growth of bacteria, especially spore-forming Gram-positive rods which are inevitably present on the beans. Failure to observe this precaution results in generation of heat and production of acids and other products of microbial action, imparting undesirable flavours to the finished product. The discarded water contains reducing sugars and ninhydrin-positive compounds (Goel, 1974) and may be assumed to contain gums and polysaccharides released by the seeds; it must possess a substantial Biological Oxygen Demand and pose an effluent disposal problem for any large soy-sauce factory.

The soaked beans are then cooked, either by prolonged boiling or in a pressure vessel; process economics and improvement of the quality of the cooked beans combine to favour the latter method, which has the additional advantage of sterilizing the beans. Typical conditions would be autoclaving at 115°C for one hour in sufficient water to cover the beans. The choice of cooking conditions is discussed by Yong and Wood (1977a,b). The cooking water is discarded, and again has a high Biological Oxygen Demand.

The seed coat or testa can be removed before or after cooking, but we find it more convenient to effect this operation before cooking since, once the beans have been cooked, it is essential to carry them forward to the next stage of the process as rapidly as possible. Traditionally, and on the laboratory scale, the seed coat is ruptured by rubbing the seeds between the hands, but on a modern industrial scale mechanical removal may be employed. The testa is conveniently removed from the cotyledons by water classification, and can be used as animal feed. I am not convinced that this dehulling is an essential process, particularly if the beans are cooked in the autoclave, so sterilizing them, and my research group has produced apparently accetable soy sauce using intact beans, although it is rather easier to work with the naked cotyledons in the subsequent stages. A proper economic assessment would be interesting. Discussions with members of the Whole Food and 'Macrobiotic' foods trades reveal a marked preference for the use of intact beans in these markets. In practice, whole beans seem to be widely used throughout South-East Asia, despite suggestions to the contrary in the literature.

2. Wheat

It seems that soy sauce was traditionally made exclusively or mainly with soybeans in China. In Japan, on the other hand, mixtures containing up to 50% wheat are favoured. As I understand it, legumes and grains are associated with opposite poles of the dualism of Nature (central to certain Oriental philosophies) names for which (Yin and Yang) have become more familiar in the West as a result of the interest in Zen and similar philosophies during the 1960s. It is a curious fact that the amino-acid make-up of legume and grain seeds, both of which exhibit nutritional deficiencies for the human diet, are complementary; for example, grains tend to be deficient in lysine whereas legume seeds are rich in this amino acid. The result is that blending of the two types of seed makes a nutritionally better food than does either alone (Siegel and Fawcett, 1976). Although it is unlikely that this is very significant for soy sauce, which typically contains about 1.0 to 1.4 g of total organic nitrogen per 100 ml (Onaga *et al.*, 1957), even at the consumption rate of 25 ml per head per day, which is found in Japan and parts of South East Asia, it could be important for *miso* (soy paste also used at about 25 g per head per day in Japan) and other fermented soy products.

Soy sauce made entirely or almost entirely from the bean is known as '*Tamari*'; unfortunately this name has been applied without discrimination to all types of soy sauce by devotees of some food cults (Wood, 1977); this is a great pity because true *tamari* would be acceptable in the very restricted diet of, for example, the unfortunate victims of coeliac disease, who exhibit a drastic intolerance to wheat glutenin. According to

Yokotsuka (1960) [cited by Hesseltine (1965)], *tamari* accounts for most of the soy sauce produced in China, but is of rather small importance in Japan, where *Koikuchi* (a blend of fermented soy sauce and acid hydrolysates as already described above) and *Usukachi* (made from soy beans, wheat flour and a little rice) account for the overwhelming part of the total Japanese trade in soy sauce.

As with every other aspect of this subject, it is difficult to make definitive statements but, from my observations of labels, it seems that most of the soy sauce sold in the United Kingdom contains wheat.

The wheat may be whole wheat, which is first roasted and then coarsely crushed, or wheat flour which is generally steamed before use although it can be roasted (Yong and Wood, 1977). A low-protein wheat is suitable. Although most of my group's studies have been conducted with white flour, some unpublished experiments suggest that whole-wheat flour gives a darker sauce with a fuller aroma and flavour. The studies of Asao and Yokotsuka (1957) suggest that aromatic compounds (such as vanillin, vanillic acid and 4-ethylguaiacol) are produced on cooking wheat and make a significant contribution to the aroma and flavour of the finished sauce.

C. Mixing

The prepared beans and wheat are thoroughly mixed, so that each bean becomes coated with wheat, thus presenting a dry surface. This is regarded as very important since it permits mould growth, but discourages growth of bacteria. This is probably less important in modern facilities with their air filters and high standards of hygiene, but even so it is probably impossible to achieve much better than clinical standards of hygiene, and the provision of conditions tending to suppress bacterial growth will be of economic advantage.

D. Koji

To the western observer, conditioned to the idea that moulds (with the exception of those found in blue-veined cheese, on the skins of certain Continental sausage and on wine grapes undergoing the 'Noble Rot') are foul contaminants of foodstuffs, the Oriental use of moulds is amazing and a little alarming. Millennia before the modern discovery of mycotoxins, it seems that Oriental civilizations had learned to select and use moulds for production of foodstuffs (Groff, 1919). The species of *Aspergillus* employed (e.g. *A. oryzae*, *A. sojae*, *A. tamarii*) belong to the same group (Murakami, 1971), characterized by the absence of mycotoxins. It is difficult to see how this was achieved in the absence of any knowledge of chemistry, biochemistry or microbiology.

A great deal could be written about the production and use of mould inocula but, for the present account, it must suffice to note that specially produced spores (*tane-koji*) or a portion of koji from a satisfactory previous run are incorporated with the raw materials, and the whole mass is then incubated at 30°C. Traditionally, the mixture was spread as shallow layers in woven bamboo baskets or wooden trays (preferably with perforated bases to permit good aeration). Racks loaded with such trays were incubated in rooms held at 30°C under conditions of high humidity. Today, although the traditional practice is still followed in many factories, more modern methods are also found. These include mounting trays on conveyor belts, or using rotating horizontal cylinders partially filled with the raw material. Deep-tank methods, involving powerful forced aeration through a perforated false bottom, have also been tried (Aidoo, 1979).

Under these warm, moist and well-aerated conditions, germination of the mould spores is rapid, becoming evident about 20 hours after the commencement of incubation. Biochemically, one of the first changes to become evident is an increase in the amount of reducing sugars consequent on release of sucrase by germinating spores, and hydrolysis by this enzyme of sucrose present in the soybeans. Thereafter, mould growth is rapid and, even in shallow trays, the metabolic heat which is released may elevate the temperature in the koji to around 35°C. In deeper layers, the temperature may exceed 40°C (B.J.B. Wood, unpublished observations; Aidoo, 1979) and harm the mould.

If the koji is left undisturbed, the mycelium binds the whole into a solid cake, and in practice it is essential to stir at intervals. This mixing, in addition to keeping the koji fairly open to penetration of air, ensures a more even distribution of water through the koji, preventing excessive drying and consequent early sporulation in the material on the outside of the mass. It also secures a more uniform temperature through the mass, thereby preventing local over-heating. Traditionally, this stirring is accomplished by hand, but mechanical stirring can be used to good effect. Published information suggests that continuous stirring is undesirable, resulting in damage to the delicate mould mycelium. Accordingly, it is stated that horizontal cylinder fermenters should only be rotated occasionally, but we find that excellent and rapid mould growth can occur in koji contained in horizontal cylinders which are rotated or which are stirred by internal blades, in both cases operating continuously (Aidoo, 1979).

Incubation is complete after three days. During that time the fermenting mass gives off a mild, sweetish, musty odour but, if incubation is further prolonged, sharp, harsh odours develop, coincident with the onset of sporulation by the mould. A part of the art of koji fermentation is making the correct judgement of when to terminate the process, since stopping it too early will result in insufficient hydrolysis of the proteins and polysaccharides in the beans and wheat, whereas permitting sporulation to occur will result in the development of unacceptable mouldy flavours, giving rise to a poor-quality product. The dilemma is analogous to that of the maltster in Western brewing.

Indeed, the whole process of making koji bears comparison with malting, and it is no surprise to find that Japanese writing in English sometimes refer to the finished koji as 'malt' (for example see Togo, 1977). In both processes, warm, moist and well aerated conditions are provided to encourage even and rapid germination of the mould or barley, respectively. Development of the organism concerned results in the elaboration of numerous enzymes which hydrolyse various polymers present in the system. By timing the process carefully, maximum concentrations of the necessary enzymes are obtained. Subsequently, mixing the product of this first stage with water permits further hydrolysis of proteins and polysaccharides present in the material, so providing a rich medium for the growth of selected anaerobes.

A considerable number of enzymes are released by the mould. Those identified by various workers include sucrase, proteinases (acid, alkaline and neutral), lipase(s), phosphatase(s), cellulase(s), amylases and maltase. Some evidence obtained by Aidoo (1979) suggests that extracellular deaminases may be present in the last stages of the koji fermentation. Surprisingly, our efforts to demonstrate pectinase activity in koji have so far proved unrewarding, despite ample evidence for production of enzymes of this class by *Aspergillus* species including *A. sojae* (Fogarty and Ward, 1974), *A. saito* and *A. niger* (Rombouts and Pilnik, 1972). Although my group has not so far examined soy sauce moulds for hemicellulases, β-glucosidases (other than the cellulase complex) or α-galactosidases, their presence would be significant. Smiley *et al.* (1976) have reported production of α-galactosidase by *A. awamori* grown on moist wheat bran, and its subsequent use in a hollow-fibre reactor to remove galacto-oligosaccharides from soy milk. Similarly, information on the presence or otherwise of nucleases would be valuable because of the potential importance of the products of their hydrolytic activities in flavouring.

Among enzymes attacking oligomers of lower molecular weight, peptidases would be of particular significance in determining the quality of the finished soy sauce. For a more detailed study of the sequence of enzyme release and the resulting changes in levels of reducing sugars and protein hydrolysis products, the reader is referred to the papers by Yong and Wood (1974, 1976, 1977a,b), and Goel and Wood (1978) and to Aidoo's work (Aidoo, 1979). These workers have not examined the proteinase complex in any detail, and it would be interesting to know more about the specificity of the enzymes concerned, and if they are released concurrently or sequentially. The descriptions acid, neutral and alkaline proteases seem a little uninformative, but the review by Morihara (1974) makes it clear that fungal enzymes belonging to the categories characterized by the three pH optima exhibit distinct specificities. Sekine (1976) reports that *A. sojae* possesses two neutral proteases. These are type I, stable to 50°C with a molecular weight of 42,000 and active against casein; and type II, thermostable (100°C for 60 minutes) with a molecular weight of 16–18,000 and active against basic proteins, e.g. histone. *Aspergillus oryzae* only possessed the type I neutral protease. In general, neutral proteases are specific for hydrophobic amino-acid residues on the amino side of the bond being broken. Kundu and Manna (1975) report on formation by *A. oryzae* of both neutral and alkaline proteases.

Alkaline proteases have pH optima around pH 10. Their specificity is for aromatic or hydrophobic amino-acid residues located at the carboxyl side of the bond which is undergoing hydrolysis. Acid proteases are also commonly found among the enzymes liberated by *Aspergillus* species. Tsujita and Endo (1976) reported on the presence of acid protease activity in Takadiastase, a bran culture of *A. oryzae*. These enzymes have optima around pH 3 to 4 and are specific for aromatic or hydrophobic amino-acid residues at both sides of the bond being hydrolysed. The enzymes obtained from *Aspergillus* species are described

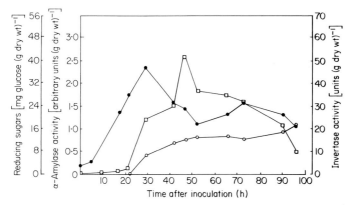

Fig. 1. Development of α-amylase (○) and sucrase (□) activities and changes in the contents of reducing sugar (●) in an experimental soy-sauce koji made with wheat flour and soybeans, and inoculated with spores of *Aspergillus oryzae* strain NRRL 1989 and incubated at 30°C. After Yong (1971).

as 'pepsin-like' (as opposed to 'rennin-like') in their activities. It seems reasonable to surmise that these acid proteases will be the chief source of proteolysis in the Moromi stage, and that the neutral proteases will dominate in the Koji stage of soy sauce and *miso* production. The significance of the alkaline proteases is less clear. Figures 1 to 5 summarize some of the changes in enzymes and metabolite levels found in our studies with a typical soy-sauce mould. It should be stressed,

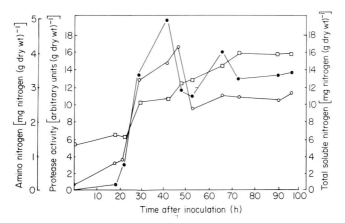

Fig. 2. Appearance of protease activity (○) and changes in contents of amino nitrogen (□) and total trichloroacetic acid-soluble nitrogenous compounds (●) in koji prepared under the conditions described in Figure 1. After Yong (1971).

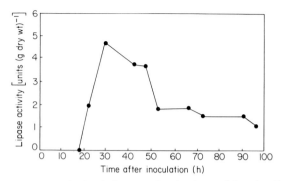

Fig. 3. Appearance of lipase activity in a koji prepared under the conditions described in Figure 1. After Yong (1971).

however, that moulds isolated from a variety of soy-sauce kojis will exhibit a range of concentrations of enzymes. There is unlikely to be a 'best' mould strain for all purposes since it would seem reasonable to postulate that different moulds would contribute to the differences observed in organoleptic qualities between soy sauces produced in different areas and by different manufacturers.

E. Moromi

The salt-mash fermentation is even more complex than the mould stage

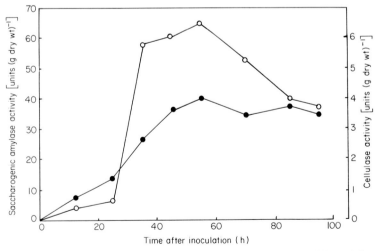

Fig. 4. Appearance of saccharogenic amylase activity (O) and cellulase activity (●) in a koji prepared under the conditions described in Figure 1. After Goel (1974).

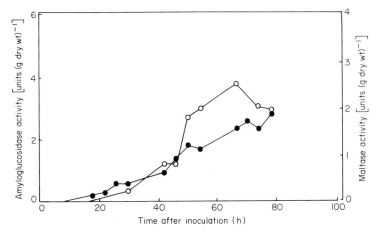

Fig. 5. Appearance of maltase (●) and amyloglucosidase (○) activities in a koji prepared under the conditions described in Figure 1. After Aidoo (1979).

of soy-sauce production. In essence, the finished koji is mixed with brine and then stored for a period of a few months to several years. During this time an initial period of active microbial growth and a subsequent period of maturation produce a subtly aromatic brew which is then decanted from the solid residues, clarified, pasteurized and finally packaged for distribution and sale. Traditionally, much importance has been attached to long fermentations and to exposing the mash to the seasonal cycles of temperature. Even today, manufacturers who stress the natural and traditional nature of their process wax lyrical on the changes in their fermentation in response to the weather. The scientific basis of much of this would be hard to establish, but in the economics of this industry much depends on the fact that not only a product is being marketed. The chemically assayed qualities of the product and its organoleptic worth are intimately integrated into the consumer's philosophical and even religious views.

Traditionally, the koji, sea-salt and water were mixed and the fermentation was allowed to proceed as it would, perhaps aided by the addition of a portion of a previous batch of moromi which had yielded a good fermentation. Under these conditions, many microbes developed in the salty mash, but it seems that two groups, namely yeasts and lactic acid-producing bacteria (lactobacilli and pediococci), are of paramount importance. The work of several groups (e.g. Yong and Wood, 1976; Yong *et al.*, 1978) has demonstrated that the yeasts found in soy sauce (typically *Saccharomyces rouxii*) can only grow in the very salty mash (18% sodium chloride) when the pH value has dropped to 5 or below. Rather remarkably, the pH value of a mixture of koji and brine prepared under aseptic conditions, and in which no bacteria can be detected, will drop from its initial value of about 6.4 to 4.4 over several days (Yong and Wood, 1976), and will then sustain vigorous growth of the soy yeast. In practice, however, lactic-acid bacteria (for example strains of *Lactobacillus delbruekii*) gain access to the newly prepared mash and rapidly ferment it, so giving the desired pH value; thereafter these bacteria die out. By analogy with obervations made in cheese (Piatkiewicz and Kasperkiewicz-Jamroz, 1976), it may be that on dying they release proteases, but attempts to demonstrate this in *miso* were not very rewarding, in part at least due to the complex nature of the system (Abiose, 1980; S.H. Abiose and A. Piatkiewicz, unpublished observations). In laboratory experiments, the lactobacilli attain between 10^6 and 10^7 organisms per ml at the peak of their growth (Goel, 1974; Yong and Wood, 1976) and it seems not unreasonable to suggest that products of their dissolution, such as proteins and nucleic acids, may contribute to the properties of the finished soy sauce. In experimental fermentations (Yong and Wood, 1977b), soy sauce of good quality was made from moromis acidified with lactic acid. However, it did not have quite such good organoleptic appeal as soy sauce made with lactobacilli although, as Goel's (1974) careful studies have demonstrated, it is rather difficult to obtain statistically significant results in taste-panel experiments. These results may suggest a role for lactic-acid bacteria other than simple acidification of the mash.

Once the correct pH value has been attained, yeast growth is vigorous and marked production of carbon dioxide indicates that an alcoholic fermentation is taking place. Typically, a 'fully brewed' soy sauce will contain between 1 and 2% (v/v) ethanol and now has spirit duty imposed on it in the United Kingdom. The yeast's role in production of soy sauce is complex, but its effect is clear; in experimental moromis, the typical soy sauce aroma only develops if yeast fermentation takes place.

Space, time and labour are increasingly expensive throughout the World, and it will be appreciated that the ancient ways of moromi fermentation are extravagant of all three. The necessary considerations of commercial confidentiality make it difficult to obtain reliable data, but anecdotal and hearsay evidence suggest that there is an increasing trend toward shortening the period of fermentation. This can be achieved by controlling temperature and using larger inocula of appropriate organisms. Replacement of part at least of the whole beans by defatted soybean meal seems to be fairly widely practised. In addition to giving more rapid extraction of the desired nutrients, use of this preparation, by lowering the amount of oil present in the mash, simplifies the subsequent pressing and clarification of the sauce. Another tempting prospect is supplementing the enzymes present in the koji with readily available, powerful and comparatively inexpensive purified enzymes which are now marketed. However, such evidence as is available seems to suggest that these are of somewhat limited utility in practice; for example, excessive proteinase activity can give rise to bitter off-flavours. The use of acid hydrolysates of soybeans and wheat flour to supplement or replace koji has already been adverted to.

Not surprisingly, given the nature of the trade, there has been a strong reaction in some areas to intrusion of these modifications to the traditional process but, if the increasing World market for reasonably priced soy sauce is to be satisfied, it seems reasonable to argue that an acceptable compromise between tradition and science must be found.

As to the biochemistry of the changes that occur during moromi fermentation, the nature of the traditional process makes it difficult to offer any generally applicable statements. Fermentations, carried out by members of my group at 40°C with controlled inocula of a selected yeast and bacterium, yielded the results shown in Figures 6, 7 and 8. It is surprising how little difference the presence or absence of microbes makes to the contents of reducing sugars and amino-nitrogen compounds. One possible explanation, activity of hydrolytic enzymes elaborated by yeasts and bacteria, does not seem to be substantiated by these results, and the need for further and more detailed research is demonstrated (Yong and Wood, 1977b).

In addition to the development of flavour and aroma, the colour of the moromi also deepens during fermentation and maturation. Yong (1971) found no evidence for tyrosinase action, and the colour is probably due to non-enzymic reactions.

Thus, at the end of the fermentation, the moromi possesses all the essential attributes of soy sauce: flavour, aroma, saltiness and colour. In addition, despite the abundance of reducing sugars and amino-nitrogen compounds present in the liquid, it seems to be fairly resistant to further microbial action, a product of its salinity, pH value and alcohol content.

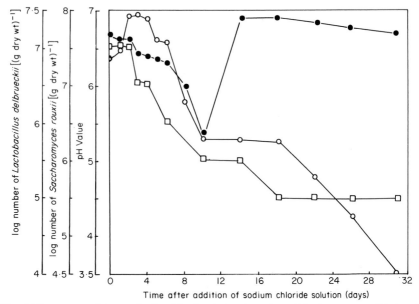

Fig. 6. Time-course of changes in the pH value (□) and numbers of *Saccharomyces rouxii* (○) and *Lactobacillus delbrueckii* (●) in an experimental soy-sauce moromi incubated with pure cultures of the two organisms previously trained to grow in the presence of 18% (w/v) sodium chloride. The moromi was incubated at 40°C. After Yong (1971).

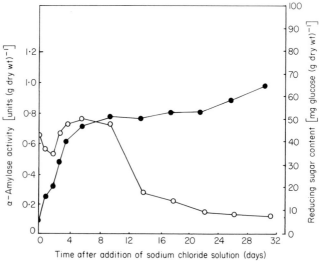

Fig. 7. Time-course of changes in the contents of reducing sugar (●) and α-amylase activity (○) in a moromi prepared under the conditions described in Figure 6. After Yong (1971).

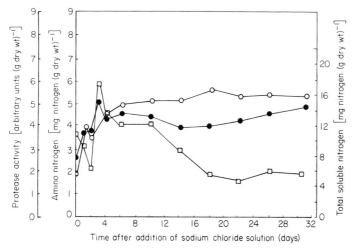

Fig. 8. Time-course of changes in the contents of total soluble nitrogen (●), amino nitrogen (○) and protease activity (□) in a moromi prepared under conditions described in Figure 6. After Yong (1971).

Preservatives may be added, however, in order to increase further the stability of the final soy sauce.

In traditional practice, the finest soy sauce is decanted from the fermentation, with the liquid extracted by pressing the residues being regarded as having inferior qualities. The solid residues have no culinary value and go for animal feed. The liquid can be clarified and freed of heat-coagulable protein by heating with kaolin or similar filter aid, then filtering off the solid material. The clarified liquid, after introduction of any desired additives (preservatives, colouring and molasses), is pasteurized and packaged.

Some Indonesian types of soy sauce are concentrated by evaporating some of the water.

III. OF BEANS, MICROBES AND MISO

A. Beans

The soybean (*Glycine max*) is an ancient food legume of the East. The ancient Chinese character for the soybean clearly shows the root nodules (Wolf and Cowan, 1975). Now, these structures are known to be the site of nitrogen fixation, but it is remarkable that these rather small underground nodules should be deemed worthy of inclusion by the people of perhaps 3,000 years ago.

The bean can only be grown economically in the warmer parts of the World, and production of fermented products from other types of bean is an attractive idea. Siegel and Fawcett (1976) show that the soybean accounted for 49% of World legume production in 1972, and this suggests that there is considerable scope for development of products analogous to soy sauce and *miso* from, for example, the field bean and other temperate-climate legumes. Unpublished work by my group suggests that the field ban makes acceptable sauce and 'Miso'. Robinson and Kao (1974) have reported successful substitution of soybeans by chickpeas and horse beans in *Tempeh*.

The raw soybean has a range of undesirable constituents. Siegel and Fawcett (1976) list trypsin inhibitors, haemagglutenins and saponins, to which Wolf and Cowan (1975) add evidence for other, as yet unidentified, anti-nutrients, and also point out that the sugars stachyose and raffinose which are normally present in beans 'have been implicated as causative factors in digestive disturbances such as flatus'. Furthermore, Siegel and Fawcett (1976) point out that protein from beans is somewhat difficult to digest. However, they go on to state that fermentation improves digestibility and increases the concentration of anti-oxidants. Moreover, the fermented products are free from the various anti-nutritional factors listed above, and may show increases in vitamin concentrations. In addition, they cite substantial evidence that various types of beans show increases in contents of ascorbic acid, tocopherol, B-group vitamins, choline and available iron during 2- to 4-day germination, although how much change might be expected during the 16 or so hours soaking to which beans are exposed in preparation for soy sauce and *miso* production, is not clear.

B. *Miso*

Miso is a paste. It is produced by growing selected strains of *Aspergillus* on grain. The grain is then blended with soaked, cooked soybeans and salt. After a fermentation, which involves yeasts and bacteria in much the same way as in soy sauce, the aromatic paste is used as a condiment and cooking aid. Although less well known in the Occident than is soy sauce, it is probably as important in the Orient as soy sauce, again in a bewildering array of names and local varietal differences. The work of Hesseltine's group has done much to introduce this interesting fermentation to the West, and we have recently completed a programme of biochemical and microbiological work which parallels our studies on soy sauce (Abiose, 1980).

Among the grains used in *miso* production, rice is easily the most important, but barley also accounts for a substantial amount of production. A further variety is made from soybeans alone. Furthermore, each of these groups is divided into a number of types depending on the ratio of beans to grain, amount of salt added, and duration of the moromi (salt mash) fermentation. Thus there is a range from light, sweet *misos* high in rice, low in salt and fermented for only a short period, to dark salty, richly flavoured beans-only *misos* fermented for two years or so.

Traditionally, *misos* were sold 'live', that is to say they contained viable yeasts and detectable enzyme activity, and such *misos* are readily available from specialist food suppliers even in the Occident. However, there is a preference for conveniently packeted *misos*, and it was soon found that sealed plastic pouches had a disconcerting tendency to swell and burst when fill with live *miso*. In consequence, the manufacturers and wholesalers now offer convenient plastic pouches of pureed and pasteurized *miso*.

The mould used in *miso* production, although drawn from the same *Aspergillus oryzae* group as those used for soy sauce, is richer in amylases and poorer in proteinase activity than its soy sauce counterpart. This is a little surprising despite the fact that it is initially grown on grain, since clearly hydrolysis of proteins from the beans and grain seeds will be just as important for full flavour production in both cases.

Typically, mould growth is allowed to continue for only 50 or so hours, rather than the 72 hours traditionally used for soy sauce. The grain, having previously been cleaned, soaked overnight, cooked and cooled, is inoculated with mould spores (*Tane-Koji*), spread into trays or baskets, and incubated in a warm humid room (30°C). Mould growth is rapid and the koji must be turned and mixed a couple of times during incubation. Meanwhile, the beans are soaked and cooked as usual. Moulded grain, beans and salt are then blended in appropriate proportions. In laboratory studies, pure cultures of yeasts and bacteria are also added, but in the more traditional of the commercial producers, inoculation by the old method of adding a portion of a sound previous run is still considered to be the better procedure. Traditionally, sequential incubation of the salty mash at different temperatures was employed, and laboratory studies seem to show that this more laborious procedure gives more abundant microbial growth and a better-flavoured product. The fermentation is rather slower than that found in soy-sauce moromi, doubtless reflecting a lower water availability in the paste than in liquid preparations. Even so, it is complete in a few weeks in the lightest *miso*.

About 25 g of *miso* is consumed per head per day in Japan, roughly comparable with *shoyu* utilization. The more concentrated nature of *miso* means that this represents a significant intake of calories and protein, unlike soy sauce, where even Japanese consumption rates will contribute but tiny amounts of these dietary factors. Since the whole of the paste is consumed, there are no solid wastes for disposal. The liquid which exudes from the paste is sometimes collected separately. One school of thought holds that this is true 'Tamari' and it has even been regarded as ancestral to soy sauce, although others consider soy sauce to be the older fermentation.

C. Microbes

1. Koji

Some discussion of the *Aspergillus* species found in soy sauce is contained in Wood (1977). The moulds belong to the *flavus-oryzae* subdivision of the genus. Various specific names have been applied, e.g. *A. oryzae*, *A. sojae* and *A. tamarii*. How real these distinctions are is not clear, and I suspect that they merge into each other fairly imperceptably. From time to time there are suggestions that bacteria contribute to koji. Specifically, Sakaguchi (1959b) has suggested that *Bacillus subtilis* in moderate amounts prevents development of a turbidity in the finished soy sauce and also helps in production of the desired aroma.

An interesting recent development is the report by Hayashi *et al.* (1977) that adding a small amount of acetic acid to the koji (preferably 0.2 to 0.8% by weight, based on the water content of the koji) is beneficial in that it inhibits growth of bacterial contaminants. The inhibition seemed to be a specific attribute of acetic acid among the various organic and inorganic acids which they tested. Within the specified range of concentration, the acid did not have any adverse effect on mould growth or mould protease production, although higher concentrations were inhibitory. In koji that had been experimentally inoculated with typical bacterial contaminants, mould growth and protease formation were better in the presence of acetic acid than in its absence.

2. Moromi Bacteria

Yong and Wood (1976) followed numerous earlier investigators in selecting *Lactobacillus delbrueckii* for their study of the Moromi stage of soy sauce production. However, Sakaguchi (1959a,b) has reported that strains of a *Pediococcus* species, described as *P. soyae*, is present in soy sauce. Like *L. delbrueckii* they grew well in salt concentrations up to, and even in excess of, 18% (w/v). They were homofermentative, producing only lactic acid. On transfer to media containing a higher concentration of salt than that in which they had previously been growing, they exhibited a lag before resuming growth, interpreted as representing a need for physiological adaptation to the higher concentration of salt. Again, this behaviour exactly resembles that exhibited by lactobacilli in soy sauce, as does the inability to grow in salt-containing media at pH values below 5.0. The reported temperature optimum of 20–30°C is, however, rather below that of lactobacilli. The demonstration by Sakaguchi (1959c,d) that nutritional factors derived from koji are essential for maximal growth, confirms my own unpublished observations on lactobacilli. Sakaguchi (1960a,b) has also reported the requirements by *Pediococcus* spp. for vitamins, amino acids, glycylbetaine and carnitine, as essential or stimulatory media constituents. Sakasai *et al.* (1975a,b) have further studied the effects of various environmental factors on the growth of *Pediococcus soyae*.

A particularly interesting series of papers by Noda *et al.* (1976a,b,c,d) deals with the effects of vanillin and its isomers and of guaiacol on the growth of pediococci and soy sauce yeasts. Apparently, they can exhibit antibiotic activity towards bacteria but can be either stimulatory or inhibitory to yeasts. These observations are significant since the compounds specified will be formed during roasting of the wheat used in the fermentation (Asao and Yokotsuka, 1957).

Bacillus subtilis has been reported as occurring in moromi (Sakaguchi,

1959a,b) and is considered as beneficial by some workers (Sakaguchi, 1959b), particularly in view of its ability to form proteases (Kamekura and Onishi, 1974). Abiose (1980) found *Bacillus* spp. in some samples of *miso*.

3. Moromi Yeasts

Again, numerous species of yeasts have been isolated from moromi (Yong and Wood, 1974). In our own experimental work, we have confined our attention to *Sacch. rouxii*, but it can be argued that a consortium of yeasts is desirable for producing the best quality soy sauce, exactly as the same case may be made for wine brewing. A complicating factor in attempting to discuss the species of yeasts found in moromi arises from the use by different workers of the names *Sacch. rouxii* and *Zygosaccharomyces major* seemingly to refer to the same organism. Other yeasts reported as present in soy sauce include *Zygosacch. soja*, *Zygosacch. sulsus*, *Zygosacch. japonicus* and species of *Torulopsis* and *Monilia* (Onishi and Suzuki, 1970). Hesseltine (1965) used *Hansenula subpelliculosa* in production of soy sauce.

In general, yeast growth is essential to production of soy sauce's characteristic flavour and aroma as noted earlier, where it was also pointed out that the desirable yeast species can only grow in the salty mash when the acidity has been increased as the result of bacterial formation of lactic acid. Yokotsuka (1960) found that the yeasts were able to grow at higher temperatures (up to 40°C) in saline media than they would tolerate in normal low-salt media. He also observed that glycerol formation is enhanced in the presence of high concentrations of salt. If the yeast to be used in fermentation of soy sauce has been maintained on normal laboratory media appropriate for yeasts, it is necessary to 'train' the organisms by passaging through media of progressively higher concentrations of salt. At each step there is a marked lag before normal growth is resumed, suggesting that physiological adaptation is taking place. I have also observed that growth is more sparse at higher concentrations of salt if normal yeast media with added salt are used. If, however, koji extracts are used, growth seems to be profuse as assessed by the production of carbon dioxide. This suggests that, at the higher concentrations of salt, the organisms are more demanding in their nutritional requirements. In their interesting study of the relationship between pH value, salt concentration and yeast growth, Yong *et al.* (1978) demonstrated that yeast grows significantly better in media containing a high concentration of salt acidified with lactic acid than it does in otherwise similar media where hydrochloric acid was used for acidification. The physiological and nutritional relationships between the yeast and its medium are obviously complex.

Not all yeasts are desirable in soy-sauce fermentations. Some species of *Torulopsis* form pellicles or rings on the surface of the sauce, which is not acceptable in a commercial product (Onishi and Suzuki, 1970). On the other hand, Yokotsuka *et al.* (1967a,b) found that 4-ethylphenol, 4-ethylguaiacol and 2-phenylethanol, all important components of soy-sauce aroma, were only formed by *Torulopsis* spp.

Many of these microbiological matters are more fully discussed by Shibani (1978) on whose review I have drawn extensively.

4. The Microbiology of Miso

In general, microbes associated with miso are very similar to those found in shoyu. As has already been noted, the miso producer seems to prefer strains of *A. oryzae* which are more profuse amylase producers than those used for shoyu. In her review, Abiose (1980), in stating that *Sacch. rouxii* is again the dominant yeast, points out that *Sacch. soya*, *Zygosacch. soya*, *Zygosacch. japonicus* and *Zygopichia japonica*, all of which have been isolated from miso, are but synonyms for *Sacch. rouxii*. Matsumoto and Imai (1977a,b) isolated *Hansenula subpelliculosa*, *Torulopsis versatilis* and *T. etchellsii* from various *misos*, apparently in association with *Sacch. rouxii*. Abiose (1980) notes that many lactic acid-producing bacteria are said to develop in *miso*, although it is not certain that they are essential for the fermentation. Numerous aerobic bacteria have also been observed, notably *Bacillus* species, although their role (if any) is equally obscure.

IV. TRADE IN SOY SAUCE

A. Introduction

In general, it has proved remarkably difficult to obtain statistics. I am very grateful to the Japanese Embassy for statistics from the Japanese Ministry of Agriculture; to the United States Embassy for American

Customs information, to the Food and Agriculture Organisation (F.A.O.) of the United Nations Organisation for official F.A.O. trade returns, to the Kikkoman Shoyu Company of Japan and Amoy Sauce Company of Hong Kong for statistics relating to their two countries, and also in the Kikkoman Company's case for information concerning their own operations in Japan and the U.S.A.

Most of the figures relate to exports and imports between countries, and only Japan furnished data on total national production. It is certain that other countries, for example Indonesia, produce soy sauce-like condiments, and it seems reasonable to assume that there will be some sort of export trade in these commodities also. The Japanese figures suggest that exports are only a tiny part of the total soy-sauce production, and it must also be remembered that soy sauce is widely made on a domestic scale for private consumption. These considerations strongly suggest that it is impossible to produce any sort of reliable estimate of World soy-sauce consumption, and any figures claiming to represent the 'World trade' in soy sauce should be approached with considerable reserve, to say the least. It would, however, seem reasonable to conclude that the total trade in this commodity is very substantial and, as will be apparent subsequently, is exhibiting steady growth in real terms, with the increasing European and American markets making a significant contribution to this growth.

Inevitably perhaps, the greatest enigma is the very home of soy sauce, mainland China. It will be seen later that such import figures as are available show that the Peoples' Republic of China has a considerable export market. Inspection of the labels on bottles in specialist food shops in the United Kingdom suggests that there is a useful Chinese export trade here and presumably also to the rest of Europe.

Soy-sauce imports to Europe attract customs duty at the full rate of 18% *ad valorem*, but there is provision for a decrease or exemption from duty in the case of importations from certain countries that are not members of the European Economic Community but which enjoy preferential arrangements with the community. *Miso* is treated similarly, with the rate of tax being 20% *ad valorem*. In the United Kingdom, there is a further excise duty based on the concentration of ethanol in the products; currently this is levied at the rate of £10.47 per litre of ethanol. I am indebted to H.M. Customs and Excise Statistical Office, Bill of Entry Section and the International Customs Branch, for this information. This is the position in one country; other provisions will apply elsewhere, and trading in soy sauce and *miso* is evidently not a simple matter. It is not clear to me what happens in countries where dietary laws prohibit use of alcohol, yet which import substantial amounts of soy sauce. Obviously no-one is going to be aware of any effects from the small amounts (1–2%) of alcohol in the products, since their strong flavours and high contents of salt impose constraints on the amounts that can be accepted in foods, but none-the-less they do, for example, attract spirit duty in the United Kingdom, so that, to the official mind at least, this is not an insignificant matter.

B. Statistics

The statistics I have succeeded in obtaining are collated in Tables 1 to 8. Results have all been converted into metric tonnes. Returns quoted in litres have been converted to weight by assuming a specific gravity of 1.2. This is rather arbitrary, but is a reasonable average. Rapid fluctuations in the relative values of currencies create special problems in trying to compare the value of the trade in various countries. The data for 1978 have been converted into pounds sterling by using exchange rates current at the time of writing (December 1979 to January 1980) but the results from earlier years have been left in their original form.

In general, the correlation between one country's export figures and another's import figures is quite good; differences can be accounted for by different recording procedures, the use of various units and the inevitable minor errors which occur in making returns. Japanese statistics are particularly detailed, showing amounts of soy sauce exported in cans and bottles (in litres) and 'other soy (sauce)' or 'soy (sauce) n.e.s.' (in kilograms). For only two countries, the Korean Republic and Japan, were figures on soy paste (*miso*) available to me.

Table 1 cumulates detailed figures of exports of soy sauce from Hong Kong, the Korean Republic, Singapore and Japan to a total of 97 countries in 1978, grouping the countries of destination into geographical regions and employing the names and groupings on the original lists. Table 2 lists the total export figures of those four countries for 1976, 1977 and 1978. Table 3 gives Japanese export figures for the same three years broken down by the types of containers used. Table 4 gives the total soy sauce and *miso* production by Japan over a 13 year period. Table 5

Table 1

Exports of soy sauce 1978 (tonnes)

		Export volume (in tonnes) from				
Population (millions)	Destination	Hong Kong	Korean Republic	Singapore	Japan	Total
1. *North America*						
23.5	Canada	683.4	40.0	2.4	117.3	843.1
251.9	U.S.A.	2,750.7	281.7	11.6	2,165.4	5 209.4
						6 052.5
2. *South and Central America: West Atlantic*						
23.4	Argentina	—	—	—	32.8	32.8
4.7	Bolivia	—	—	—	13.0	13.0
108.0	Brazil	—	—	—	3.1	3.1
11.0	Chile	—	—	—	25.3	25.3
2.1	Costa Rica	—	—	—	2.9	2.9
6.5	Ecuador	—	1.2	—	—	1.2
4.5	El Salvador	—	—	—	13.0	13.0
6.8	Guatamala	10.0	—	—	12.9	22.9
0.7	Guyana	0.5	—	—	1.7	2.2
3.1	Honduras	2.7	—	—	0.4	3.1
61.0	Mexico	22.2	—	—	31.5	53.7
2.4	Nicaragua	1.5	—	—	0.4	1.9
1.8	Panama	0.8	—	—	—	0.8
2.5	Paraguay	—	—	—	0.7	0.7
0.4	Surinam	—	—	—	2.7	2.7
13.5	Venezuela	121.6	—	—	15.0	136.6
						315.9
0.1	Grenada	0.7	—	—	—	0.7
2.1	Jamaica	18.7	—	—	—	18.7
2.0	Trinidad and Tobago	363.0	—	—	—	363.0
						382.4
	Former Dutch West Indies	348.1	—	—	—	348.1
3. *Europe*						
7.5	Austria	12.9	—	—	5.3	18.2
9.8	Belgium	1.8	—	—	160.2	162.0
15.1	Czechoslovakia	—	—	—	7.7	7.7
5.1	Denmark	2.8	—	2.0	57.7	62.5
4.8	Finland	—	—	9.5	25.0	34.5
53.2	France	42.7	13.2	11.9	161.4	229.2
61.3	German Federal Republic	61.0	13.4	15.7	346.6	436.7
8.8	Greece	—	—	—	0.8	0.8
56.0	Italy	3.2	—	—	59.1	62.3
13.9	Netherlands	202.5	6.9	21.8	242.1	473.3
4.1	Norway	8.2	—	—	16.8	25.0
9.5	Portugal	7.4	—	—	8.9	16.3

Table 1 (*cont.*)

Population (millions)	Destination	Export volume (in tonnes) from				
		Hong Kong	Korean Republic	Singapore	Japan	Total
36.2	Spain	14.1	23.4	4.8	40.4	82.7
8.3	Sweden	35.5	—	0.3	141.7	177.5
6.3	Switzerland	15.3	—	0.01	35.2	50.5
55.5	U.K.	703.0	—	271.8	173.5	1 148.3
—	U.S.S.R.	—	—	—	30.2	30.2
						3 017.7
4. *Near and Middle East*						
0.2	Bahrain	1.5	94.9	4.9	—	101.3
38.0	Egypt	—	—	—	12.7	12.7
606.2	India	—	—	0.5	0.8	1.3
28.4	Iran	—	14.9	—	113.0	127.9
12.4	Iraq	—	16.8	—	19.6	36.4
2.8	Jordan	—	11.9	0.8	—	12.7
1.1	Kuwait	1.6	52.5	3.4	16.5	74.0
0.8	Oman	—	—	0.6	—	0.6
0.2	Qatar	9.3	4.8	—	1.5	15.6
9.2	Saudi Arabia	28.2	609.0	6.5	149.7	793.4
0.9	United Arab Emirates	1.9	5.9	1.4	5.0	14.2
7.0	Yemen Arab Republic	1.0	1.4	—	1.0	3.4
						1 193.5
5. *Far East and Western Pacific*						
0.2	Brunei	8.1	—	177.3	—	185.4
4.7	Hong Kong	—	—	1.0	327.2	328.2
129.0	Indonesia	106.0	—	—	36.5	142.5
111.9	Japan	2.3	6.0	—	—	8.3
36.4	Korean Republic	—	—	—	217.9	217.9
0.2	Macao	9.5	—	—	—	9.5
12.9	Malaysia	12.3	—	354.1	52.2	418.6
42.8	Philippines	0.2	—	—	12.6	12.8
*	Sabah	7.9	—	332.9	98.7	439.5
*	Sarawak	12.0	—	110.5	—	122.5
2.3	Singapore	19.5	—	—	168.5	188.0
17.2	Taiwan	—	—	—	16.8	16.8
42.0	Thailand	—	—	—	49.4	49.4
						2 139.4
6. *Pacific and Australasia*						
14.3	Australia	222.3	5.5	196.2	400.0	824.0
0.02	Cook Islands	5.6	—	—	—	5.6
0.003	Christmas Islands	—	—	5.5	—	5.5
0.6	Fiji	20.4	—	3.0	14.1	37.5
0.1	Guam	—	—	—	372.2	372.2
0.007	Nauru	0.5	—	—	9.2	9.7
0.1	New Caledonia	—	—	—	68.7	68.7
0.1	New Hebrides	—	—	—	6.8	6.8
3.1	New Zealand	46.0	—	1.3	8.1	55.4

(*Included in Malaysia)

Table 1 (*cont.*)

Population (millions)	Destination	Export volume (in tonnes) from				
		Hong Kong	Korean Republic	Singapore	Japan	Total
—	Oceania n.e.s.	133.4	—	—	—	133.4
2.8	Papua New Guinea	4.5	—	1.0	22.1	27.6
0.6	Portuguese Timor	—	—	2.5	—	2.5
1.2	Samoa and Tonga	6.1	6.9	2.1	60.4	75.5
0.2	Solomon Islands	0.1	—	—	6.4	6.5
0.007	Tuvalu (Ellicels)	—	—	—	9.2	9.2
—	U.S. Oceania	7.4	—	—	—	7.4
						1 647.5
7. *Africa*						
—	Africa (not elsewhere specified)	—	—	39.8	—	39.8
18.2	Algeria	—	—	—	37.4	37.4
1.2	Canary Islands	—	15.5	—	17.2	32.7
29.4	Ethiopia	1.8	—	—	—	1.8
0.05	Gambia	0.4	—	—	—	0.4
8.5	Ghana	1.4	—	—	1.7	3.1
12.9	Kenya	0.4	4.5	—	5.8	10.7
2.9	Libya	—	3.4	—	—	3.4
8.0	Malagasy	16.2	—	—	—	16.2
5.6	Malawi	1.1	—	—	—	1.1
0.9	Mauritius	1.3	—	64.6	—	65.9
79.8	Nigeria	5.1	—	1.8	8.5	15.4
23.9	South Africa Republic	58.8	—	—	25.8	84.6
19.5	Sudan	2.2	—	—	—	2.2
0.5	Réunion Islands	—	—	45.7	2.2	47.9
16.0	Tanzania	2.0	—	—	—	2.0
21.6	Zaire	—	—	—	1.1	1.1
						365.7
	Totals	6 192.8	1 233.5	1 713.6	6 591.6	15 731.5
	Value	HK$ 17,249,393	US$ 953,508	S$ 1,784,000	Yen (000) 1,255,987	
	Value in pounds sterling[a]	1,547,031	421,906	365,574	2,369,786	4,704,297
	Value in pounds sterling per tonne	249.0	342.0	213.3	359.3	

[a] The following exchange rates (per pound sterling) were used: Hong Kong $, 11.15; Singapore $, 4.88; U.S.A. $, 2.26; Japanese yen, 530.

Table 2

Soy sauce exports by four major producing countries for 1976 to 1978

	Hong Kong	Korean Republic	Singapore	Japan
1978				
Tonnes	6,192.8	1,233.5	1,713.6	6,591.7
Value[a]	17,249.4	953.5	1,784.0	1,256.0
1977				
Tonnes	5,583.6	681.8	1,640.0	7,879.0
Value[a]	15,452.8	343.7	1,639.8	1,542.1
1976				
Tonnes	5,056.5	350.3	1,159.1	7,184.2
Value[a]	13,633.7	151.2	1,121.0	1,406.1
Re-Exports				
1978				
Tonnes	2,945.3	N.A.	109.5[b]	N.A.
Value[a]	7,515.5		88.6	
1977				
Tonnes	2,234.3		533.2	
Value[a]	5,147.3		418.7	
1976				
Tonnes	1,401.0		814.0	
Value[a]	2,762.0		568.3	

N.A. indicates not available

[a] In local currency except Korean Republic ('000 U.S.A. $). Units '000 HK$; '000 S$; '000,000 Yen.

[b] Obtained by subtracting 'Domestic Exports' from 'Total Exports'. Other figures all relate to 'Domestic Exports'.

Table 3

Total soy sauce exports from Japan

Type	Unit	Amount	Value in Yen ($\times 10^{-3}$)
1976			
Bottles	l	1,880,064	601,528
Cans	l	2,721,528	480,805
Other	kg	1,662,277	323,752
Totals	kg	7,184,187	1,406,085
1977			
Bottles	l	2,150,775	682,737
Cans	l	3,255,488	564,504
Other	kg	1,391,458	294,894
Totals	kg	7,878,974	1,542,135
1978			
Bottles	l	2,011,595	633,720
Cans	l	3,121,906	531,561
Other	kg	431,408	90,706
Totals	kg	6,591,609	1,255,987
			(£2,365,324)[a]

[a] Calculated as described in the footnote to Table 1.

Table 4

Soy sauce and *miso* production in Japan over a 13-year period

	Production in tonnes of	
Year	Soy sauce[a]	*Miso*
1965	1,235,905	492,650
1970	1,345,903	552,207
1975	1,354,800	561,000
1976	1,472,400	573,000
1977	1,380,699	572,460
1978	1,434,000	567,000

[a] These values are calculated from kilolitres by multiplying by 1.2

Table 5

Soy sauce and *miso* production and export in Japan for the years 1976–1978

	Soy sauce				Miso				Miso (as percentage of soy sauce)		
		Exports				Exports				Exports	
Year	Production (tonnes)	Quantity (tonnes)	Value (Yen·10^{-3})	Exports (percentage of total production)	Production (tonnes)	Quantity (tonnes)	Value (Yen·10^{-3})	Exports (percentage of total production)	Production	Amount	Value
1976	1,472,400	7,184	1,406,085	0.49	573,000	865	221,510	0.15	38.9	12.0	15.75
1977	1,380,699	7,879	1,542,135	0.57	572,460	1,012	260,314	0.18	41.5	12.8	16.9
1978	1,434,000	6,592	1,255,987	0.46	567,000	1,081	278,121	0.19	39.5	16.4	22.1

Table 6

Imports of soy sauce

Origin	Import (tonnes) into			
	Hong Kong	Singapore	United States of America	
1978				
Taiwan	2.0	—	228.9	
Indonesia	0.6	—	—	
Japan	288.8	133.5	2,280.3	
China Peoples Republic	13,588.6	841.2	837.7	(total 15,237.5)
Malaysia	10·4	35·5	—	
Hong Kong	—	67·3	2,256.8	
Korean Republic	—	1.8	191.1	
Thailand	—	0.5	—	
United Kingdom	—	1.7	—	
United States of America	—	0.05	—	
Mexico	—	—	138.3	
Others	—	—	99.0	
Total	13,890.4	1,576.0	6,040.1	
Value	24.8×10^6 Hong Kong dollars	1.22×10^6 Singapore dollars	—	
Value (£ sterling)	2.22×10^6	0.25×10^6	—	
1977				
Total	12,028.4	1,657.4	5,224.9	
(China Peoples Republic	11,573.3	635.9	498.8)	
Value	$20.1 \cdot 10^6$ Hong Kong dollars	$1.20 \cdot 10^6$ Singapore dollars	—	
1976				
Total	10,104.7	1,919.4	4,791.9	
(China Peoples Republic	9,872.4	945.6	523.8)	
Value	$13.9 \cdot 10^6$ Hong Kong dollars	$1.48 \cdot 10^6$ Singapore dollars	—	

Table 7

Re-exports of soy sauce from Hong Kong and Singapore in 1978

Destination	Re-export (tonnes) from	
	Hong Kong	Singapore
1. *North America*		
Canada	828.6	0
U.S.A.	858.5	0
2. *South and Central America; West Atlantic*		
Costa Rica	2.4	0
Dominican Republic	9.0	0
Guatemala	6.3	0
Honduras	13.6	0
Mexico	34.4	0
Nicaragua	0.2	0
Panama	101.2	0
Venezuela	31.1	0
Trinidad and Tobago	62.5	0
Former Dutch West Indies	35.9	0
3. *Europe*		
Belgium	25.8	0
Denmark	12.0	0
France	89.2	0
German Federal Republic	2.2	0
Netherlands	44.5	0
Sweden	1.8	0
United Kingdom	229.9	0
4. *Near and Middle East*		
India	0	0.5
Pakistan	2.3	0
Saudi Arabia	4.2	0
United Arab Emirates	0	0.3
5. *Far East and Western Pacific*		
Brunei	22.8	7.1
Indonesia	177.0	0
Japan	178.6	0
Malaysia	54.2	69.1
Philippines	44.3	0
Sabah	65.3	5.2
Sarawak	0	4.3
Singapore	291.0	0
Sri Lanka	0	0.1
Thailand	0	0.3

BRIAN J. B. WOOD

Table 7 (*cont.*)

Destination	Re-export (tonnes) from	
	Hong Kong	Singapore
6. *Pacific and Australasia*		
Australia	783.1	11.0
Fiji	27.2	0
Nauru	10.5	0
New Zealand	13.7	0
Oceanea n.e.s.	52.4	0
Oceanea (U.S.)	12.9	0
Papua New Guinea	19.4	0.2
Samoa and Tonga	15.2	0
Solomon Islands	2.0	0
7. *Africa*		
Ghana	5.2	0
Malagasy Republic	41.4	0
Nigeria	13.0	0
South Africa Republic	39.5	0
Togo	2.3	0
Total	2,945.3	109.5
Value	$7.52 \cdot 10^6$ Hong Kong dollars	$0.09 \cdot 10^6$ Singapore dollars
Value in pounds sterling[a]	674,439	18,443

[a] Calculated as described in the footnote to Table 1.

brings together Japanese figures from previous tables, comparing production and export of soy sauce and miso for 1976, 1977 and 1978. Table 6 lists imports of soy sauce into Hong Kong, Singapore and the U.S.A. from some 12 countries. Table 7 shows re-exports of soy sauce by Hong Kong and Singapore in 1978 to 47 countries, again grouped by geographical regions. Table 8 gives details of soy paste exports by Japan and the Korean Republic in 1978. Table 9 gives details of total exports of soy paste by these two countries in 1976, 1977 and 1978.

In examining data in these tables, it is essential to bear in mind that they represent only a part of the World trade in soy sauce and soy paste. It is quite certain that these products (and other similar ones) will be extensively traded between all of the countries of South East Asia and the Western Pacific, but only a few countries present the kind of

Table 8

Exports of soy paste (*miso*) by the Korean Republic and Japan in 1978

Destination	Export (tonnes) by	
	Korean Republic	Japan
1. *North America*		
Canada	1.5	42.4
U.S.A.	72.4	550.9
Total	667.2	
2. *Central and South America*		
Brazil	—	0.4
Chile	—	3.1
Guyana	—	0.5
Mexico	—	4.0
Surinam	—	3.1
Venezuela	—	1.0
Total	12.1	
3. *Europe*		
Austria	—	1.9
Belgium	—	31.8
Denmark	—	0.5
France	3.0	25.6
German Federal Republic	3.6	19.6
Italy	—	10.8
Netherlands	1.3	36.7
Portugal	—	6.9
Spain	19.9	2.3
Sweden	—	0.4
Switzerland	—	1.6
United Kingdom	—	57.1
U.S.S.R.	—	1.0
Total	224.0	
4. *Near and Middle East*		
Bahrain	68.4	—
Bangladesh	0.8	—
Egypt	—	8.6
Iran	2.7	26.2
Iraq	12.6	16.6
Jordan	8.8	—
Kuwait	21.2	6.6
Qatar	21.7	0.8
Saudi Arabia	336.3	16.6
United Arab Emirates	6.8	5.3
Yemen Arab Republic	—	1.0
Total	561.3	

Table 8 (*cont.*)

Destination	Export (tonnes) by	
	Korean Republic	Japan
5. *Far East and Western Pacific*		
China	—	0.1
Hong Kong	—	22.6
Korea Republic	—	10.3
Malaysia	—	1.3
Philippines	—	2.5
Sabah	—	1.2
Singapore	—	66.5
Total	104.5	
6. *Pacific and Australasia*		
Australia	—	25.7
Fiji	—	0.6
Guam	—	13.5
Indonesia	—	14.2
New Hebrides	—	0.5
Papua New Guinea	—	1.3
Samoa	10.3	—
Solomon Islands	—	1.0
Total	67.1	
7. *Africa*		
Algeria	—	16.5
Canary Islands	11.1	6.0
Ghana	—	1.6
Kenya	—	2.0
Libya	14.4	0.5
Mozambique	—	0.2
South Africa Republic	—	8.3
Zaire	—	0.6
Total	61.2	

statistical data needed for the present survey. The largest problem is China; as usual no trade figures are directly available, but the remarkable amount of soy sauce which Hong Kong imports from China (Table 6) hints at an extensive Chinese trade in this product. Inspection of the shelves in specialist food shops in the United Kingdom suggests a substantial annual importation of soy sauce from China, but unfortunately H.M. Customs and Excise, despite levying spirit duty on the product, has no separate record of United Kingdom trade in it.

Table 9

Exports of soy paste (*miso*) by the Korean Republic and Japan in 1976–1978

	Exports (tonnes) from	
	Korean Republic	Japan
1978		
Quantity	624.1	1,081
Value	$749.6 \cdot 10^3$	$278.1 \cdot 10^6$
	U.S. dollars	Yen
Value in pounds sterling[a]	331,681	524,717
1977		
Quantity	358.4	1,012
Value	$296.6 \cdot 10^3$	$260.3 \cdot 10^6$
	U.S. dollars	Yen
1976		
Quantity	190.7	865
Value	$117.8 \cdot 10^3$	$221.5 \cdot 10^6$
	U.S. dollars	Yen

[a] Calculated as described in the footnote to Table 1.

Again, it must be stressed that these data relate to trade between countries, and that only in Japan's case are there figures for the total annual production. In 1978, Japan produced 1,434,000 tonnes of soy sauce and 567,000 tonnes of *miso*, with less than 0.5% going for export; this corresponds to an annual production of 12.81 kg of *shoyu* and 5.07 kg of *miso* per head of population, a daily consumption of 35.1 g of *shoyu* and 13.90 g of *miso*. Although no figures are available, hearsay suggests that such rates of consumption are prevalent in many parts of South East Asia. If, for simplicity, we take the total population of the countries listed under Section 5 (Far East and Western Pacific) of Table 1, ignoring the numerous countries (e.g. Viet Nam, Laos, Cambodia) not listed therein, we arrive at a total population of 399.6 million and hence an estimated annual soy sauce consumption of $5 \cdot 10^6$ tonnes in that region. These are truly formidable figures and lend weight to the view that fermented soy products come third after alcohol and milk products in a World 'league table' of fermented foods.

The U.S.A. imports of soy sauce from the four countries listed in Table 1 constitute 86% of the total imports (Table 6). In addition, the

Kikkoman Foods Inc. Wisconsin plant produced 18 000 kl (about
21 600 tonnes) in 1978, i.e. 3.58 times the total U.S.A. imports,
suggesting a total North American annual trade of around 28,000
tonnes soy sauce.

Total and *per capita* consumption of soy sauce in South and Central
America and the West Atlantic seems to be much lower on average than
in North America. However, the inadequacy of the available data is
illustrated by comparing the listed Mexican imports of 53.7 tonnes in
1978, with its exports of 138.3 tonnes to the U.S.A. in the same year
(Table 6). Venezuela is remarkable among the South American
countries in importing about 10 tonnes per million population, over
one-third of the entire region's imports, which makes an interesting
comparison with more usual regional rates of 2–3 tonnes per million
persons per year. Grenada and Jamaica import at rates comparable to
Venezuela, but Trinidad and Tobago import a remarkable 181 tonnes
per million people per year, which would represent five times the
Canadian consumption or nearly twice the U.S.A. value, and suggests
either a heavy Oriental influence on the diet or a substantial re-export
trade.

Among the European nations, imports of about 4–5 tonnes per
million per year seem to be common, and Czechoslovakia (0.5 tonnes
per million per year) is the only Comecon country other than the
U.S.S.R to figure in these tables to any significant extent. Switzerland is
a fairly heavy importer (8 tonnes per million per year), and Sweden with
21.4 tonnes per million per year exceeds the United Kingdom (20.7
tonnes per million per year) on a *per capita* basis. Old ties between
countries can be discerned in some of the trading patterns. Thus 80% of
Singapore's European exports are to the United Kingdom and both
Singapore and Hong Kong export more to the United Kingdom than
does Japan. The Netherlands are the second most important market to
both the former countries. It is reasonable to assume that the
Netherlands will receive substantial imports from her former territories
in the Far East, but these do not appear in the data available to me.

The Arab countries, with a population of about 63 million, import
1,180 tonnes per year, compared with about 3,000 tonnes per year for
the 335 million West Europeans. Kuwait (67.3 tonnes per million per
year), Saudi Arabia (86.2 tonnes per million per year) and Bahrain,
whose 101.3 tonnes (nearly all from the Korean Republic) is equivalent
to 506.5 tonnes per million per year are major importers, although even

Bahrain's figures correspond to only 1.4 g per person per day, minute in comparison with the Japanese figure of 35 g. Nonetheless, these figures show that the Near and Middle East represents an important market for soy sauce, particularly for the Korean Republic (about whose remarkable annual growth in fermented soy exports I shall say more subsequently), for whom it is the only region where she commands a greater market share than do the other major exporters. This is mainly due to the Korean export of 609 tonnes to Saudi Arabia, but the country also has the main share of soy sauce imports of Bahrain, Kuwait and Jordan.

In the Far East and Western Pacific, the trade figures are small, demonstrating the importance of locally made soy sauce. Korea's only contribution to this market is a small export (6 tonnes) to Japan, from whom it imported 218 tonnes. Malaysia, including Sabah and Sarawak, imported 787 tonnes of soy sauce from Singapore, taking the lion's share of that country's exports in this region.

In the Pacific and Australasia region, Australia's 824 tonnes (57.6 tonnes per million per year) is by far the largest importation, but some of the smaller countries import very large proportionate amounts of soy sauce, for example (all in tonnes per million per year): the Cook Islands, 280; Christmas Island, 1,867; Fiji, 62.5; Guam, 3,722; Nauru, 1,386; New Caledonia, 687; New Hebrides, 680; Samoa and Tonga 63; Tuvalu, 1,314. These figures may represent stockpiling or, more probably, a re-export trade, but certainly add up to a very substantial market in proportion to the population of the region, being slightly more than half of the European imports with a population only about 7% that of the latter region (such population figures are not very reliable but serve as a useful guide).

Africa, on the other hand, is but little developed as a market for soy sauce, although the existence of local fermentations of various legume and other seeds suggests that it has considerable potential. Certain areas, notably the Canary Islands, Mauritius and Réunion Island represent a disproportionate share of the market relative to their population, no doubt for historical reasons.

From the summary figures it can be seen that Japanese soy sauce not only has the largest share of the market but also commands the highest price.

Turning now to the total exports over the last three years (Table 2) we see relatively little change in the period (1978, 15 732 tonnes; 1977, 15 785 tonnes; 1976, 13 750 tonnes), but a remarkable increase in the

Korean Republic's exports (which nearly quadrupled in tonnage and had a six-fold increase in revenue). Singapore and Hong Kong showed steady increases in exports, while Japan showed no clear trends, roughly following the performance of the country's total production (Table 5). It is interesting to note that while Hong Kong's re-export market doubled during the period being considered, Singapore's declined sharply, indicating a considerable increase in exports of domestic soy sauce.

The important Japanese exports, when broken down by the type of container used (Table 3) show no definite trends in exports in bottles or cans, but a definite increase in the use of other types of containers; the containers in this last group are not specified but it seems reasonable to assume that it represents the advent of the convenient and light-weight plastic containers.

Over the longer term (Table 4) it seems that Japanese production of both soy sauce and *miso* reached a plateau in the early 1970s and has remained fairly static ever since. In considering Japanese production, however, it is essential to remember that the Kikkoman Company's production facility in Wisconsin, which was built in the early 1970s, produced a massive 21,600 tonnes of soy sauce in 1978 which was equal to three times the total exports from Japan in the same year and 1.5% of the total production in Japan (4.6% of the parent company's production). When these figures are taken into account, it is seen that Japan's total share of the World soy sauce market remains very healthy indeed and is showing substantial growth. It is also necessary to appreciate (Table 5) that Japan's exports are, on average, only about 0.5% of its total annual production and that slight pricing adjustments would permit the Japanese to capture a much greater share of the World market without significant change in the scale of their industry's operation. For convenience, the same table presents the data for *miso* (soy paste), although this product will be discussed in greater detail later in this review. Both production and export of *miso* are showing a steady increase in Japan with the value of the exports rising by 25% during the period under consideration. *Miso* exports are still fairly small in comparison with soy sauce, but on a rising market Japan's exports still only represent 0.2% of its annual *miso* production; clearly there is considerable room for expansion here.

The data for soy sauce imports into Hong Kong, Singapore and the U.S.A. are in broad agreement with the appropriate export figures, the small differences probably reflecting differences in reporting and accounting procedures. The most interesting information in this table is the steady increase in exports from the China People's Republic to Hong Kong and the U.S.A. In Hong Kong's case, Chinese imports to it equal nearly twice Hong Kong's exports to other countries. During the three years examined, Hong Kong and the U.S.A. showed increases in soy sauce imports of 37.5% and 27%, respectively, whereas Singapore's imports decreased by 18% which, although substantial, is less than its decrease in re-exports. No particular trends are evident in the detailed breakdown of re-exports from Hong Kong and Singapore for 1978 (Table 7) but these figures are included for completeness.

The trade in *miso* is smaller than that in soy sauce, total exports in 1978 by Korea and Japan being worth £856,000, compared with their £2.8 million of soy sauce exports. Japanese exports showed a steady increase between 1976 and 1978. Korea increased its exports more than three-fold in the same period and increased its share of the market from 18% to 36%. The 1978 exports of *miso* were worth £531.5 per tonne to Korea, whereas the soy sauce had an average value of £343.0 per tonne. Japanese *miso* averaged £485.4 per tonne, rather closer to the value for soy sauce of £359.5 per tonne. Thus a comparison of Tables 1 and 9 will show that *miso* is of greater relative importance to Korea than it is to Japan.

Among European countries (Table 8), Belgium and Holland import the greatest amount on a *per capita* basis (about 3 tonnes per million per year) whereas the United Kingdom, which tops the list in terms of total imports, ranks third with a little over 1 tonne per million per year. Interestingly, Spain imports a fair amount of *miso*, whereas Sweden, which has the greatest consumption of soy sauce *per capita*, imports very little *miso*. The U.S.A. and Canada had total *miso* imports about 10% of their soy sauce imports but Central and South American imports were minute.

Once again, the near and Middle East provide a few surprises. Total *miso* imports were twice the European total from Japan and Korea, although the region's total soy sauce imports (from the four major exporting countries) were only one-third of Europe's imports. Bahrain showed a high relative importation of 342 tonnes per million per year of *miso*, comparable with its 506 tonnes per million per year of soy sauce. Saudi Arabia was another important market for *miso* (38 tonnes per million per year).

Africa is interesting, importing nearly as much *miso* as the Pacific and

Australasia region, although its soy sauce imports were only 22% of the latter region's.

I consider that the *miso* market is capable of considerable expansion, and that the great range of flavours readily produced by varying the fermentation conditions and ingredients should offer scope for a greater potential market than that for soy sauce, especially in Europe and the U.S.A.

In conclusion, let me stress that the necessary restrictions of space which the Editor must impose on each author mean that this review is necessarily brief to the point of being superficial. A full treatment of the topics which have been lightly touched upon herein would require at least one whole volume of this series to be devoted to soy sauce, a second to miso, and at least one more to the other related fermentations. I hope that this brief outline will tempt readers to tackle some of the primary sources listed in the bibliography.

V. A TOUR OF SOUTH EAST ASIA

A tour of some South East Asian countries during summer 1980 gave me a valuable opportunity to gather data on production of fermented soy products at first hand, and thus to update this review somewhat. This section summarizes the major points discovered during the tour which included visits to Kikkoman Co. factories in Noda, Japan and Wisconsin, U.S.A., and to smaller local producers in Bangkok, Thailand and Kuala Lumpur, Malaysia as well as to research centres in all these places and also in Singapore.

A. Technical and Scientific Aspects

Despite references in the literature to the use of dehulled soybeans for soy-sauce making, in practice many manufacturers use intact beans. My enquiries on this point were met with some surprise. I was assured that no problems were associated with using whole beans and that dehulling would impose intolerable additional burdens on the manufacturer in terms of manpower, equipment and time. The koji samples which I examined were all well covered with mould and appeared quite indistinguishable from those which my research group produce from dehulled beans. We plan to compare the biochemical parameters of kojis made with dehulled and with intact beans.

In Section II.D (p. 45), I refer to the use of deep-tank methods for koji production, and I was very interested to see this system operating in Kikkoman plants in Japan and the U.S.A., and in one plant in Bangkok. An excellent film about soy-sauce, made by the Kikkoman Company, vividly illustrates the process that they employ. Cooked beans and roasted crushed wheat are mixed, inoculated with their specially selected *Aspergillus* strain and charged into very large troughs, each of which occupies an entire room. The troughs have a perforated false bottom through which air can be blown. The charge is evenly distributed in the trough and smoothed down mechanically. Powerful compressors force a steady stream of filtered humidified air through the koji. Probes buried in the koji monitor temperature and humidity, and regulate temperature and humidity of the air blast to maintain conditions in the koji optimal for mould growth and enzyme production. The koji is mixed at intervals by a mechanical flail which traverses the length of the trough, mixing its entire contents. At completion of the koji fermentation, the trough is emptied mechanically and then is rigorously cleaned before charging with the next batch of materials. Although on a rather small scale, the deep-tank process which I saw in Thailand was generally similar to the one described above, the main difference being the use of manual labour for charging, mixing and discharging the koji. In both processes, I would estimate that the koji layer was about 300 mm deep and the importance of using probes to monitor conditions throughout the mass was stressed.

As to the raw materials, the marked differences in organoleptic qualities between soy-sauce from different regions had led me to expect substantial differences in the ratio of beans to wheat being employed, an impression apparently confirmed by statements in the rather scanty literature on the subject. In fact, when converted into a dry-weight basis, the ratio of beans to wheat was consistently around 50:50 in all cases. Outside Japan, white flour seems to be mainly used, whereas Japanese producers favour whole wheat which is toasted and crushed, preferably only a short time before use. Considerable importance is attached to the way in which the beans are soaked and cooked. Soaking may be overnight, for as little as three hours, or omitted altogether. After soaking and rinsing, the beans are boiled at ambient pressure until sufficiently soft, cooked in pressure vessels at about 120°C (sometimes in rotating vessels) or (in the most modern Japanese practice) subjected to very high pressure, short-time cooking in a continuous cooker. There seems to be general agreement that the way in which the beans are cooked is very important, but some diversity of views as to which is the best way. Japanese arguments for the high-pressure short-time process are supported by evidence that this gives a substantial increase in the percentage of bean protein which is solubilized in the subsequent fermentations, but this method is probably only appropriate for large companies where heavy capital outlay is rewarded with process economies.

Systems used for securing satisfactory inoculation with mould provided interesting comparisons. As would be expected, major Japanese manufacturers have elaborate systems to ensure consistent inocula of very high quality, coupled with active research into strain selection and improvement. At the other extreme, some producers reject entirely the use of pure cultures, insisting that the inoculum provided by koji trays which have been used regularly for long periods is essential if top-quality sauce is to result. Under these latter conditions, the koji stage is allowed to continue for up to seven days, resulting in a product which is dry and hard, but apparently without development of the harsh and ammoniacal aromas normally observed at the onset of mould sporulation. When I enquired about this, it was suggested that perhaps the koji dries out too quickly for sporulation to occur because of the high ambient temperatures and the shallow trays used, but no experimental work seems to have been undertaken on this matter.

The work of Mrs Napha Lotong and her associates in Bangkok on methods of spore-inoculum production appropriate for the smaller independent soy-sauce brewers deserves special mention. Appropriate amounts of rice and water are placed into autoclavable plastic bags which are then closed with a specially designed air-filter assembly and autoclaved. After cooling, the rice is inoculated with a pure culture of mould. The mould proliferates rapidly and sporulates profusely in ambient conditions typical of the region. When sporulation is complete, the filter is removed to permit the mass to dry thoroughly, in which condition it will keep for several weeks. Because of the high spore yield, only a small amount of moulded rice is needed to inoculate a batch of koji. The system is so designed that everything, including fabricating the air-filter assemblies, can be undertaken by even a small soy-sauce business and I saw it being employed by such a small and otherwise rather primitive brewery whose owner was delighted with the improvement in his operations. This development can contribute substantially to the survival of smaller individual producers at a time when they are beginning to come under pressure from application of stricter standards for safety and hygiene.

It is in the moromi stage that the greatest contrast between Japan and the other countries occurs. Japan has a very marked seasonal cycle with hot summers and cold winters and it is considered that exposure of the developing moromi to this cycle, or a simulation of it, is essential for production of high quality soy sauce. Furthermore, the use of pure-culture inocula of desirable yeasts and lactic-acid bacteria finds favour, although I understand that this is by no means universal and some manufacturers continue to prefer 'natural' inocula. In Thailand and Malaysia (and, I am informed, in many Singapore breweries), the moromi stage is carried out in small earthenware pots rather than the larger fermentation vessels which I saw in Japan (but again I stress that I cannot generalize about Japanese practice). These pots stand in direct sunlight and must reach daytime temperatures will in excess of $40°C$. The fermentation takes about three months and relies on inoculation with organisms remaining in the pot from previous fermentations. The Japanese breweries that I visited recover the soy sauce by pressing it from the fermented mash, leaving a fairly dry, solid residue. The other breweries, however, syphon ('draw') off liquid from the mash, replace it with an equal volume of brine, leave for a few days, make a second drawing and so on, resulting in about five drawings of gradually diminishing quality before the mash is exhausted. After blending, the first and second drawings are retailed as the 'superior' qualities of soy sauce, with little or nothing added to them. The later drawings go for blending with molasses and/or caramel to a rather different product with distinct culinary uses; again different quality sauces result from the different blendings.

The solid residues from soy-sauce production may be used as animal feed or be processed into a kind of *miso* for human consumption. I was told in Japan that, despite the high content of salt, cattle so like the taste that once they have been fed on a diet containing these residues they refuse formulations that omit them.

B. Trade Aspects

In Section IV (p. 61) of this review, I refer to the difficulties of assembling complete statistical data. This can be illustrated by reference to Australia, where I show (Table 1, p. 66) imports of 824 tonnes in 1978. While I was in Australia, Steetley Chemicals (Division of Steetley Industries Ltd) very generously obtained more complete data for me. This shows that in 1978–1979 Australia imported a total of 1,343,560 litres (1612 tonnes) and in 1979–1980 imported 1,201,602 litres (1201 tonnes), about 71% more than I had estimated. Their data give a breakdown of consumption by State which shows that, in general, this equates to the population of each State. In each year the trade was worth slightly more than $1.2 \cdot 10^6$ Australian dollars, making the cost F.O.B. $A per litre 0.90 in 1978–1979 and 1.02 in 1979–1980.

In Thailand, there are about 50 soy sauce factories, the majority of which are small, producing less than 100 kilolitres per year, although it should be noted that most of them also produce soybean paste and soybean cheese. The total annual consumption of soy sauce in Thailand is estimated at about 6,000 kilolitres (about 7200 tonnes). These data were compiled from information generously provided by Mrs N. Lotong (Biology Department, Kasetsart University, Bangkok) and M. Sundhagul and colleagues (Applied Scientific Research Corporation of Thailand, Bangkok).

In 1979, Taiwan imported 84 tonnes of soy sauce and exported 833 tonnes, figures which seem to be about average for the period 1976 to 1980 according to information very kindly supplied by Mr H. C. Tso of the Farmers' Bank of China, Taipei.

In Malaysia, there are about 140 soy sauce factories producing in total an estimated $5.5 \cdot 10^6$ gallons of soy sauce per year according to the proprietor of a leading brewery in Kuala Lumpur. This is about 21,000 tonnes per annum. I was unable to obtain information on exports of soy sauce from Malaysia although total imports in 1978 (Table 1, p. 65) were about 1000 tonnes.

VI. ACKNOWLEDGEMENTS

To Yong Fook-Min, Shatish Goel, S. B. Shibani, Kofi Aidoo and Sumbo Abiose for their parts in increasing my knowledge and understanding of the topics covered in this review. To Jean Winter and Margaret Provan for converting an untidy pile of badly written pieces of paper into a neat manuscript. To Isabella Docherty for producing elegant diagrams at short notice.

REFERENCES

Abiose, S.H. (1980). Ph.D. Thesis: University of Strathclyde, Glasgow.
Aidoo, K.E. (1979). Ph.D. Thesis: University of Strathclyde, Glasgow, U.K.
Asao, Y. and Yokotsuka, T. (1957). *Bulletin of the Agricultural Chemical Society of Japan* **21,** 622.
Fogarty, W.M. and Ward, O.P. (1974). *Progress in Industrial Microbiology* **13,** 59.
Goel, S.K. (1974). Ph.D. Thesis: University of Strathclyde, Glasgow.
Goel, S.K. and Wood, B.J.B. (1978). *Journal of Food Technology* **13,** 243.
Groff, E.H. (1919). *Phillipines Journal of Science* **15,** 307
Hayashi, K., Noda, T.M. and Nagarayana, T.Y. (1977). United States Patent 4,028,470.
Hesseltine, C.W. (1965). *Mycologia* **57,** 149.
Kamekura, M. and Onishi, H. (1974). *Applied Microbiology* **27,** 809.
Kundu, A.K. and Manna, S. (1975). *Applied Microbiology* **30,** 507.
Matsumoto, I. and Imai, S. (1977a). *Journal of the Society of Brewing, Japan* **72,** 74.
Matsumoto, I. and Imai, S. (1977b). *Journal of the Society of Brewing, Japan* **72,** 147.
Minor, L.J. (1945). *Food Industries* **17,** 758.
Morihara, K. (1974). *Advances in Enzymology* **41,** 179.
Murakami, H. (1971). *Journal of General and Applied Microbiology, Tokyo* **17,** 281.
Noda, F., Sakasai, T. and Yokotsuka, T. (1976a). *Journal of Food Science and Technology, Tokyo* **23,** 59.
Noda, F., Sakasai, T. and Yokotsuka, T. (1976b). *Journal of Food Science and Technology, Tokyo* **23,** 67.
Noda, F., Sakasai, T. and Yokotsuka, T. (1976c). *Journal of Food Science and Technology, Tokyo* **23,** 74.
Noda, F., Sakasai, T. and Yokotsuka, T. (1976d). *Journal of Food Science and Technology, Tokyo* **23,** 80.
Onaga, D.M., Luh, B.S. and Leonard, S.J. (1957). *Food Research* **22,** 83.
Onishi, H. and Suzuki, T. (1970). *Reports of the Noda Institute of Scientific Research* No. 14, 71.
Piatkiewicz, A. and Kasperkiewicz-Jamroz, T. (1976). *Acta Alimantaria Polonica* Vol. II (XXVI) No. 4, 321.
Robinson, R.J. and Kao, C. (1974). *Cereal Science Today* **19,** 397.
Rombouts, F.M. and Pilnik, W. (1972). *Chemical Rubber Company Critical Reviews in Food Technology* **3,** 1.
Sakaguchi, K. (1959a). *Bulletin of the Agricultural Chemical Society of Japan* **23,** 22.
Sakaguchi, K. (1959b). *Bulletin of the Agricultural Chemical Society of Japan* **23,** 100.
Sakaguchi, K. (1959c). *Bulletin of the Agricultural Chemical Society of Japan* **23,** 428.
Sakaguchi, K. (1959d). *Bulletin of the Agricultural Chemical Society of Japan* **23,** 443.
Sakaguchi, K. (1960a). *Bulletin of the Agricultural Chemical Society of Japan* **24,** 489.
Sakaguchi, K. (1960b). *Bulletin of the Agricultural Chemical Society of Japan* **24,** 638.

Sakasai, T., Noda, F. and Yokotsuka, T. (1975a). *Journal of Food Science and Technology, Tokyo* **22**, 474.

Sakasai, T., Noda, F. and Yokotsuka, T. (1975b). *Journal of Food Science and Technology, Tokyo* **22**, 481.

Sekine, H. (1976). *Agricultural and Biological Chemistry* **40**, 703.

Shibani, S.B. (1978). MSc. Thesis: University of Strathclyde, Glasgow.

Siegel, A. and Fawcett, B. (1976). 'Food Legume Processing and Utilisation'. International Development Research Center Technical Studies, I.D.R.C., Ottawa, Canada.

Smiley, K.L., Hensley, D.E. and Gasdorf, H.J. (1976). *Applied and Environmental Microbiology* **31**, 615

Togo, K. (1977). 'Rice Vinegar; An Oriental Home Remedy'. Kenko Igakusha Co. Ltd., Tokyo.

Tsujita, Y. and Endo, A. (1976). *Biochimica et Biophysica Acta* **445**, 194.

Wolf, W.J. and Cowan, J.C. (1975). 'Soy Beans as a Food Source'. C.R.C. Press, Cleveland, Ohio.

Wood, B.J.B. (1977). *In* 'Genetics and Physiology of *Aspergillus*' (J.E. Smith and J.A. Pateman, eds.), p. 481. Academic Press, London.

Wood, B.J.B. and Yong, F.M. (1974). *In* 'The Filamentous Fungi, vol. 1; Industrial Mycology' (J.E. Smith and D.R. Berry, eds), p. 262. Edward Arnold (Publishers), London.

Wood, B.J.B., Cardenas, O.S., Yong, F.M. and McNulty, D.W. (1975). *In* 'Lactic Acid Bacteria in Beverages and Food (Fourth Long Ashton Symposium, 1973)' (J.G. Carr, C.V. Cutting and G.C. Whiting, eds.), p. 195. Academic Press, London.

Yokotsuka, T. (1960). *Advances in Food Research* **10**, 75.

Yokotsuka, T., Sakasae, T. and Asao, Y. (1967a). *Journal of the Agricultural Chemical Society of Japan* **41**, 428.

Yokotsuka, T., Asao, Y. and Sakasae, T. (1967b). *Journal of the Agricultural Chemical Society of Japan* **41**, 442.

Yong, F.M. (1971). MSc Thesis: University of Strathclyde, Glasgow.

Yong, F.M. and Wood, B.J.B. (1974). *Advances in Applied Microbiology* **17**, 157.

Yong, F.M. and Wood, B.J.B. (1976). *Journal of Food Technology* **11**, 525.

Yong, F.M. and Wood, B.J.B. (1977a). *Journal of Food Technology* **12**, 163.

Yong, F.M. and Wood, B.J.B. (1977b). *Journal of Food Technology* **12**, 263.

Yong, F.M., Lee, K.H. and Wong, H.A. (1978). *Journal of Food Technology* **13**, 385.

4. Breadmaking

ERKKI OURA*, HEIKKI SUOMALAINEN* and RISTO VISKARI†

* Research Laboratories of the State Alcohol Monopoly (Alko), P.O. Box 350, SF-00101 Helsinki 10, Finland
† Oy Karl Fazer Ab, Bakery Group, P.O. Box 40, SF-15101 Lahti 10, Finland

I. INTRODUCTION

A. The Early History of Making Leavened Bread

The history of breadmaking covers a period of some 6000 years. The discovery of using microbes to leaven bread was, of course, made by chance when primitive man was able to prepare various food products from grain, including unleavened bread. Possibly some dough was forgotten and left to stand so that it underwent spontaneous fermentation. When this dough was then used by the 'baker', the loose and light bread he obtained must have surprised and delighted him to the extent that he tried to repeat and perpetuate the method: leavened bread was discovered. It may have first occurred in Egypt since the earliest records of the history of fermentation show that breadmaking as an art began with the ancient Egyptians, who appear to have been the first to use leaven to make their bread light. The Egyptians are also credited with the invention of the oven for baking bread.

The leaven used at that time consisted in all probability of a microbial population of lactic acid-forming bacteria with some kind of yeast. The method spread outside Egypt, for example with the Jews during their Exodus and later to Greece and Rome. The Romans developed the breadmaking technique nearer to its modern level. For instance, Pliny in his Naturalis Historiae (Liber XVII: 12, 26, 77 A.D.) states that 'the bread is there (in Gallia and Hispania) lighter than elsewhere' because foam from beer fermentation was used to raise the dough. The yeast skimmed from grain-malt wort primarily fermented for the alcohol yield was used for nearly 2000 years in the production of fine bread, whereas the raising of ordinary bread, especially rye bread, was performed using sours.

From Roman times, through the Middle Ages, right up to the middle of the Nineteenth Century, there was no production of yeast especially intended for use in breadmaking. Even in 1847, Balling stated that in the Austrian Empire 14,400,000 pounds of thick brewer's yeast suspension were used in breadmaking and estimated the total yeast production in Europe to be 150 million pounds. At the beginning of the Nineteenth Century, a beer process in which bottom yeast was used started to spread from Bavaria, and this method gradually displaced the older method based on top-fermenting yeast. Bottom brewer's yeast was not suitable for baking and this put the wheat-bread bakers into a difficult situation from which they were rescued only when production of baker's yeast was started.

B. Discovery of Yeast in Fermentations

Up until the 1860s, the biological and biochemical nature of yeast and other microbes and their relation to fermentation were completely obscure, even though Anthony van Leeuwenhoek using his own microscopes had in 1680 observed and sketched some bodies that were later interpreted as being yeast cells. Gay-Lussac formulated an equation to represent alcohol fermentation (which is still used in its original form and known as the Gay-Lussac equation) as early as 1810, but without the slightest inkling of the forces behind the process. Moreover, although Cagniard-Latour, Schwann and Kützing (1835–1837) convincingly showed that fermentation was the result of vegetative growth of yeast, and that these bodies had even been named *Saccharomyces* by Meyen (1838), the role of yeast in fermentations and also in breadmaking was doubted and not generally accepted. The position was cleared up by Louis Pasteur in his works of 1857–1863 where he demonstrated decisively the role of yeast (and other microbes) in fermentations and described its anaerobic and aerobic nature. In the 1880s, Hansen of the Carlsberg Laboratory in Copenhagen initiated the study of pure cultures and, finally, in 1897 Büchner produced fermentation using cell-free extracts of yeast. These are a few of the early events in the history of yeast fermentations. The controversies that existed were settled by these works (and others similar to them) and eventually yeast obtained the recognition it deserves as the generator of fermentations, and the sciences of microbiology and biochemistry could be founded.

II. PRODUCTION OF BAKER'S YEAST

A. Developments in the Field of Baker's Yeast Production

In the Netherlands, distiller's yeast had been delivered to bakers in pressed form since at least 1781 (Kiby, 1912). A more decisive step in starting production of a special baker's yeast was the development of the

Vienna Process in about 1846 by the Viennese Mautner. In this process, 10–12 kg of yeast and 28 litres of alcohol were obtained from 100 kg of grain, often barley (Olsen, 1961). The new bakery yeast was an important advance in bringing the very irregular process of dough fermentation (as it was during the use of sour or brewery yeast) under control and in producing bread of higher volume. The process was used for production of baker's yeast as long as alcohol prices were favourable and supplies of grain abundant (Peppler, 1967). The following stage in the development was the Air Process, introduced by Eusebius Bruun of Copenhagen in 1877. The method was based on Pasteur's observation that yeast growth was greatly stimulated by blowing air through the fermentation medium, and in about 1900 it was possible to obtain some 20–22 kg of yeast and the same amount of alcohol from 100 kg of grain.

Modern production of baker's yeast begins with the improvements of the production process during 1915 to 1920. They are covered by several German, Hungarian and Danish patents that describe a fed-batch Process (Zulauf or Z process, also called the Danish method) based on incremental feeding of wort that was adjusted to growth. The yield of yeast was originally about 75% and alcohol was not formed in the final production stage. The high yield and improved storage stability were the main advantages of this method. The production process was further developed by a change from malted-grain to beet-molasses wort supplemented with ammonium salts. In the development of the new techniques, the names of the German Hayduck and the Dane Sak are prominent. Those interested in the lively patent race connected with this development are referred to Wagner (1936). Although some other technological ideas have been suggested or even implemented, e.g. continuous fermentation (Olsen, 1960, 1961; Sher, 1969), they have not displaced the fed-batch method which is still commonly used. Naturally, the development of fermentation technology, such as better aeration systems and instrumentation, and the development of improved yeast strains have greatly influenced the modern production process for making baker's yeast.

B. Technology of Baker's Yeast Production

Although *Candida* (*Torula*) yeast has been occasionally used for baking (Burrows, 1970) and some *Saccharomyces carlsbergensis* strains have been patented for use as baker's yeast (Koninklijke Nederlandsche Gist- en Spiritusfabriek, 1967), pure strains of *Saccharomyces cerevisiae* are almost universally employed.

Several descriptions of the process for production of baker's yeast, including active dried yeast, have appeared recently (Burrows, 1970; Harrison, 1971; Reed and Peppler, 1973), even in this *Economic Microbiology* series (Burrows, 1979), and therefore the topic will not be discussed here.

C. The Requirements Presented for Baker's Yeast

The following requirements can be drawn up for a baker's yeast strain, its handling and for the entire industrial process.

The yeast used in the production process must be suitable for aerobic cultivation on molasses, grow well and give a good yield. Yields as high as 56.7 parts of yeast dry substance per 100 parts of hexose sugar (i.e. nearly 59.7 g of yeast per 100 g of sucrose) have been quoted (Peppler, 1960). During industrial production, the cultivation proceeds in several stages, where the yield in the first stages is not maximum. Because other losses will occur, the practical yield will be of the order of 48 g of yeast dry substance from each 100 g of sucrose.

Yeast cultivation should result in a baker's yeast of a light colour and fresh smell. The cells should be separate and of even size, and thus easy to filter.

The yeast should have a good leavening activity that is not rapidly lost on storage. These two requirements are easily satisfied individually by choosing correct culture conditions. When the product yeast has a high content of protein, its overall enzyme activity is increased but this simultaneously implies rapid autofermentation and breakdown reactions, i.e. bad storage ability. On the other hand, a low content of protein results in a somewhat decreased activity for fermentation but stability increases. High protein contents are easily obtained by cultivating yeast under substrate-rich conditions. Generally speaking, keeping quality decreases as fermentative activity increases, assuming other factors remain unchanged (Bergander and Bahrmann, 1957). Because fermentative activity and stability are inversely related, the production process must be balanced to give a satisfactory compromise. In fact, there exist two different types of baker's yeast: one very active but with only reasonable stability, and the other with somewhat lower activity but good stability. When there is refrigeration between the yeast factory and the bakeries, and when little fresh yeast is used by the small-scale consumers, the first type of yeast is often produced. However, when a considerable part of the pressed yeast is to be sold in grocery stores, the time from the producer to consumer can be several days or even 2–3 weeks, and a yeast with better stability is needed, which can be produced only at the expense of activity. Some other factors, such as local climate and use of active dried yeast, influence which of the two yeast types will be produced.

The industrial production process must always lead to a yeast of consistent performance, or in Frey's words 'in the highly mechanized baking industry the processes are so timed that it is necessary for the yeast to act almost as a chemical compound and to have a constant and uniform fermentation rate from day to day' (Frey et al., 1936).

The genetic stability of the strain used must be good and the stock cultures must be stored so that its original properties remain unchanged. As Harrison (1971) stresses, this is a problem because all of the important requirements are of a quantitative nature and quite small changes can be significant. Because variations always exist, the original strain must be stored under conditions where only little growth occurs and must be maintained by reselection (Lincoln, 1960).

D. Production Values for Baker's Yeast

The International Union for Pure and Applied Chemistry (IUPAC, 1966) charted the World production of baker's yeast for 1963 and later in 1971 revised and extended the values with data for 1967. According to those countries responding to the enquiries, World production was somewhat over 700 000 tonnes of fresh yeast and about 16 000 tonnes of active dried yeast. Bronn (1976) has extended the IUPAC data using various sources and with his own information, and concluded that total production of fresh baker's yeast for 1974 was 970 000 tonnes. A recent estimate made at the State Alcohol Monopoly (Alko) in Finland (K. Edelmann, unpublished observations) was based partly on collected production figures and partly on estimates. According to this, baker's yeast production in 1980 was as high as 1 430 000 tonnes of yeast (Table 1). This includes a value for the Soviet Union that was estimated to be

Table 1

Estimated World production of baker's yeast in 1980. Unpublished work of K. Edelmann.

	Production (tonnes fresh yeast, 28–32% dry matter)
Western Europe (European Economic Community)	315 000
Western Europe (non-Community countries)[a]	128 000
Eastern Europe[b]	394 000
North and Central America	330 000
South America	75 000
Africa	47 000
Asia	132 000
Australia	10 000
Total	1,430,000

[a] Includes the Asiatic part of Turkey
[b] Includes the Asiatic Soviet Union

about 270 000 tonnes using the figures given in the article by Makarenko *et al.* (1978).

The data in Table 2 show that only a few countries in the Third World reach the consumption of European countries. However, when customs become westernized, consumption of leavened bread and hence baker's yeast increase, as exemplified by Morocco and Syria. The food authorities in Syria estimate that consumption will increase to almost 2 kg per person in the future. Bearing this in mind, K. Edelmann has projected that the annual yeast consumption in the World could eventually reach 1 kg per head. The final figure reached, however, will depend greatly on what happens in the populous countries of China and India.

The annual consumption of yeast per person is highest in Finland, at 2 kg (Table 2). Consumption in other Scandinavian countries is nearly the same, about 1.8–1.9 kg, in Central and South European countries it is 1.5–1.7 kg and 1–1.5 kg, respectively and in the U.S.A. and Canada it is 1.1–1.2 kg. In the Soviet Union, it was estimated to be 1.1 kg with an additional 0.4 kg of yeast coming from distilleries (Makarenko *et al.*, 1978). Finnish bread consumption is nevertheless exceeded by several South European countries (Italy, Greece, Yugoslavia, Roumania and Hungary; Pelshenke, 1961), and despite Sweden's high yeast consump-

Table 2

Recent annual consumption of baker's yeast per person in different countries of the World. Unpublished results of K. Edelmann.

	Annual consumption per person (kg)
Finland	2.0
Sweden	1.9
Austria	1.9
Denmark	1.8
West Germany	1.6
Greece	1.4
Hungary	1.3
Poland	1.2
U.S.A.	1.2
Canada	1.1
England	~1.1
Spain	1.1
Soviet Union	1.1
Chile	~1.0
Morocco	1.0
Syria	0.8
Zambia	0.6
Argentina	0.6
Kenya	0.3
Japan	0.3
Ethiopia	0.1
South Korea	0.1
India	~0.01

tion, it and the Netherlands have the lowest bread consumption in Europe. The large amount of yeast used in Finland can be explained by the prevalence of home baking and the number of small bakeries where the use of yeast is not as economical as in larger bakeries. A small bakery that starts using, for example, a yeast slurry tank can decrease its yeast needs by about 20%. As for Sweden, there is a considerable export of bread (Swedish knacke bread) and a tradition for sweet bakery goods. Even a fairly sweet bread involves an increased use of yeast during production. These two examples show that the demand for baker's yeast and bread consumption do not necessarily move together, and that there are several factors that can affect the ratio of these two figures.

E. Measuring the Activity of Baker's Yeast

The leavening ability of yeast and its behaviour in the bread are determined using various methods ranging from simple measurements of gas evolution to simulating the whole breadmaking process in a test bakery. Because there is good agreement between breadmaking tests and measuring gas generation in the fermentation stage, several yeast factories have discontinued the breadmaking test or use it only at intervals for the sake of comparison (Parisi, 1970).

Fermentation tests can be done with either flour or one or more kinds of sugar. The quantity of carbon dioxide evolved can be measured by its volume (either total gas evolution or gas retention) at constant pressure or by pressure changes at constant volume.

The oldest and most classified method although seldom used now is that of Hayduck-Kusserow (Kretzschmar, 1955), which measures the amount of carbon dioxide formed during sucrose fermentation. A second classical method, the Berliner method, in which the time needed for the dough to reach a given height (gas retention is measured), is, because of its simplicity, still in use. It does, however, give discordant results with different grades of flour. Mitterhauzerová and Sedlárová (1966) found a strong correlation between dough volume and gas volume in comparative tests. More expensive fermentographs have been developed. The first of these, the zymotachegraph (Kent-Jones and Amos, 1967), was a relative of alveographs and farinographs. More commonly used are the fermentographs of Brabender and the well known Swedish SJA (Fig. 1). In both of these, total carbon dioxide evolved during yeast fermentation in a normal dough is measured and registered. In the Brabender equipment a rubber balloon with the fermenting dough is suspended by a chain in a thermostatically regulated water tank. As fermentation proceeds, the balloon expands and its rise in the water is recorded. In the SJA method, a pan bread is made and enclosed in the apparatus. The amount of gas generated is collected in a gasometer which rises and the movement is registered. The machine allows measurement of both total carbon dioxide evolved and that retained by the dough, the latter being indicated by the volume of the dough itself (Kent-Jones and Amos, 1967; Parisi, 1970). The Chefaro Balance instrument is claimed to record automatically both gas production and its retention during the whole course of fermentation (Elion, 1940; Kent-Jones and Amos, 1967).

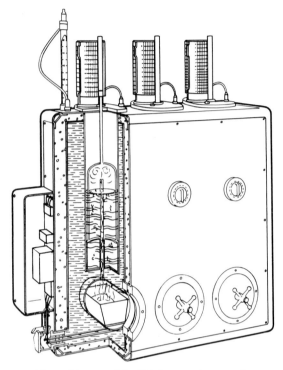

Fig. 1. Diagram of the SJA fermentation recorder.

Yeast activity tests based on fermentation in the dough are affected by inconsistencies in the quality of the flour. To overcome this, Schulz (1962, 1965) returned to using chemically defined media. His method employs a non-fermentable base of starch and bread flour, to which is added either maltose alone or, in separate tests, one only of the following sugars: maltose, sucrose, glucose or fructose. The increase in volume of the dough is measured and registered in simple graduate cylinders over three hours. Unfortunately, the method is difficult to quantify (Parisi, 1970), and in the absence of a medium as complex as flour an accurate relationship between fermentative power and breadmaking tests cannot be relied on. A better correspondence is possible with the Burrows and Harrison method (1959), which employs a simple fermentometer to measure the gas evolved from a thin dough suspended in fermentation bottles (Fig. 2). Methods measuring changes in pressure include the Sandstedt-Blish pressure meter system (Sandstedt and Blish, 1934).

The International Union of Pure and Applied Chemistry (1970)

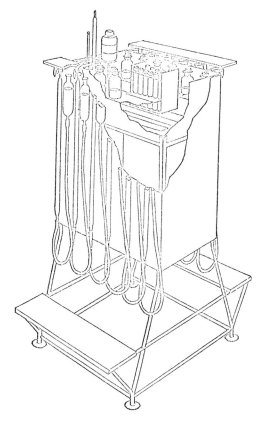

Fig. 2. Cut-away drawing of the Burrows and Harrison fermentometer. From Burrows and Harrison (1959).

initiated a study in which yeast-leavening activity using different lots of flour was measured in several laboratories around the World. It was found that the same method with the same flour and yeast gave different results in different laboratories because each experimenter, probably unconsciously, made small variations in the method. It seems that a method for evaluating baker's yeast was not treated as a precise determination in the same way as it would be in other fields. Some determinations based on the total gas generated gave far better agreement. This was chiefly those carried out with SJA fermentation recorder in the Nordic countries, where the apparatus is very much used, and probably uniformly because of the easy reciprocal contacts (Parisi, 1970). A broadly similar result was obtained with the Burrows and Harrison (1959) fermentometer. Equivalent results for some of the other test methods could not be obtained, mainly because they were evaluated at only one or very few laboratories.

III. SOME PROPERTIES OF BAKER'S YEAST

A. Yeast and the Leavening Gas

1. *Yeast Fermentation*

Dough can be leavened by carbon dioxide formed in the fermentation of sugars by the yeast. This process involves a sequence of 12 enzymic reactions (when starting from glucose), the sum of which can be represented as follows:

Glucose $\longrightarrow$ 2 Ethanol $+ 2$ CO$_2$

2 ADP $+ 2$ P$_i$ 2 ATP $+ 2$ H$_2$O

According to this equation, about 0.49 g of carbon dioxide (270 ml at NTP) is evolved for each gram of glucose fermented.

During alcohol fermentation, some side-reaction compounds are always produced, the main ones being glycerol and succinic acid (Oura, 1977a), and small amounts of a vast number of other components are also formed. Glucose is consumed to produce these components so that the practical yield of carbon dioxide from glucose is less than the value presented above. Furthermore, yeast cells multiply during dough fermentation, and sugar is used for synthesis of the new cell material. When yeast is transferred from its package to a medium rich in fermentable sugars (as dough is), it uses sugars not only as described above but also in a very simple way. Cells take up sugar and store it in the form of glycogen. This phenomenon is of negligible importance in brewery and distillery fermentations where the ratio of yeast to fermentable sugars is low, but it can be significant during dough fermentation. It is characteristically of short duration and requires a much higher yeast concentration than other fermentations (Reed, 1974). The relatively high amount of yeast in doughs is often able to use a substantial part of the fermentable sugars for glycogen synthesis.

Production of side-reaction components and, especially, the increase of cell material and/or cell number can consume a considerable amount of sugar. Practical measurements have shown that, during dough fermentations lasting some hours, only 70–75% of the glucose consumed has been fermented (Schultz *et al.*, 1943; van Niel and Anderson, 1941). The rate at which yeast produces carbon dioxide depends on the physical and chemical environment (pH value, temperature, osmotic pressure and concentration of sugar) and is discussed in Section III.B (p. 101). Naturally, this rate can be dependent on the amount of yeast, as indeed it is in a lean dough in the first part of the fermentation (during the second part, the mobilization of maltose from starch is the rate-limiting step). The fermentation rate at the enzymic level is limited by the slowest step in the reactions involved in the formation of alcohol and carbon dioxide. The potential activity of glycolytic enzymes does not limit the rate of fermentation, as can clearly be seen by comparing the rates of glucose consumption in fermenting yeast and in yeast growing anaerobically (Table 3). The enzymes of the glycolytic path are able to function 2 to 3 times as fast during anaerobic growth as they do during anaerobic fermentation. From the equation already presented, it can be seen that fermentation involves phosphorylation of ADP to ATP and would stop if the cellular ADP were completely used up in this process. Adenosine triphosphatase (s) (ATPase) in yeast are able to hydrolyse excess ATP to ADP and inorganic phosphate. The rate of anaerobic carbon dioxide production can be increased by encouraging

Table 3

The rates of glucose utilization by yeast during anaerobic glucose fermentation and growth

Condition	Glucose utilization [μmol min^{-1} (g dry yeast)$^{-1}$]
Anaerobic fermentation	135[a]
	80–120[b]
Anaerobic growth	310[c]
	290[b]

[a] Calculated from the value [400 mm^3 CO_2 h^{-1} (mg dry yeast)$^{-1}$] presented by Hoogerheide (1975).
[b] Calculated from values presented by Oura (1974).
[c] Calculated from value [maximal uptake of glucose: 18.5 mmol h^{-1} (g dry yeast)$^{-1}$ presented by Schatzmann and Fiechter (1974).

fermentation with yeast growth (Table 3), glycogen synthesis (E. Oura, unpublished observation) or some active-uptake process (Pena *et al.*, 1967). Each of these processes requires ATP and they thus all result in a facilitated ADP regeneration. Regeneration of ADP seems to be of importance in yeast glycolysis and may even be the rate-limiting step. Racker (1976) has stressed the importance of ATPase as a necessary participant in yeast glycolysis by including it among the glycolytic enzymes.

In laboratory cultivations, yeast is able to produce 32 mmol of carbon dioxide (g dry wt)$^{-1}$ h^{-1} (continuous cultivations by Schatzmann and Fiechter, 1974), corresponding to a consumption of 2,880 mg of hexose (g dry wt)$^{-1}$ h^{-1}. Atkin *et al.* (1945) reported a rate of 730–850 mg of hexose utilized h^{-1} (g fresh wt)$^{-1}$ during the last (third) hour of fermentation in a suspension of flour in water containing asparagine. With modern compressed yeasts, this value can be approached, or even attained already during the first hour of panary fermentation, and is later exceeded.

Baker's yeast has a constitutive enzyme system for fermentation of glucose and fructose. Although it is affected by the degree of catabolite repression, the saccharase content of yeast is, in all cases, active enough to hydrolyse sucrose to glucose and fructose faster than these components are fermented. Besides the functioning of the normal glycolytic pathway, the fermentation of maltose also needs the operation of α-glucosidase and maltose permease. These are normally inducible, and are very sensitive to catabolite repression. In many present-day strains of baker's yeast, some of the α-glucosidase isoenzymes are constitutive, but nevertheless a vigorous fermentation of the main sugar in dough, maltose, starts at full strength after a more or less prolonged induction period following consumption of glucose and fructose. Milk or defatted dried milk are frequently included in a bread formula, and baker's yeast is not able to ferment lactose present in milk. Baker's yeast is unable to do so because β-galactosidase induction and even adaption of yeast to galactose fermentation (which is, besides glucose, the second sugar component in milk sugar) take a fairly long time for which a normal dough fermentation is insufficient.

2. Carbon Dioxide

The leavening gas, carbon dioxide, weighs 1.977 g l^{-1} at NTP.

Corrected for water vapour and expansion to 25°C, the volume obtainable from 1 g of sugar is theoretically 281 ml (Atkin *et al.*, 1946). Its solubility in the water at 25°C is 0.1449 g in 100 g of water when the total pressure (i.e. the sum of the partial pressure of the gas plus the aqueous tension) is 760 mmHg (100 KPa). This solubility is far greater than those for oxygen and nitrogen (0.0039 and 0.0018 g 100 g^{-1} respectively; Lange, 1967). Over 99% of carbon dioxide in pure water is in the form of dissolved carbon dioxide, with the remainder existing as H_2CO_3, H^+ or HCO_3^-.

Carbonic acid is a weak, unstable acid which is only very slightly ionized and, hence, contributes only minimally to lowering the pH value of dough. In the absence of materials that can combine with the acid, its solubility (neglecting molecular carbon dioxide) is in essence no different from the solubilities of most gases (Umbreit *et al.*, 1972). Indifferent acids and salts decrease the solubility of carbon dioxide only slightly (Karwat and Kruis, 1957). When the pH value of the solution increases to alkaline, however, the situation changes, although at the pH value of normal doughs (pH 5 to 6) the amount present in a bound form can be considered negligible for most purposes (Matz, 1972).

The water phase of dough is very complex, containing sugars, ethanol and mineral salts. It has been estimated that about 55 mg of carbon dioxide dissolve in 100 g of dough or about 7.5% of the total carbon dioxide evolved in one hour (Reed and Peppler, 1973). As the water phase heats up during baking, it loses much of its ability to retain carbon dioxide which is released in the form of bubbles.

B. Effect of Environment on Baker's Yeast

1. *Effect of Temperature*

Baker's yeast is able to grow at temperature up to about 42°C (Stokes, 1971), and its fermentative activity increases by a factor of 1.6 to 2.8 for each 10°C increase in temperature starting in the range 20–35°C (Atkin *et al.*, 1946; Reed and Peppler, 1973). Temperature has an effect on the texture of dough, and temperatures below the maximum are normally used for leavening, usually 35–37°C, occasionally temperatures of even 50°C can be used. As the temperature rises further, a critical point is reached at 54°C (Wood, 1956), when only about 5% of the yeast population in a suspension at pH 7 can survive for 15 minutes (Cerny,

1980). During baking, yeast does not have much time to benefit from the elevated temperature because it takes only 10–12 minutes before the temperature even in the centre of the loaf is too high (55°C) for yeast to tolerate (Walden, 1955; Marston and Wannan, 1976).

Compressed baker's yeast freezes at about $-1.5°C$ (Buckheit, 1971), and Bailey et al. (1940) have recommended that it be stored at a temperature of $-1°C$. This temperature, however, is not in general use. Yeast cells, although frozen, are not killed. By carefully adjusting the conditions, the speed of freezing, and the temperature and time used for thawing, most cells can be kept alive. When cells are cooled rapidly to $-30°C$ or below, fewer than 0.01% survive, whereas when they are cooled slowly, up to 50% survive (Mazur, 1961a,b). The death of yeast cells at sub-zero temperatures is a consequence of intracellular ice crystals. During freezing, ice is formed first in the extracellular medium. The intracellular water will initially be supercooled and, because of its relatively high vapour pressure, it will flow out of the cells and freeze outside. Consequently, if cooling is slow, leakage will equilibrate the vapour pressures and no intracellular ice will form. The cells will in fact be dried.

In contrast, the best thawing process to obtain a maximum number of survivors should be rapid (Lorenz, 1974). Survivals after freezing and storage above $-10°C$ are high regardless of the cooling and warming temperatures (Mazur, 1961a,b; Mazur and Schmidt, 1968; Hsu et al., 1979).

2. Effect of pH Value

The internal pH value of the yeast cell lies in the range 5.1 to 6.3 (Barton et al., 1980), the most quoted value in the older literature being pH 5.8. The cell membrane of baker's yeast efficiently isolates it from the surroundings, and the intracellular pH value, for example, remains in the physiological range even when the extracellular pH value varies considerably. This becomes apparent when yeast is grown in a medium containing large amounts of sugar and no attempt is made to control the pH value. The pH value of the medium may fall as low as pH 2.2 before growth ceases (Hartelius, 1934). Additions of acids to a medium containing sucrose have shown that the pH value must fall to pH 2.3–2.7 before growth of yeast is prevented (Weldin, 1925). These limits, however, hold only if the acidity is due to strongly dissociated compounds, such as inorganic acids. The undissociated forms of weak acids, such as aliphatic carboxylic acids (but not for example lactic, pyruvic or succinic acids) can easily penetrate into the cell interior (Oura *et al.*, 1959).

Although surprisingly resistant to an increase in the hydrogen ion concentration of the medium, yeast is more sensitive to alkaline medium; it is well known that the amounts of glycerol and acetic acid are substantially increased if an alkaline fermentation is carried out (Neuberg and Hirsch, 1919). This mode of metabolism is called Neuberg's third form of fermentation. The effect can be seen even at pH 6 when yeast is fermenting glucose, and growth, although it does proceed in alkaline media, is very poor above pH 7.7–8.0 (Neish and Blackwood, 1951).

The optimum acidity for leavening action in dough is between pH 4 and 6 (Cooper and Reed, 1968), but, in rye sour dough or in San Francisco sour dough (French bread), the yeasts *Saccharomyces cerevisiae* and *Sacch. exiguus*, respectively, have to operate for at least a couple of hours at a pH value of less than 4 (Mathason, 1978) in the presence of considerable amounts of acid, e.g. acetic acid (Galal *et al.*, 1978).

3. Effect of Osmotic Pressure

Baker's yeast is fairly sensitive to high osmotic pressures generated by sugar or salt, or both, in dough. The yeast internal pressure is most often quoted to be equivalent to 0.8 M sugar, which corresponds, for instance, to about 27% (w/v) sucrose, 14% glucose, or 2.3% NaCl (calculated as fully dissociated). A dough of 280 g of flour and 160 ml of water containing 44 g of sucrose would have a sugar concentration of 0.8 M. On addition of yeast to this system, hydrolysis of sucrose starts immediately, rapidly proceeding to completion. Because yeast saccharase splits sucrose at a rate 300 times as rapid as the rate of hexose fermentation (Demis *et al.*, 1954), the osmotic pressure nearly doubles from the original value. Moreover, during the hydrolysis, some of the water is bound in the newly formed hexoses, which further increases the molarity of the sugar solution. The amount of sucrose in this example is small compared with the amounts used in sweet doughs, about 35 to 50 g per 100 ml of water in the dough. Then the molarity of sucrose is 1.9–2.7 at the beginning and increases from this value.

The adverse effect of high osmotic pressure is counteracted by using two to three times as much yeast as when there is little added sugar. The use of dried yeast with low saccharase activity also helps (Reed and Peppler, 1973). Osmotolerant yeasts for leavening high sugar and fat doughs have been isolated and patented (Windisch and Steckowski, 1971a), even though osmotolerance as such was found not to be the only important factor in good leavening (Windisch and Steckowski, 1971b). Subsequently, Windisch *et al.* (1976) demonstrated that non-osmotolerant hybrid baker's, brewer's and distiller's yeast can have good leavening ability even in doughs where the osmotic pressure is high. Their idea was to use these yeasts as substitutes for chemical leavening agents.

4. Effect of Some Other Physical Factors

The volume of the dough after leavening largely depends on the volume of carbon dioxide evolved. Atmospheric pressure can vary between 720 and 800 mm Hg (96 and 107 KPa), which can cause a difference of about 10% in the volume of the carbon dioxide evolved. In test bakings, the leavening value obtained is often corrected to normal atmospheric pressure.

Background radiations from nature are not expected to have much effect on yeast in dough. When yeast is exposed to ultraviolet radiation or X-rays, lower doses are mutagenic and higher doses are lethal. Although microwaves (2450 megacycle) are bactericidal (Napleton, 1976), frozen dough can be carefully thawed in a microwave oven. Should the yeast become inactivated, it is presumably because of excessive heating and not the action of microwaves.

Against ultrasound, yeast cells are more resistant than bacterial cells. Ultrasound together with 0.4–0.8 MPa pressure has been used in bread-making (Auerman *et al.*, 1977). The treatment was claimed to lower process time and improve bread quality.

5. Effects of Organic Acids and Ethanol

Various species of lactic-acid bacteria are used to make sour dough breads. Lactic acid is the dominant acid produced. In San Francisco sour dough (French bread), the acids comprise approximately 70% lactic, 25% acetic and 1% C_3–C_5 volatile organic acids, the rest being unidentified (Galal *et al.*, 1978).

Acids at their usual concentration in sour dough (about 80 $\mu eq/g^{-1}$) are more inhibitory to baker's yeast than to sour dough yeast fermentations (Sugihara et al., 1970). Only at high concentrations does lactic acid have an effect on the fermentation of various sugars by yeast (Schulz, 1972), possibly because of impermeability (Oura et al., 1959). Non-dissociated acetic acid easily enters the cells and it, in particular, inhibits fermentation (Sugihara et al., 1970). According to Samson et al. (1955), acetic acid inhibits the uptake of phosphate, glucose fermentation and respiration of ethanol and, further, the fermentation of fructose 1,6-bisphosphate and 3-phosphoglycerate in cell-free extracts. Decarboxylation of pyruvate (Samson et al., 1955) and utilization of endogenous reserve carbohydrates (Guiraud et al., 1972) are stimulated. Efforts have been made to circumvent the deleterious effect that aliphatic organic acids at low pH value can have on the leavening action of baker's yeast by using separate sub-doughs with lactobacilli and yeast in the making of, for instance, Russian wheat bread (Shcherbatenko et al., 1972).

As the concentration of ethanol rises during fermentation, both fermentation and growth of yeast gradually cease. *Saccharomyces* species are in general quite tolerant to ethanol, although their growth slows down at 2% alcohol in the medium and completely stops at about 10%. Fermentation is more resistant to alcohol, the corresponding values being approximately 5% and 12% (Ingram, 1955). Assuming that 3.2–5.5% of 100 g of flour is fermented (p. 116) during leavening, the ethanol produced will be about 1.6–2.8 g. If the water:flour ratio is 160:280, the content of ethanol in water can rise to 2.8–4.8 % (w/v). The upper part of this range is definitely inhibitory to yeast, which means that the leavening process slows down towards the end.

Nagodawithana et al. (1977) found that ethanol inhibits non-competitively the first enzyme (hexokinase) in the Embden-Meyerhof-Parnas pathway of sugar fermentation. α-Glycerophosphate dehydrogenase is inhibited as well, which could be of importance in dough fermentation, for it would lower the amount of glycerol formed and so channel more sugar in the desired direction of ethanol and carbon dioxide production (Oura, 1977a).

Concerning the inhibitory effect of ethanol, it has been reported that, in fermenting yeast, the concentration of ethanol is higher inside the cells than outside (Navarro and Durand, 1978; Rose, 1978).

6. Effect of Mineral Salts

Besides suitable sources of nitrogen, phosphorus and sulphur, yeast requires a supply of the following minerals in order to grow: potassium, magnesium, iron, copper, zinc and manganese (Oura and Suomalainen, 1977). Chlorine appears to be essential for growth but sodium seems to be unnecessary, and although apparently not necessary, calcium does stimulate growth of yeast cells (Oura and Suomalainen, 1977). Sufficient concentrations of iron, copper, zinc, manganese and potassium are available in molasses used for cultivation of baker's yeast. The only elements that have to be added to molasses wort are nitrogen, phosphorus and magnesium. Yeast suspended in dough is able to ferment the sugars of the flour and even to grow because, although the yeast's own reserves of necessary elements may be inadequate, the flour and water mixture usually contains enough minerals. Nevertheless, prolonged fermentations may require some supplementation (Atkin et al., 1945; Reed and Peppler, 1973).

The relative concentrations of copper, zinc and iron are important in the propagation of yeast. High iron and low copper and zinc (and manganese) have been reported to lower the protein content and low iron and high copper, zinc and manganese to increase the protein content of yeast (Ringpfeil et al., 1974).

Only few salts or ions normally occur in water in sufficient concentration to exert any important effect on dough fermentation. Some of these, particularly calcium sulphate, although having no effect on yeast itself, are beneficial to doughs by strengthening gluten (Matz, 1972). Chlorine and fluorine, at the concentrations present in tap water, seem not to have any effect on yeast growth or fermentation in dough. According to White's investigations (1954), cadmium is the most poisonous element for yeast growth, followed by copper. In the classical study of Finney et al. (1949), the depressant effect of several cations on dough fermentation was reported as (expressed as the minimum inhibitory concentration in mg per 100 g of flour): cadmium chloride, 0.1; zinc chloride, 0.1; mercuric chloride, 1; cupric chloride, 10; nickel chloride, 10; lead acetate, 10; and stannous chloride, 100. The order of these minerals differs somewhat from that presented by White (1954). Matz (1972) has described a case where water treated with a copper salt to prevent growth of algae was the cause of the production of unmarketable bread, and that further investigations revealed that as little as 3 p.p.m. of copper sulphate has a profound effect on fermentation. Furthermore, cadmium entering a dough from plated mixer arms has been known to be sufficient to retard fermentation (Matz, 1972), apparently by suppressing yeast metabolism. A similar effect has been proposed for vanadium. Thus, the trace elements present in water may play a more important role in breadmaking than is ordinarily believed.

7. Effect of Flour and Dough Components

Flour contains proteins and amino acids, polysaccharides and sugars, minerals and vitamins, all of which are beneficial to the leavening action of yeast. The free sugars in flour are glucose, fructose, sucrose, maltose and glucofructosans. They make up about 1–2% of the flour (see p. 115). The gassing activity of yeast in dough is mainly dependent on fermentation of glucose, fructose, maltose and, if added to the dough, sucrose (Cooper and Reed, 1968). Maltose fermentation as such, for instance, is not a prerequisite for good leavening action, the yeast may be able to ferment maltose only very poorly and still raise the dough excellently (Suomalainen et al., 1972). This is somewhat surprising since starch from flour is mainly degraded to maltose by β-amylase during dough making. It could be that glucose in the dough inhibits maltose transport (Hautera and Lövgren, 1975).

The presence in dough of glutathione, an oxidizing–reducing agent, can cause slackening by reducing disulphide bonds in dough proteins. Addition of 2% inactive dry yeast to enhance the flavour of the bread causes a definite slackening effect because of the presence of glutathione at 7 to 12 mg per g of yeast (Reed and Peppler, 1973).

Ascorbic acid, cysteine, potassium bromide and azodicarbonamide are other additives that react with the disulphide and thiol groups in the gluten chains of proteins causing oxidations and reductions (Wood, 1980). Here again the effect is restricted to the dough rheology; these compounds at the concentrations they are used in dough have no effect on yeast metabolism.

Mould inhibitors are routinely added to some commercial bakery products. Because yeast is a unicellular mould, additions like calcium or sodium propionates, sorbic acid or sorbates and benzoates are expected to affect yeast too. Sorbic acid is used principally in chemically leavened baking goods, normally at 0.025–0.10% of the weight of the flour.

Because it can be used over a broader pH range than propionate and benzoate it has also been used in yeast-leavened bread. It has been observed that sorbic acid at the concentrations used in some bread varieties substantially inhibits the fermentation if the yeast strain used is not specifically selected for its resistance to sorbate. It is a stronger inhibitor of fermentation than is propionic acid, and it has been recommended that it should be used only for treating the bread surface or the packing material (DeSa, 1966).

8. Yeast Stimulants and Yeast Food

The following vitamins are beneficial for yeast fermentation: biotin, pantothenic acid, inositol, thiamin, pyridoxine and niacin (Peppler, 1960), although all baker's yeast strains have an absolute requirement only for biotin (Oura and Suomalainen, 1978).

Yeast requires biotin for some carboxylation reactions, e.g. in biosynthesis of fatty acids and aspartate. Panthothenic acid is a structural component of coenzyme A, which participates in fatty acid synthesis. Inositol in the cell is required as a structural component of phosphatidylinositol and phosphoinositol-containing sphingolipids, and may additionally play a direct role in the synthesis or secretion of yeast glycans (Hanson and Lester, 1980). Thiamin in the form of thiamin pyrophosphate acts as the coenzyme for pyruvate decarboxylase and some other enzymes. Pyridoxine, as pyridoxal phosphate, has an important role in the metabolism of amino acids. Niacin, which some baker's yeast strains synthesize via tryptophan (Oura, 1977b), is used to synthesize nicotinamide adenine dinucleotide and its phosphate, which are required for a multitude of dehydrogenation and reduction reactions in the cell.

In 1945, Atkin et al. started to study the effect of different additives on dough fermentation rate. Starting from the well-known fact that fermentation in dough is more vigorous than fermentation of a sugar solution, they showed that some of the substances commonly considered to be necessary for growth also have an accelerating effect on gas production by yeast fermenting a synthetic medium. Although fermentation in the dough proceeds faster than in synthetic media (Hoseney et al., 1969) even here it is possible to stimulate gas production by using the above-mentioned compounds. Potassium, magnesium, ammonium, sulphate and phosphate are the principal inorganic ions required.

The term 'yeast food' is primarily understood to mean the nitrogenous food and the mineral salts taken up by yeast in order to stimulate its fermentation and growth. Other salts may be present in yeast food, which act as dough conditioners by being oxidizing agents or water correction salts, or both. As in doughs, the only nutrient that has any effect on gassing rate is the ammonium salt (Maselli, 1959). For short-term fermentation, yeast needs no exogenous nutrients because of its own reserves, as has already been mentioned. Further, most, if not all, of the accelerating substances used in yeast food are present in dough in sufficient amounts to support fermentation at optimum rates for a time. For prolonged leavening and prefermentations, the addition of minerals and suitable nitrogen sources increases the gassing rate. Thus, adding ammonium chloride at 0.15% of the flour weight shortens the proof time from 65 to 55 minutes, the effect being similar with both compressed and active dried yeast (Peppler, 1960). The principal mechanism of this action is (as discussed on p. 100) the ADP level. When growth is possible, as it is in the presence of assimilable sugar and nitrogen source, it couples regeneration of ADP to ATP usage and the growth rate is not limited by a deficit of available ADP. Thiamin (vitamin B_1) can markedly increase the rate of fermentation. When yeast is used in preferments, which contain no flour and hence no thiamin, the thiamin must be added separately. For this reason, baker's yeast in U.S.A. contains more than 50 mg of thiamin per g yeast dry matter (Reed and Peppler, 1973).

Some calcium salts are normally included in yeast nutrients. They have no effect on fermentation but, in doughs made with very soft water, addition of these salts somewhat lowers proof time because of their gluten-strengthening function (Reed, 1972; Matz, 1972); with water of normal hardness the effect of calcium in yeast nutrients is insignificant.

IV. YEAST IN BREADMAKING

A. Leavening Agents

There are various materials that can provide the leavening or 'raising' effect in baking. Most baking products are leavened by the fermentative action of yeast, with the carbon dioxide produced acting as the leavening gas. In some cases, for instance in doughs rich in added sugar, yeast cannot function as intended, because it is sensitive to high osmotic pressures. Air incorporated during whipping is a commonly used and very effective aerator of sweet cakes based on egg-white foam. A number of chemical leaveners are also in use. The most common alternative to yeast-leavening systems is based on chemical production of carbon dioxide from the reaction of sodium bicarbonate with an acid salt. This reaction is initiated by moisture when mixing into the dough and is completed by heat during baking. Since sodium bicarbonate is soluble in water, the leavening acid determines the conditions during which carbon dioxide is released. In some cases, ammonium bicarbonate can be used which, during the heat of baking, decomposes to ammonia gas and carbon dioxide, which together form the leavening gas. The use of this salt is limited to products baked to a low moisture content, since the odour of ammonia will remain if the water content remains high in the finished product. Baking powders based on sodium bicarbonate are themselves restricted by their characteristics, and the usual maximum is 6% on the weight of flour (Lawson, 1962). The total carbon dioxide evolution from this amount, 214 ml, is less than that obtainable from yeast (2.5% on dough weight yields 350 ml of carbon dioxide in an hour; Reed and Peppler 1973), and using higher levels of baking powder can lead to undesirable aromas in bakery products. The use of chemical leaveners in frozen doughs or instant-bread mixes reveals a second disadvantage. When bread is made by chemical leavening, as is desirable for sportsmen, the military and in all kinds of emergencies and rescue operations, the result, although resembling a yeast-leavened loaf in grain and texture, has a flat flavour that is cereal-like and lacks the aroma of fermented dough. The flavour has been improved by adding suitable aroma substances to the dough, e.g. lyophilized flavour broth prepared by fermenting glucose in the presence of non-fat milk solids (Matz *et al.*, 1958).

Oxygen generated from peroxide by the action of catalase has been proposed as a leavening gas (Selman, 1953), or even small pieces of ice can be mixed into the dough just before transfer to a very hot oven, where the water vapour formed acts as the leavening gas. This kind of method can be used in the baking of crisp bread (knacke bread). In all doughs, water vapour makes a considerable contribution to the volume increase of the carbon dioxide bubbles in bread, and ethanol formed in the leavening process may play a role in expanding the vesicles formed in other ways. Dissolved carbon dioxide in dough becomes available when the dough is heated during baking. Further, the size of carbon dioxide bubbles can be increased by vacuum.

Although it is possible to replace yeast in the production of some bakery goods, and its role in altering wheat gluten structure (see p. 117) can be partly played by mechanical work, yeast is still needed to produce flavour and aeration. It seems at present that there is no ready substitute in view for yeast (Axelsen, 1974).

B. Methods for Making Dough

Detailed accounts of the science and technology of baking include that by Pyler (1973). There are four basic methods for making the dough of wheat bread, two of which, the traditional sponge and dough process and the straight-dough process, are used in modern bakeries. The other two, which represent newer technology, are the continuous dough-mixing and the liquid ferment systems. The two major objectives of the dough-mixing process in breadmaking are the thorough and uniform dispersion of ingredients to form a homogeneous mixture and inducing physical development of the gluten structure in the dough.

The straight-dough method is a single-step process, in which all the ingredients are mixed together in a single batch. In the sponge and dough method, the major fermentative action takes place in a preferment (referred to as a sponge), in which normally more than one half of the total dough flour is subjected to the action of an active yeast fermentation. Several variations and modifications of both methods have evolved over the years.

A recent innovation is the accelerated dough-development method based on ultrahigh-speed dough mixing. Doughs developed by this method are treated as no-time doughs, i.e. they are not subjected to bulk fermentation. The necessary fermentative flavour production and carbon dioxide generation are achieved in the intermediate and final proofing stages. This principle has been applied, for example, to the Chorleywood Bread process in Great Britain (Axford et al., 1963). The method is also characterized by the use of ascorbic acid as an oxidant and an increase in the yeast level by 50–100%. This no-time dough principle can be successfully applied to the production of bread with continuous mixers. Special interest in the development of this kind of method has been shown in U.S.A., e.g. Do-Maker and Amflow systems (Snyder, 1963).

The liquid-ferment process has been modified to simplify the sponge and dough method by replacing the sponge stage with a liquid, pumpable ferment, the basic formula of which normally includes water, yeast and yeast nutrients, and other ingredients such as salt, sugar or non-fat dry milk. Nowadays, the liquid-ferment system has been widely adapted to continuous dough mixing processes especially in the U.S.A.

In contrast to the American systems, the ingredient incorporation and mechanical dough development in European continuous systems are in most cases done by a single machine. This usually cannot provide full mechanical dough development, and the mixed dough has to be subjected to bulk fermentation of variable duration to attain proper maturity.

Production of rye bread is normally based on the use of sour dough in methods that may be considered as modifications of the sponge and dough method. In this case, only the sponge consists of sour ferment. The dough-mixing process is different from that used for wheat flours; rye doughs are mixed slowly without great energy input to prevent their becoming too tough through development of the very viscous properties of pentosans. As the proportion of wheat flour in a mixed rye–wheat bread increases, the gluten-forming properties of the wheat flour become more important (Drews and Seibel, 1976).

After the bulk or panary fermentation, the dough goes to the so-called make-up stage, which can include such operations as dividing, rounding, intermediate proof, moulding, panning and final proof. After the proof periods, the final stages of the fermentation occur as the temperature of the loaf is raised in the oven, but above about 60°C the yeast is inactivated and fermentation ceases.

C. Role of Yeast in Breadmaking

Yeast performs three main functions in a panary fermentation (Bennion, 1967). It produces carbon dioxide, in sufficient quantities and at the right time, to inflate the dough and produce a light spongy texture that will result in a palatable bread when correctly baked. The yeast also helps bring about essential changes in the gluten structure, known as 'maturing' or 'ripening' of the bread. Moreover, it produces a complex mixture of chemical compounds that contribute to the flavour of the bread.

In addition to producing carbon dioxide, the lactic acid-forming bacteria also produce acids which, by lowering the pH value of the dough below the optimum for rye amylase, decrease the amount of additional maltose formed by amylases, thus preventing the bread from becoming too sweet. Moreover, the acids contribute to the flavour of the finished bread and enhance the storage properties.

1. Dough Fermentation

Flour is a varied and complex mixture which, however, is quite stable when kept dry. The formation of dough initiates a series of chemical and physical changes which do not cease even when the dough has been baked into bread (Atkin et al., 1946). Although any interpretation of the changes in concentrations of different sugars in bread doughs is complicated by production of maltose from starch and of fructose (and sucrose) from glucofructosans by the action of flour enzymes, the following values for wheat flour are generally accepted.

The sugar content of wheat flour varies from 1–2% depending on the kind of flour used. The sugars in flour are glucose (about 0.01% of the flour), fructose (0.02%), sucrose (0.10%), maltose (0.07%), raffinose, and a series of oligosaccharides composed of D-fructose and D-glucose residues, called glucofructosans (Colin and Belval, 1935a; Koch et al., 1951; Williams and Bevenue, 1951; MacKenzie, 1958; Täufel et al., 1959; Vaisey and Unrau, 1964; Tanaka and Sato, 1969; Saunders et al., 1972; Suomalainen et al., 1972; Varo et al., 1979). An important point is that 60–75% of the free sugars in flour consist of fructose-containing sugars other than sucrose, i.e. glucofructosans and raffinose (Colin and Belval, 1935a; MacKenzie, 1958; Tanaka and Sato, 1969; Saunders et al., 1972). The term levosine is often used to describe these sugars in this connection. They were originally isolated from cereals by Tanert (1891a,b), and were reported by Colin and Belval (1935b) to yield 1 mol of glucose and 9 mol of fructose. The term levosine is frequently used quite loosely; here the expression 'glucofructosans' will be used instead of the word 'levosine' used in the cited references.

That the fermentation of the mixture of sugars in dough proceeds in several phases was elegantly demonstrated by Larmour and Bergsteinsson in 1936 by following carbon dioxide production during panary fermentation. Changes in the levels of different sugars during the proof period have been more clearly shown in the studies of Koch *et al.* (1954), MacKenzie (1958) and Suomalainen *et al.* (1972). According to these workers, the first change is a rapid disappearance of sucrose. It has always been held that the saccharase activity of yeast is not a limiting factor during panary fermentation, and it is certainly true that sucrose is hydrolysed faster than its products are fermented. Of the hexoses present as free sugar in the flour, or formed by inversion of sucrose (or by hydrolysis of glucofructosans), glucose is consumed more rapidly than fructose, particularly at the beginning of the fermentation period. During the time needed for fermentation of these sugars, the content of maltose in the dough increases through the action of flour amylase which begins to work when the flour is mixed with water. When the easily fermentable sugars have nearly disappeared, gas production from maltose as substrate will start. Between these two fermentation phases there exists a more or less clear induction period (an interlag phase), as can be seen in the characteristic 'camel-back' appearance of the zymotachegram where the first hump lasts about 1 hour and the second begins after a little less than a further hour (Larmour and Bergsteinsson, 1936).

Changes in sugar concentrations during fermentation of a straight dough are shown in Figure 3. The dough consisted of 280 g of wheat flour and 160 ml of water containing 4% salt, which was leavened with 5 g of yeast. Tanert (1891a,b) described levosine as an easily hydrolysable but non-fermentable carbohydrate, and Geoffroy (1935) showed it to be practically unfermentable in pure solutions. Colin and Belval (1935a,b) and Geoffroy (1935) agree that most of it disappears during fermentation, and Guillemet (1935) found that, in a yeast–water suspension, 30% of the glucofructosans disappeared in 24 hours. Later it was shown that glucofructosans are rapidly hydrolysed during the fermentation period in breadmaking (de Grandchamp-Chaudun, 1950). MacKenzie (1958) reported that glucose-difructose could disappear from the dough during the proof time nearly as quickly as sucrose. Hydrolysis (and fermentation) of the fraction of higher glucofructosans and raffinose, however, occurred more slowly, and was dependent on the yeast strain used. He demonstrated also that baker's yeast was able to ferment an isolated levosine fraction that probably contained more than five hexose units. Colin and Belval (1935a) showed that enzymes in the flour are inactivated if the flour is boiled in 95% ethanol: the residue then yielded only 0.15% alcohol. This amount came from the small proportion of

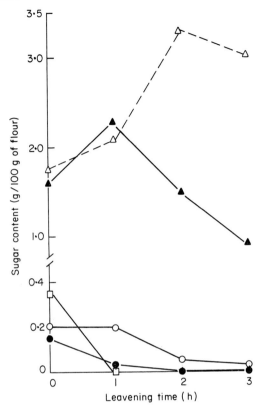

Fig. 3. Time-course of changes in the contents of sucrose (□), glucose (●), fructose (○) and maltose (△ in non-yeasted dough and ▲ in yeasted dough) in a dough during test leavening and of maltose in non-yeasted dough. From Suomalainen *et al.* (1972).

fermentable sugars (glucose, fructose, sucrose and maltose) originally present in the flour, which indicates that hydrolysis of glucofructosans requires the action of enzymes in the flour. The results presented by Saunders *et al.* (1972) also give the impression that flour enzymes are responsible for hydrolysis of these oligosaccharides.

Free sugars in 100 g of flour (typically 1–2%, of which at least 60% consists of hexoses, sucrose and utilizable glucofructosans) can theoretically produce 165–330 ml of carbon dioxide, or, assuming a practical yield of 75% (see p. 98), 125–250 ml. It appears that, in a straight-dough fermentation, the bulk of the fermentation occurs mainly at the expense of free sugars in the flour, as has been shown by comparing gas production in a dough of normal flour and flour with inactivated amylase (Fig. 4). Moreover, measurements of gas evolution in the fermentation of individual sugars in synthetic nutrient solutions (Atkin *et al.*, 1946; Koch *et al.*, 1954) led to the same conclusion.

Changes occurring in the sugar composition of dough during sponge and dough fermentations have been reported by Koch *et al.* (1954), Lee *et al.* (1959), and in a preferment dough system by Piekarz (1963). Koch *et al.* (1954) have followed changes in sugars of a dough containing added sugars.

From the observed volume of carbon dioxide evolved during leavening, it is possible to estimate the decrease in dry matter content (fermentation loss) during leavening. In a straight dough fermentation, it is 3.2% and, in a sponge dough, about 5.5% of the total solids. According to Fransson (1977), during leavening of French bread 1.4 g of sugars (calculated as glucose) are lost in an hour from 100 g of flour. In the Chorleywood process, the yield of bread from flour is about 4% higher than in bread made by traditional methods, the difference being largely due to a substantial decrease in fermentation losses (Elton, 1969).

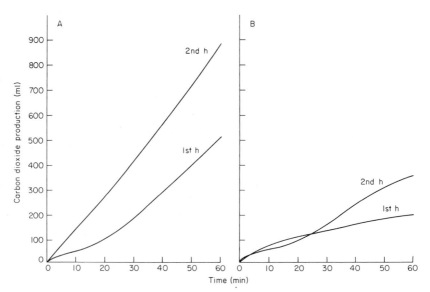

Fig. 4. Time-course of carbon dioxide production during the first and second hours of a test leavening using normal dough (A) and dough made from autoclaved flour (B).

2. Ripening of the Dough

One of the principal functions of flour is to form an elastic dough able to retain the gas produced by yeast fermentation. Bread, however, is not made from a simple gas-charged dough, for the dough has to attain a definite degree of 'maturity' or 'ripeness' before it can produce good bread. Changes occurring in the dough during this maturation period are closely connected with certain changes in protein structure, resulting in a stage where the balance between extensibility and the resistance of the protein structure lies between fairly close limits. At this point, the protein network in dough is able to retain carbon dioxide as small gas cells and is elastic enough to expand without disruption when the fermentation proceeds and again when the gas volume increases in the heat of the oven. This conditioning is believed to be brought about by chemical changes in the bonds of gluten and other wheat proteins. Protein helices in the dough are originally tightly coiled. The mechanical work done during mixing of the dough breaks some of the hydrogen and disulphide bonds linking neighbouring protein chains, and the new bonds formed are in different positions. Carbon dioxide produced during fermentation is held within this three-dimensional network. As more gas is produced, the gas bubbles covered with a thin protein film expand, and this mechanical work on the protein network causes a further change of bonding to take place. When the dough is ripe, the position of the bonds between adjacent protein chains is such that the relationship between rigidity of the protein network and its extensibility is optimal (Stafford, 1967). Allowing the leavening to continue beyond this stage leads to a 'dropping' of the dough at the point when the gluten mesh collapses and gas leaks away (Bennion, 1967).

The bulk fermentation stage necessary for this ripening can be shortened or even eliminated by intense mechanical working of the dough. For instance, the Chorleywood bread process brings about mechanically in a few minutes changes in the elastic properties of the dough that in normal breadmaking processes require several hours to occur by the action of yeast fermentation. Even the Chorleywood process, however, still requires yeast for aeration and to produce the flavour.

3. Aroma Formation

In addition to any salty, sweet or sour taste, the flavour of bread is also the result of the concerted effect of the aromas originating from the microbial fermentation and the thermal reactions during baking. The components of the dough formula itself, except for sugar, salt and mould inhibitor, normally contribute relatively little to the flavour of white bread sold today (Peppler, 1960).

Suomalainen and Nykänen (1972) have shown that the aroma components produced by yeast are qualitatively almost identical in a nitrogen-free sugar solution and in alcoholic beverages, such as beer, wine, whisky and cognac. These same compounds are also produced during the fermentation of dough, ethanol and carbon dioxide being the main products. Only a little ethanol remains in the bread (Rothe and Thomas, 1959; Smith and Coffman, 1960), most of it in fermented dough escapes with the oven gases. Some other alcohols, namely propanol, butanol, isobutanol, pentanol and isopentanol, although present in dough in only trace amounts, are still detectable in the finished bread presumably because their higher boiling points prevent complete vapourization in the hot oven. Bacteria associated with yeast and flour are responsible for formation of some organic acids during fermentation (Robinson et al., 1958a). Acetic and lactic acids are the major products, but traces of other acids are also formed. Organic acids in normal white bread subtly affect the flavour (Johnson and Sanchez, 1972), whereas in sour breads their influence is more pronounced. The organic acids present react with ethanol to produce the ethyl esters that are found in minor quantities in bread. Minor quantities of carbonyl compounds, including acetaldehyde, propanal, butanal, pentanal and furfural, are produced during fermentation, and become more concentrated in the bread crust produced during the baking process (Linko and Johnson, 1963).

Some aroma components, including propan-1-ol, acetaldehyde, propanal and furfural, are common to both bread leavened with yeast and bread baked using baking powder, while others such as ethanol, isobutanol (2-methylpropan-1-ol), isopentanol, optically active pentanol (2-methylbutan-1-ol), butanal, pentanal, acetic acid and ethyl acetate have been detected only in yeast-leavened bread (M. Mäki-Esko and E. Oura, unpublished observations). Yeast is most probably responsible for many more aroma components than those listed here; at least 211 different ones have already been identified in different breads (Rothe, 1974).

The effects of yeast metabolism on dough are not limited to the aroma compounds produced during yeast fermentation. Some fermentation byproducts appear to be intimately involved in the development of the bread aroma complex by enhancing the rate of browning reactions and formation of melanoids and caramel polymers, in the bread crust during the baking process (El-Dash, 1971). Fermentation in dough results in a considerable depletion of the free amino acids formed in the dough, and these amino acids are possible reactants in the Maillard reaction, which is a condensation between a reducing sugar and an amino acid. It appears that the pale coloured and insipid bread resulting from over-fermented dough may be largely caused by a vast depletion of amino acids during the long fermentation process (El-Dash, 1971). Moreover, bakers customarily add sufficient sugar to the dough to assure an excess for reactions with free amino groups during the baking process and, in any case, yeast can use much of the maltose produced during leavening and thus produce a pale coloured and insipid bread (Johnson and Sanchez, 1972).

Yeast itself has a characteristic odour and flavour, which is diluted beyond recognition by other ingredients of the formula. Brasch (1951) attributed this odour to yeast fat. Bacteria accompanying the yeast as well as other ingredients, such as flour and defatted milk, are similarly flavourless in bread. Bacteria that are always present in compressed yeast number from 100 to 200 million per g (Reed and Peppler, 1973). When compressed, yeast contains about $20 \cdot 10^9$ cells in a gram (Vraná *et al.*, 1968) and a yeast cell weighs about 50 times more than one bacterial cell (assumed to have $10 \cdot 10^{-13}$ g fresh matter; Mandelstam and McQuillen, 1973). The total active material in the form of yeast cells is is 10,000–20,000 times the amount present as bacterial cells. The comparatively small number of bacteria derived from wheat varies from less than 100 to over 100,000 per g (Pomeranz, 1971), and can be neglected in this evaluation. Nevertheless, bacterial activity seems to be as important as that of yeast for optimum flavour development, and Carlin (1958) and Robinson *et al.* (1958a,b) claim that, besides yeast, certain bacteria are essential for a good bread flavour.

D. Active Dry Yeast

Prior to 1920, a large number of patents were issued in Europe and the U.S.A. for production of active dry yeast, but none of these fulfilled the demands for a yeast with good fermentation rate combined with stability over a period of 6 to 8 months (Frey, 1957). A completely acceptable product was not made until the Second World War. Nevertheless, use of active dry yeast has been at a minimal level in countries with temperate climates and short distribution lines. In many other countries, however, it has largely replaced compressed yeast in home baking, but less so in wholesale bakeries that can easily be supplied with refrigerated compressed yeast. Areas where bakeries are remote from yeast production plants can especially benefit from its easy shipment and storage. Exports to tropical and subtropical countries are generally in the form of active dry yeast.

The most important advantage of active dry yeast over fresh yeast is its better stability at normal temperatures. A useful storage life of 6 and 21 months at 32°C and 21°C, respectively, has been reported (Felsher et al., 1955). Although high concentrations of sugar and some fungicides, such as propionate, do inhibit its activity, active dry yeast tolerates them better than fresh yeast. For production of sweet goods, in particular, active dry yeast is beginning to replace compressed yeast. Its disadvantages are a higher price and the necessity for mixing the dried yeast granules with water in a narrow temperature range. During the drying process, the cell membrane is partly damaged and, on rehydration, it takes some time before its properties are restored. At this time, the cell is not able to retain all intracellular components, and some cellular material leaks out (Suomalainen et al., 1965). The colder the rehydration temperature the more the cell loses components and the weaker the fermentation activity becomes. It can be seen from Figure 5 that the temperature range in which active dry yeast maintains its activity is quite narrow. A further disadvantage of active dry yeast is that it supplies more reduced compounds, such as reduced glutathione, which may lead to slacker doughs unless the formula is changed to contain more oxidants than are usual in baking with compressed yeast.

Active dry yeast is normally made from a yeast that has a lower nitrogen content than highly active compressed yeast. Normally at least about half of the amount of compressed yeast in a formula is used. To give the same dry matter content, the conversion ratio should be 32–33%. When sugar is added to the dough, a lower conversion ratio can be used; Thorn and Reed (1959) give a value of 25–30% for sweet dough containing 20% sugar.

Several years ago, a new type of active dry yeast was introduced

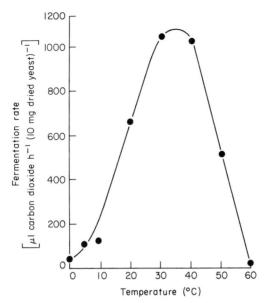

Fig. 5. Influence of rehydration temperature on the ability of laboratory dried baker's yeast to ferment glucose.

(Langejan, 1974; Khoudokormoff and Langejan, 1976). It was prepared from a yeast with a high protein content using rapid drying at fairly high temperatures and a fluidized bed system. Addition of some emulsifiers helps in the rehydration process, and the yeast does not have to be dissolved in water before use. Consequently, because of this property, this kind of active dry yeast is called 'instant' yeast. Its activity is comparable to compressed yeast (at equivalent solid contents). Table 4 compares some properties of active dried yeasts, dried by different methods, with those of compressed yeast.

E. Yeast-Leavened Frozen Dough

Frozen yeast-leavened dough has proved a promising candidate for a more efficient method of producing bakery goods (Rollag and Enochian, 1967). The dough can be prepared up to the stage of forming rolls and loaves, but then, instead of leavening completely and baking, it is frozen for extended storage. This permits preparation of fresh bread in places where it is possible to leaven and bake the dough but which lack

Table 4

Fermentative activity and other parameters of some active dried yeasts and compressed yeast. From Oszlanyl (1980)

Form	Method of drying	Protein[a] (%)	Gas production[b] (ml)
Compressed yeast	None	52	390
Active dried yeast			
irregular particles	Belt	40–42	140–160
irregular spheres	Drum	41–43	160–185
powder	Spray	40–44	95–160
threads	Fluidized bed	52	341

[a] Expressed as the nitrogen content multiplied by 6.25.
[b] Carbon dioxide produced in 165 minutes per 300 mg yeast dry matter in dough.

fermentation and make-up equipment. Moreover, the baker can more economically use the capacity of the oven and other implements and plan his production schedule for different kinds of products. The housewife, however, still finds it quite frustrating at times trying to prepare 'home baked' goods from frozen dough that does not always behave in the correct way although she carried out all of the steps as stated on the package (Lorenz, 1974). In this case, the yeast did not act as a chemical compound (see p. 92) as normal compressed baker's yeast does. On the other hand, whether yeast is the only guilty part in this will be revealed only when the behaviour of frozen unbaked dough has been more adequately investigated.

Studies have been carried out on improving the stability of frozen dough by incorporating different ingredients at various levels, and by choosing the best dough-processing techniques, proper packaging materials and optimum and efficient storage conditions, but the possible effects of the freezing and defrosting rates on the viability of yeast appear to have been neglected. The first problem to be solved with frozen, unbaked dough is the retention, during frozen storage, of sufficient yeast viability and gassing power to avoid excessive proofing times, or even complete loss of proofing power, after thawing (Merritt, 1960). As has been discussed on p. 102, the sequence of slow cooling and rapid thawing produces more surviving yeast cells than does the sequence of rapid cooling and slow warming, but which is presently practised in the manufacture of frozen dough (Lorenz, 1974). The viability data are based mainly on results using cells of *Sacch. cerevisiae* rather than on actual yeast performance in a frozen-dough. Nevertheless, most probably the results apply to yeast in a frozen-dough product as well.

The level of yeast in frozen dough should be increased to 4–5% (Merritt, 1960; Lorenz and Bechtel, 1964; Javes, 1971); lower amounts prolong proofing times and higher amounts lead to a bread with less desirable flavour (Lorenz and Bechtel, 1964). The stability of frozen dough is related to yeast quality. It is believed that doughs frozen after fermentation give poorer bread than non-fermented or only partially leavened doughs (Meyer *et al.*, 1956; Merritt, 1960; Sugihara and Kline, 1968). The greater stability of frozen yeast in the dough is probably due to the dormant state of the yeast (Merritt, 1960). During fermentation, yeast reaches an incipient state of growth when it is vulnerable to freezing damage (Lorenz, 1974).

V. SOUR DOUGHS

A. Introduction

Sours are doughs in which the flour, most usually rye, is fermented by lactic-acid bacteria rather than by baker's yeast. The methods used are still largely empirical, and have evolved over long periods of time (Pyler, 1973). Even today, the starter for the sour dough is frequently produced simply by creating optimum conditions for acidification and leavening by growth of indigenous bacteria and yeasts in the flour–water mixture.

Present methods of sour-dough fermentation in rye-bread production are designed principally to attain the degree of acidification needed to render rye flours suitable for baking, to control the development of different flavour components (until now mainly the ratio of lactic acid to acetic acid), and to achieve dough leavening by yeast growth in the various stages of sour-dough propagation. This last function is frequently augmented by adding baker's yeast to the final dough (Schulz, 1966).

Rye flour is difficult to bake without souring, and the same is true for mixtures of flour containing more than 20% rye. Doughs containing rye swell during souring, becoming suitably elastic and able to retain gas (Spicher and Möllemann, 1975).

Besides the effect on gelatinization of starch, acidification also changes the baking properties of flour pentosans and proteins, and it is evident that the micro-organisms in sour dough do considerably more to influence the rheological properties of the dough than by simply decreasing the pH value (Wood *et al.*, 1975).

The enzymes from sour-dough micro-organisms strongly degrade pentosans during souring, which lowers the viscosity of rye dough (Rehfeld and Kraus, 1961). The proportion of water-soluble proteins, peptides and amino acids in the dough also increases through the action of lactic-acid bacteria (Kosmina, 1977).

Sprouting in the ear is a common harvest defect of rye grown in areas where the harvest can be wet. During sprouting, the concentration of starch-degrading α-amylase activity increases exponentially, the rate of increase being particularly rapid in the case of rye. The enzymic activity of proteases and pentosanases also increases on germination. Excessive amounts of α-amylase in rye flour produce a sticky crumb and, in severe cases, a wide-open grain, a diminution in loaf volume, and cavitation of the loaf (Westermarck-Rosendahl, 1978; Barret, 1975). With rye flours, optimum baking results are often obtained only at relatively low pH values that inactivate α-amylase. Thus, whereas the pH value of normal wheat bread is in the range 5.3–6.2, baking with rye is usually kept at pH 4.0–4.5. Besides its effect on the structure and flavour of bread, acidification also serves to extend the shelf life of bread by preventing mould growth and retarding staling (Brümmer, 1974; Drews and Seibel, 1976).

Areas continuing to sour-dough bake are Central and Eastern Europe, the Scandinavian countries and North America. In Finland, more than a third of all of the bread consumed is rye bread, mostly from sour doughs consisting of wholemeal ryeflour (Salovaara, 1979). Flours with lower ash content are widely used in Germany and the Soviet Union (Drews and Seibel, 1976), as is a mixture of rye and wheat, with most of the rye flour generally being used for making the sour, and wheat flour being added to the sour to make the final dough. Wolter (1974) has developed a sour-dough procedure that uses a blend of rye and wheat flours in the sour. Sour doughs of straight wheat flour are rarer. Perhaps the best known are sour dough French bread, which has became quite popular in the San Francisco Bay area, and the soda cracker process. There are also a number of special, traditional sour methods practised in various countries. Although sour bread is still universally made by natural fermentation and starter sponges, numerous culture preparations are also commercially available to simplify production.

B. Microflora of Sour Dough

The most important sour-dough bacteria are rods belonging to the genus *Lactobacillus*, the only genus in the family Lactobacillaceae. Others that may participate include members of the genera *Leuconostoc*, *Pediococcus* and *Streptococcus* of the family Streptococcaceae (Rogosa, 1974).

In the sour-dough fermentation, the main products are lactic acid, acetic acid, ethanol and carbon dioxide. Acetic-acid formation by acetic-acid bacteria is an oxidative reaction requiring oxygen; the carbon dioxide-permeated substrate of sour dough fails to provide a conducive environment for their activity. Acetic acid is, in this case, produced by heterofermentative bacteria of sour dough (Schulz, 1966). The sum of the main metabolic pathways of sour-dough bacteria, involving sequences of enzymic reactions, are presented in Figure 6. This illustrates that lactic acid is the product of fermentation in homofermentatives whereas lactic acid, acetic acid, ethanol and carbon

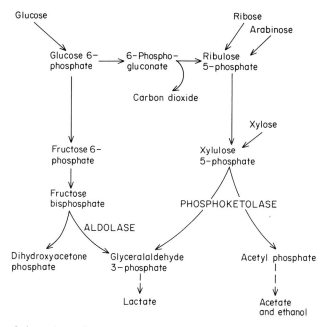

Fig. 6. Glycolytic pathway for homofermentative lactic-acid bacteria *via* aldolase and the heterofermentative lactic-acid bacteria pathway for glucose and pentoses *via* phosphoketolase. From Williams (1975).

dioxide are produced from glucose by heterofermentative lactic-acid bacteria. Production of carbon dioxide and ethanol by the yeasts present is also of significance.

Different species of lactic-acid bacteria can vary greatly in their ability to use different substrates. Some bacteria also ferment pentoses to form lactic acid and acetic acid (Fig. 6), which may be relevant in sour doughs. According to Schlegel (1976), such typical homofermentative lactic-acid bacteria as *Lactobacillus plantarum* and *L. casei* ferment glucose by the Embden-Meyerhof-Parnas pathway homofermentatively, whereas ribose (*L. casei*) and other pentoses (*L. plantarum*) are heterofermented to acetic and lactic acids *via* the phosphoketolase pathway. However, if these bacteria are first cultivated on pentoses, such as ribose, and are then transfered to a glucose-containing medium, they will then ferment glucose heterofermentatively as well. The metabolism of lactic-acid bacteria is obligatorily fermentative even though, in air, growth generally occurs; only some species are strict anaerobes (Rogosa, 1974).

1. Rye Sours

a. Lactic-acid bacteria. It has been known from the beginning of the Century that organisms producing lactic acid are of prime importance in sour-dough fermentations (Holliger, 1902; Spicher, 1959). Initially, the investigation of lactic-acid bacteria was badly obstructed by a lack of consensus concerning taxonomy and research methods. The classification system for lactic-acid bacteria was introduced in Bergey's manual in the 1950s, and ever since studies of sour-dough bacteria have been based on that system, especially in the work done by Spicher in the Federal Republic of Germany.

The members of the genus *Lactobacillus* are Gram-positive, asporogenous, rod-shaped bacteria and characteristically sucroclastic; with few exceptions only sugars and some sugar derivatives can be used as a carbon source. At least half of the end-product carbon is lactate, and lactate is not fermented. Additional products may be acetate, ethanol, formate, succinate and carbon dioxide. Volatile acids with more than two carbon atoms are not produced. They have complex nutritional requirements for vitamins, amino acids, peptides, nucleic acid derivatives, fatty acids or fatty acid esters, salts and fermentable carbohydrates, which are generally characteristic for each species. Their growth temperature range is 5–53°C, the optimum generally being

30–40°C. The optimal pH value is usually 5.5–5.8 or below, and they usually occur growing at pH 5.0 or less. At neutral or initial alkaline pH values, the lag phase may be lengthened or growth yield decreased. Pathogenicity is rarely encountered (Rogosa, 1974).

In his earlier investigations, Spicher (1959) obtained 120 isolates of lactobacilli from 17 sour doughs and four commercial so-called pure-culture sour-dough starters (Reinzuchtsauer) and characterized eight lactic-acid bacteria. Four of them were homofermentative (*Lactobacillus delbreückii, L. plantarum, L. leichmannii* and *L. casei*) and four heterofermentative (*L. brevis, L. fermentum, L. pastorianus* and *L. büchneri*). The homofermentative species occurred in 16 samples and the heterofermentative in 13, while eight samples (38%) contained both types. Different sour doughs vary greatly in their content of lactic-acid bacteria, as do commercial starters. This diversity in turn has an effect on baking techniques, and acidification and quality of bread can vary according to flora in the sour dough.

In a later investigation of the bacterial composition of German pure-culture sour-dough starters of varying origin, 245 isolates were obtained belonging to the genus *Lactobacillus* (Spicher and Schröder, 1978a). In these, the identified lactic organisms varied in number and proportion. According to the morphological, physiological and biochemical characteristics of the isolates, they were classified into three subgroups (in parentheses some species of each subgroup): *Thermobacterium* (*Lactobacillus acidophilus*), *Streptobacterium* (*Lactobacillus casei, L. plantarum, L. farciminis, L. alimentarius*) and *Betabacterium* (*Lactobacillus brevis, L. brevis* var *lindneri, L. büchneri, L. fermentum, L. fructivorans*). Of these three subgroups, members of the genus *Thermobacterium* are homofermentative, lactic acid being the major product from glucose (generally 85% or more), with relatively high optimum and minimum growth temperatures, and maximum growth temperatures generally in the range 45–50°C. Members of the *Streptobacterium* are also homofermentative, but usually have lower maximum and minimum growth temperatures and a wider variety of fermentation reactions. In most cases, ribose is also fermented and yields lactic and acetic acids without gas. The heterofermentative *Betabacterium* species produce only 50–65% of DL-lactic acid and relatively large amounts of other products in fermentation of carbohydrates, namely acetic acid and carbon dioxide, and minor products such as ethanol. They ferment ribose and produce mannitol from fructose (Rogosa and Sharpe, 1959). A more detailed study of the properties of members of the genus *Lactobacillus* and a detailed description of the 27 species in the genus can be found in Bergey's manual (Rogosa, 1974).

The three lactic-acid bacteria species that are considered of major significance in sour-dough fermentation are the homofermentative *Lactobacillus plantarum* and the heterofermentative *L. brevis* and *L. fermentum* (Schulz, 1966), with the heterofermentative lactic-acid producers believed to be more important than the homofermentative types. In fact, only when heterofermentative bacteria are present is the typical flavour of sour rye bread obtained (Spicher and Stephan, 1960).

Best quality bread is obtained by using mixed cultures of homo- and heterofermentative bacteria (Kosmina, 1977). Stegemann and Rohrlich (1958) observed that when the ferment is acidified with a pure culture of heterofermentative *L. brevis*, the resultant bread exhibits a desirable aromatic character but a deficient crumb elasticity, whereas when the homofermentative *L. plantarum* is used, the bread will be lacking in aroma but possess a suitably elastic crumb. Consequently, to obtain a product that is satisfactory as regards both aroma and crumb structure, a mixture of two bacterial species must be employed in preparation of sour dough.

Research (Spicher and Schröder, 1979a,b) has also been carried out on the vitamin and amino acid requirements of sour-dough bacteria. The only marked difference observed between the three subgroups was for the thiamin requirement; *Betabacterium* species required this vitamin whereas representatives of the other two subgenera did not. It is striking that lactic-acid bacteria with a very high requirement for amino acids predominate in sour doughs and pure-culture starters.

b. Yeasts. Although some of the lactic-acid bacteria possess an adequate leavening capacity, the leavening action of a sour dough is essentially determined by the fermentative action of its yeasts. Spicher *et al.* (1979) isolated 44 yeast cultures from various sour-dough starters. These could be subdivided into four groups by their morphological and physiological characteristics and assigned to the species *Candida krusei* (27 strains), *Saccharomyces cerevisiae* (11), *Saccharomyces exiguus* (4) and *Pichia saitoi* (2). The presence of different species yeast in sour doughs varies greatly according to the raw materials used and fermentation conditions.

Indigenous sour-dough yeasts are usually more acid- and temperature-resistant than commercial compressed baker's yeast (Schulz,

1966). Spicher and Schröder (1978b) investigated the effects of the four species of yeasts already mentioned on lactic-acid fermentation of sour dough. The sour-dough yeasts were able to proliferate in the temperature range 20–50°C, with an optimum at 35°C, and were active at pH values between 3.6 and 5.6, with the lower pH values shifting the optimum temperature towards the lower end of the range. Activity of the yeasts was more strongly influenced by addition of acetic acid than by the presence of lactic acid. Depending on the species of *Lactobacillus* involved, the yeasts had either a stimulatory, inhibitory or neutral effect on lactic-acid fermentation. It was shown that various yeasts can shorten the generation time of *L. brevis* var. *lindneri* by 10–30 minutes depending on the strain of yeast. Spicher and Schöllhammer (1977) made comparative studies of how yeasts isolated from spontaneous sour doughs and so-called pure cultures influence bread flavour. The former group resulted in flavours that varied through yeasty, fruity, alcoholic and sour, whereas the latter group produced only sour flavours.

c. Spontaneous fermentation. Bacteria rather than yeasts cause most of the fermentation in unyeasted doughs of flour and water. Initially the microflora of a dough is about the same as that on the surface of its constituent grain. Bacteria of the family Enterobacteriaceae, often belonging to the so-called coli-aerogenes group, are dominant in rye doughs (Kosmina, 1977; Neumann and Pelshenke, 1954). They are typical gas-formers. The products they form in breaking down sugars include carbon dioxide, hydrogen and organic acids. As the acidity of the dough increases, its microbial composition gradually alters so that Gram-negative gas producers diminish and disappear while Gram-positive lactic-acid bacteria become dominant (Fig. 7). A further typical change is the appearance of yeasts, which in turn can retard proliferation of coliform bacteria and which, as acidity increases, are successful only if they have good acid tolerance. In this spontaneous fermentation phase, dough is automatically purified of irrelevant microbes and is technologically acceptable as sour-dough starter.

Spicher (1966a) isolated 140 strains of lactic-acid bacteria from spontaneously fermenting doughs. The three main species were *Pediococcus acidilactici*, *Lactobacillus fermentum* and *L. casei* subsp. *pseudoplantarum*.

2. Wheat Sours

Bacteria responsible for the souring action in San Francisco sour-dough

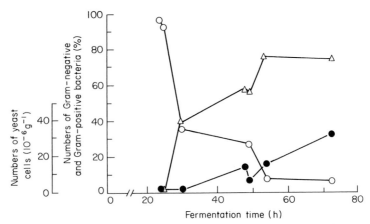

Fig. 7. Changes in the contents of yeasts (●), Gram-positive bacteria (△) and Gram-negative bacteria (○) during spontaneous fermentation of sour dough. From Kosmina (1977).

French bread process were first isolated and characterized by Kline and Sugihara (1971), who described them as a previously unreported species of heterofermentative *Lactobacillus* and proposed the name *L. sanfrancisco*. It is a non-motile bacterium of short to medium slender rods, indifferent to oxygen and does not use carbohydrates, other than maltose, as carbon source. Subsequent genetic DNA homology studies indicated that *L. sanfrancisco* is unique and not related to any known lactic-acid bacterium (Sriranganathan *et al.*, 1973). The leavening function is performed mainly by a yeast found in the starters of several bakeries in the area. Although identified as *Torulopsis holmii*, because ascospores are not produced, reference to the perfect form *Saccharomyces exiguus* is preferred (Sugihara *et al.*, 1971). Two other yeasts of lesser significance were identified as *Saccharomyces inusitatus* and *Sacch. uvarum* (Ng, 1976).

It is suggested that *Sacch. exiguus* can co-exist with sour-dough bacteria so successfully because of its tolerance to acetic acid and its resistance to an antibiotic produced by the bacteria. A further advantage for *Sacch. exiguus* is that it does not utilize maltose for which the bacteria have a high requirement. The yeast consumes virtually all of the flour glucose and fructose and larger amounts of low molecular-weight glucofructosans. In the San Francisco sour-dough process, 100 g of flour contains approximately 1.7 g of free sugars other than maltose. Approximately 5.5 g of maltose per 100 g of dry flour are produced after the flour–water mixture is prepared. The sour-dough bacterium utilizes only about 56%

of this, conveniently leaving the remainder as a necessary ingredient in crust browning (Saunders *et al.*, 1972). Thus, bacteria and yeasts do not compete for the same carbohydrate source in the dough. Moreover, both yeast and flour contribute to the unusual combination of nutritional requirements of the sour-dough bacteria for growth, which includes such factors as fresh yeast extractives, unsaturated fatty acids and carbon dioxide. The contribution of the yeast to the acidity produced is shown to be negligible. The bacteria and yeasts appear to live in a sort of symbiotic relationship (Sugihara *et al.*, 1970, 1971; Saunders *et al.*, 1972).

It is interesting that the yeast *Sacch. exiguus* (*Torulopsis holmii*) has also been isolated from German and Finnish rye sour doughs (Spicher *et al.*, 1979; Suihko and Mäkinen, 1975). In the Soviet Union, the same kind of sour-dough yeast, which does not utilize maltose, is called *Sacch. minor* (Kosmina, 1977). Successful efforts have also been made to grow *Sacch. exiguus* using a molasses-based medium (Ng, 1976).

In the soda cracker manufacturing process, starter cultures have traditionally not been used; the fermentation relies primarily on chance contamination. Sugihara (1978a) found that three species of lactic-acid bacteria have prominent roles in the fermentation process, with *Lactobacillus plantarum* as the dominant species followed by *L. delbreückii* and *L. leichmannii*. Addition of some baker's yeast to cracker sponges is essential for bacterial growth (Micka, 1955).

Everson (1972) has patented a sour-dough process for sour French bread or rye bread where a starter of *Pediococcus cerevisiae* or *P. acidilactici* is used, and Luksas (1971) used a concentrate to impart sour-dough flavour consisting of the coagulum of renneted skim-milk inoculated with cultures of *Citrobacter* or *Micrococcus* species.

C. Some Special National Uses of Sour Dough

Rye sours are used in the manufacture of certain varieties of crispbread, e.g. in Finland. The U.S.A. and German pumpernickels, which are of quite different types, are both made by the sour-dough process (Drews and Seibel, 1976).

Pannettone is a popular Italian Christmas fruit cake still being manufactured today using the traditional 24-hour natural fermentation. From this, Galli and Ottogalli (1973) isolated the sour-dough yeast *Sacch. exiguus*; lactic-acid bacteria identified were intermediate strains of heterofermentative *Lactobacillus brevis* and *L. cellobiosus* and the homofermentative *L. plantarum*, with *L. brevis* dominating. *Enterobacter* and *Citrobacter* species were also found in low numbers. Research is under way to shorten the process by using pure cultures (Sugihara, 1977).

A steamed pancake-like product called idli is common in southern India. It is prepared from rice and black gram mungo (*Phaseolus mungo*), a legume. Mukherjee *et al.* (1965) found that the micro-organism responsible for the fermentation of idli is a heterofermentative lactic bacterium, *Leuconostoc mesenterioides*. In Sangak bread, which is nutritionally important in Iran, 77% of the total bacteria found in the starter were *Leuconostoc* and *Lactobacillus* spp. (Azar *et al.*, 1977).

D. Technological Aspects of Sour-Dough Preparation

A widely used method for making rye sours, particularly among European rye-bread bakers, is the perpetuated or multiple-stage method. It is characterized by a series of preliminary sours before the final dough is obtained. The method usually consists of three steps or stages, comprising the fresh sour, basic sour, and full sour. The required starter, with its acid-forming bacteria and sour-dough yeasts, is normally taken from the preceding full sour and is mixed with rye flour and water to initiate the first stage, or fresh sour (Pyler, 1973).

Going from stage to stage, the sour is normally freshened and activated by adding flour and water to raise the pH value and to give more nutrients to the microflora. During the first stage, favourable conditions are created for development of yeasts, whereas in the second, optimum development of lactic-acid bacteria and consequent increase in acidity is encouraged. In the third stage, renewed yeast growth, formation of acid and development of consistency are all directed towards the desired final product. After a fermentation period, the full sour is used to prepare the dough with addition of the remaining flour, water, salt and special ingredients. Depending on the stage and the method used, the temperature may vary from 24°C to 35°C and absorption of water usually from 60% to 150% based on the flour. According to the product, the final pH value of the sour will be 3.5–4.0 (3.8 typical), where the functioning of microbes is already impeded. The whole multiple-stage process normally takes 16–24 hours or even longer (Doose, 1965; Spicher, 1966b). In practice, the amount of rye flour that goes through the souring is 10–40% of the total flour. Usually the higher the proportion of rye flour in the bread, the greater the amount of rye flour used in the sour. The starter sponge is most often regularly rebuilt, but it is also still common to hold back part of the sour dough from one day's production to be added to the next day's fresh sponge or dough (Pyler, 1973).

In modern sour-dough bread technology, the entire process has been simplified and automated. The multiple-stage procedure has been replaced by shorter methods based on one or two fermentation steps. These methods have been accomplished by, amongst other devices, regulating temperatures and the amount of starter. A widely used process in Europe is the Berlin short-sour process (Schulz, 1966) which concentrates on promoting growth of lactic-acid bacteria by using a suitable temperature (35°C). This temperature is too high for the sour-dough yeasts, and leavening is augmented by addition of baker's yeast to the final dough. The duration of a sour-dough fermentation may be as short as three hours. A further distinguishing feature of this method is the relatively soft dough consistency. Usually, an absorption of 90% is used. The short-sour method is essentially a sponge and dough system in which the sponge is ripened by bacteria rather than by yeast (Schulz, 1966; Drews and Seibel, 1976). Other simplified processes include the Detmold one- and two-stage methods and the Monnheim salt-sour method (Stephan, 1960, 1970; vom Stein, 1971). In the latter, acidification and preservation of the sour dough are regulated by use of salt.

Stegemann and Rohrlich (1958) described a continuous rye-sour fermentation system in which a semi-fluid rye flour ferment is fermented in a temperature-controlled tank under constant agitation with simultaneous withdrawal of fully ripe sour and compensating addition of fresh rye flour and water. Continuous growth of sour-dough bacteria requires that pH values in the fermenting liquid be maintained above the critical value of 4.0. Wutzel (1968) has patented a continuous process in which a paste of lactic-acid bacteria is continuously fed into the fermentation vessel.

The conventional San Francisco sour-dough French bread process is quite complicated with its many stages and a 24-hour production schedule. The starter sponge is rebuilt about every eight hours or at least two to three times a day, seven days a week. A very high proportion of preceding starter sponge is used in preparing a new one (Kline *et al.*, 1970).

Sugihara and Kline (1975) developed pure-culture technology for growing and stabilizing *Lactobacillus sanfrancisco*. Starter cultures are now being manufactured commercially in two different forms, namely freeze dried and frozen concentrates. The sour-dough yeast *Sacch. exiguus* proved to be very difficult to stabilize, but it was found that baker's yeast, with some modifications, can be substituted in the process to give an acceptable product (Sugihara, 1977). Stable pure-culture frozen concentrates of lactic-acid bacteria were also developed for use in fermentation of soda cracker sponge and dough. The conventional 24-hour process could in this way be lowered to eight hours, with moreover a better uniformity of product (Sugihara, 1978b).

Fermentation losses in the sour-dough process are normally in the range 1–2% based on the weight of dough. In unfavourable conditions, they may rise up to 4% (Huber, 1977). Preservation of starter sponges may be important in rationalizing the sour-dough process. To a certain extent this is possible by using lower temperatures and/or salt additions (Spicher and Stephan, 1973). It is accepted that the salt–acid process permits satisfactory storage of soured dough for 96 hours at 5–10°C. In certain cases it may be possible to freeze or freeze-dry the starter. It has been observed that freezing a fully developed San Francisco starter sponge is out of the question because the leavening power is almost completely lost in a day or two even though the souring power persists for a longer time (Kline *et al.*, 1970). A traditional way of preserving rye sour dough in Finland has involved the natural drying of the dough on the surface of wooden containers.

E. Sour-Dough Fermentation and its Regulation

There are several ways of regulating the course of the reactions in sour dough, and thus the character and quantity of reaction products. Some of the most important parameters are presented in Figure 8. The influence of different parameters naturally varies according to the microflora of the starter used.

Two examples of the acidification of sour dough are presented in Figures 9 and 10. Functioning of sour-dough microbes depends greatly

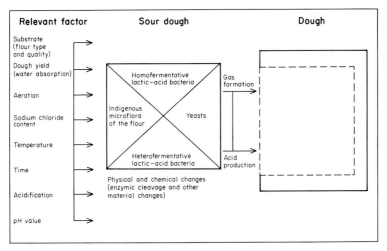

Fig. 8. Parameters in sour-dough fermentation. From Spicher (1975).

on the environmental pH value, with their acid production falling as the pH value drops. The relationship between acidity of the medium and temperature can be widely different for different species. At 15°C, 30°C and 40°C the critical pH value for *Lactobacillus plantarum* was, respectively, 4.35, 4.05 and 4.2; and the corresponding critical pH values for *L. brevis* were 4.5, 4.5 and 4.65 (Spicher, 1975). The relationship between

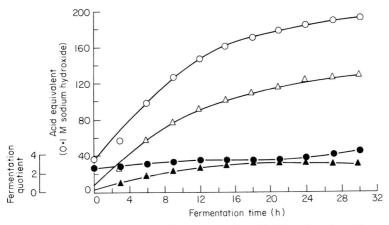

Fig. 9. Time-course of formation of total acid (O) lactic acid (△) and acetic acid (▲) and fermentation quotient (●) during the fermentation of a rye sour dough. From Kosmina (1977).

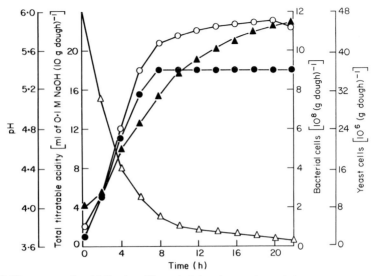

Fig. 10. Time-course of multiplication of bacteria (O) and yeasts (●) and changes in pH value (△) and acidity (▲) during the fermentation of a rye sour dough. From Rohrlich and Stegemann (1958).

incubation time and bacterial growth also varies. The lag-phase was shortest when the starter was taken from the beginning of the logarithmic phase of growth. A tendency was also seen for *L. brevis* to grow more slowly and to continue growing for a longer time than *L. plantarum* (Spicher, 1968a, 1975). For sour-dough production using mixed cultures, it is necessary to provide conditions that maintain the required balance between the bacteria.

The buffering capacity of the flour influences the quantity of acid formed and the final pH value of the sour dough. High-ash flour requires more acid to obtain a given pH value than a low-ash flour. According to Rohrlich and Essner (1960) the main buffering component in flour is phytic acid because of its residues of orthophosphoric acid. It has been shown that buffering compounds decrease acid production by *Lactobacillus fermentum* more than *L. plantarum* and *L. brevis* (Spicher and Angermann, 1965). The pH value decreases faster the greater the amylase activity of the flour (Spicher, 1968b).

Total titratable acidity is generally determined by titration of an aqueous suspension of sour with sodium hydroxide, which, however, does not differentiate between lactic acid and acetic acid. Rohrlich and Essner (1951) designated the molar ratio of lactic acid to acetic acid in sour dough as the 'fermentation quotient'. It is believed to have an important effect on the flavour of rye-sour bread. Optimum values for the fermentation quotient of a German rye bread have been quoted as 1.5–4.0 (Rohrlich, 1961) and 2.6–3.8 (80–85% lactic acid) (Büskens, 1978).

The fermentation quotient is markedly influenced by dough temperature and dough consistency. An increase in temperature or absorption favours formation of lactic acid, whereas lower temperature and stiffer doughs tend to favour acetic-acid production (Schulz, 1966). Rabe (1977) compared processes with warm/soft, warm/stiff, cold/soft and cold/stiff doughs and found that titratable acidity and the amount of L-lactate rose most rapidly in the warm/stiff doughs. Analogous results were obtained for pH values. Spicher and Rabe (1979) examined the capacity of five different heterofermentative lactic-acid bacteria to produce acetic and lactic acids, and found substantial differences. The fermentation quotient and ratio of D- and L-lactic acid for various bacteria depend on the temperature, with formation of lactic acid being more sensitive than that of acetic acid. Most of the lactic acid is usually in the L-form. Addition of certain compounds, such as sodium fluoride and calcium carbonate, to sour dough has been shown to affect acid formation and the fermentation quotient. Both increase the proportion of acetic acid (Kosmina, 1977).

The temperature optima of the most important lactic-acid bacteria according to Spicher (1968a) are listed in Table 5. Acid formation

Table 5

Temperature optima for growth and acid production of sour-dough lactic-acid bacteria. From Spicher (1968a)

	Temperature (°C)		
	Lactobacillus plantarum	*Lactobacillus brevis*	*Lactobacillus fermenti*
Fastest growth (shortest regeneration time)	35	35	40
Best proliferation (greatest cell number)	35	30–35	30–35
Fastest acid production	35	30–35	40
Maximum amount of acid	30	30	30–35

continues even below 20°C, and the upper limit for *Lactobacillus plantarum* is 40–45°C, for *L. brevis* 45–50°C and for *L. fermentum* even higher than 50°C.

According to the studies of Kazanskaya *et al.* (1977) with mixed cultures below 30°C, the dominant lactobacilli were *L. plantarum* and *L. brevis*, at 34–36°C dominant species were *L. casei* and *L. fermentum*, while at 30–32°C the ratio of the two groups became balanced. The interrelationship between sour-dough temperature and holding time has also to be taken into consideration. The cooler doughs exhibit slower fermentation rates and greater stabilities than the warmer doughs.

The rate of acidification of sours is also directly influenced by the consistency of the dough employed, the rate being at a maximum with absorptions between 80 and 90%. Stiff sours, e.g. with a 60% absorption, produce the least acidification within a given period. On the other hand, when absorptions higher than 100% are used, the rate is again retarded, probably because of the decrease in nutrient concentration resulting from increased dough dilution (Schulz, 1966; Pyler, 1973). Moreover, with a higher absorption, the final pH value is lower (Spicher, 1968b). According to Schulz (1966), the most favourable conditions for acidification of sour doughs are a temperature of 35°C and an absorption of 90%. As regards sour-dough yeasts, their growth is promoted by a high absorption and a temperature of 25–28°C (Spicher and Stephan, 1963).

Excessive additions of baker's yeast or salt to sour dough act to depress the rate of acidification and may introduce atypical flavours. Small amounts of yeast may have a positive effect (Kline *et al.*, 1970; Spicher and Stephan, 1963; Schulz, 1966). Salt also has a stabilizing effect on sour dough, which is utilized in the salt sour method. Homofermentative bacteria are usually more resistant to salt than heterofermentative (Kosmina, 1977).

It is obvious that, in sour-dough baking, the nature and proportions of the acids formed are of prime importance in giving the bread acidity and in determining whether it is pleasant. The many other compounds that are formed or liberated during fermentation can act as precursors of flavours produced during baking. Acetic acid, which has a higher vapour pressure than lactic acid, is the major acid component in the typical aroma of sour bread (Rohrlich, 1961). The sour taste of bread depends on its pH value, lactic acid being of the greater importance because it dissociates more than acetic acid. The undissociated form of acetic acid is especially inhibitory for growth of moulds on bread (Kline *et al.*, 1970). Analyses by Lück *et al.* (1975) have shown that no appreciable amounts of propionic acid are formed during sour-dough fermentation and that bread does not contain propionic acid; nor could pure cultures of propionic-acid bacteria produce this acid in doughs. Butyric-acid bacteria do not usually exist in sour doughs (Doose, 1965). Non-volatile acids in flour, such as malic, citric and fumaric acids decompose during acidification to form other acids, mainly lactic and acetic acids (Marková and Hampl, 1969).

Radler and Gerwarth (1971) have investigated formation of volatile byproducts of fermentation by several strains of lactic-acid bacteria. The homofermentative species *L. plantarum*, *Pediococcus pentosaceus* and *P. cerevisiae* were found to form small amounts of acetaldehyde, acetoin, diacetyl and traces of 2- or 3-methylbutan-l-ol and isobutanol (2-methylpropan-1-ol). In culture solutions of the heterofermentative species *L. brevis* and *Leuconostoc oenos*, additional compounds could be detected. These bacteria formed besides ethanol and products of the homofermentative organisms, small amounts of propanol, propan-2-ol, ethyl acetate, hexanol, 2,3-butanediol and octan-2-ol.

Lactic and acetic acids account for most of the total titratable acidity of San Francisco French bread, the proportion of acetic acid in the sour dough usually being in the range 20–35% although values up to 50% have been reported (Ng, 1972; Kline *et al.*, 1970). The proportion of acetic acid has been found to depend almost entirely on the degree of aeration. This effect of aeration is known to be common with heterofermentative lactobacilli (Ng, 1972). Galal and coworkers (1978) using gas–liquid chromatography revealed the presence of six other minor acids, namely propionic, isobutyric, butyric, methyl-n-butyric, isovaleric and valeric acids, that is acids with more than two carbon atoms. These acids contributed, respectively, 1.19, 0.64 and 0.59% to the total titratable acidity of the fully fermented starter sponge, the fully proofed bread dough, and the baked bread. Baking insignificantly increased the pH value but decreased the total titratable acidity by 9% mainly due to the loss of acetic acid.

Commercial chemical flavours, mostly acids or their mixtures, are also available for sour-bread production, but these will not give the full bodied flavour of microbiologically fermented bread and the mould resistance of the bread may also be lower (Jäckl, 1978; Brümmer *et al.*, 1977; Sugihara, 1977). For improved aroma maintenance, a combined method is also used, for instance the dough is first acidified only slightly using smaller amounts of sour dough and then organic acids are added (Drews and Seibel, 1976).

REFERENCES

Atkin, L., Schultz, A.S. and Frey, C.N. (1945). *Cereal Chemistry* **22,** 321.
Atkin, L., Schultz, A.S. and Frey, C.N. (1946). *In* 'Enzymes and their Role in Wheat Technology' (J.A. Anderson, ed.), p. 321. Interscience Publishers, Inc., New York.
Auerman, L.Y., Rusanova, T.V., Machikhin, S.A., Uvarov, A.P., Dimitriev, V.V. and Uvarova, I.I. (1977). Russian Patent 563,950.
Axelsen, C. (1974). *Bakers and Millers Journal* **77** (4), 7, 13.
Axford, D.W.E., Chamberlain, N., Collins, T.H. and Elton, G.A.H. (1963). *Cereal Science Today* **8,** 265.
Azar, M., Ter-Sarkissian, N., Chavifek, H., Ferguson, T. and Ghassemi, H. (1977). *Journal of Food Science and Technology* **14,** 251.
Bailey, L.H., Bartram, M.T. and Rowe, S.C. (1940). *Cereal Chemistry* **17,** 55.
Barret, F.F. (1975). *In* 'Enzymes in Food Processing' (G. Reed, ed.), 2nd edition, p. 301. Academic Press, New York.
Barton, J.K., den Hollander, J.A., Lee, T.M., MacLaughlin, A. and Shulman, R.G. (1980). *Proceedings of the National Academy of Sciences of the United States of America* **77,** 2470.
Bennion, E.B. (1967). 'Breadmaking, Its Principles and Practice', 4th edition. Oxford University Press, London.
Bergander, E. and Bahrmann, K. (1957). *Lebensmittelindustrie* **2,** 296.
Brasch, J.F. (1951). *Journal of the American Oil Chemists' Society* **28,** 396.
Bronn, W.K. (1976). *Branntweinwirtschaft* **116,** 211.
Brümmer, J.-M. (1974). *Getreide, Mehl und Brot* **28,** 45.
Brümmer, J.-M., Stephan, H. and Morgenstern, G. (1977) *Getreide, Mehl und Brot* **31,** 186.
Buckheit, J.T. (1971). *Baker's Digest* **45** (1), 46, 60.
Burrows, S. (1970). *In* 'The Yeasts' (A. H. Rose and J. S. Harrison, eds.), vol. 3, p. 349. Academic Press, London.
Burrows, S. (1979). *In* 'Economic Microbiology' (A.H. Rose, ed.), vol. 4, p. 31. Academic Press, London.
Burrows, S. and Harrison, J.S. (1959). *Journal of the Institute of Brewing* **65,** 39.
Büskens, H. (1978). 'Fachkunde für Bäcker', vol. 1, 7th edition. Verlag W. Girardet, Essen.
Carlin, G.T. (1958). *Proceedings of the American Society of Bakery Engineers,* p. 56.
Cerny, G. (1980). *Zeitschrift für Lebensmittel-Untersuchung und -Forschung* **170,** 173.
Colin, H. and Belval, H. (1935a). *Comptes Rendus Hebdomadaires des Séances de l'Academie des Sciences, Physiologie Vegetale* **200,** 2032.
Colin, H. and Belval, H. (1935b). *Bulletin de la Société de Chimie Biologique* **17,** 1040.
Cooper, E.J. and Reed, G. (1968). *Baker's Digest* **42** (6), 29.
Demis, D.J., Rothstein, A. and Meier, R. (1954). *Archives of Biochemistry and Biophysics* **48,** 55.
DeSa, C. (1966). *Baker's Digest* **40** (6), 50.

Doose, O. (1965). 'Arbeitskunde für Bäcker'. Gildeverlag H.-G. Dobler, Alfeld.

Drews, E. and Seibel, W. (1976). *In* 'Rye: Production, Chemistry, and Technology' (W. Bushuk, ed.), p. 127. American Association of Cereal Chemists, Inc., St. Paul, Minnesota.

El-Dash, A.A. (1971). *Baker's Digest* **45** (6), 26.

Elion, E. (1940). *Baker's Digest* **14**, 143, 151.

Elton, G.A.H. (1969). *Royal Society of Arts Journal* **117**, 317.

Everson, C.W. (1972). United States Patent 3,681,083.

Felsher, A.R., Koch, R.B. and Larsen, R.A. (1955). *Cereal Chemistry* **32**, 117.

Finney, K.F., McCammon, J.F. and Schrenk, W.G. (1949). *Cereal Chemistry* **26**, 140.

Fransson, G. (1977). *Livsmedelsteknik* **19**, 435.

Frey, C.N. (1957). *In* 'Yeast, its Characteristics, Growth and Function in Baked Products' (C.S. McWilliams and M.S. Peterson, eds.), p. 7. Quartermaster Food and Container Institute for the Armed Forces, Chicago.

Frey, C.N., Kirby, G.W. and Schultz, A. (1936). *Industrial and Engineering Chemistry* **28**, 879.

Galal, A.M., Johnson, J.A. and Varriano-Marston, E. (1978). *Cereal Chemistry* **55**, 461.

Galli, A. and Ottogalli, G. (1973). *Annali di Microbiologia ed Enzimologia* **23**, 39.

Geoffroy, M.R. (1935). *Bulletin de la Société de Chimie Biologique* **17**, 848.

de Grandchamp-Chaudun, A. (1950). *Comptes Rendus Hebdomadaires des Séances de l'Academie des Sciences, Chimie Biologique* **231**, 1082.

Guillemet, R. (1935). *Comptes Rendus Hebdomadaires des Séances de l'Academie des Sciences, Chimie Biologique* **201**, 1517.

Guiraud, J.P., Galzy, P. and Albert, J. (1972). *Annales de l'Institut Pasteur, Paris* **122**, 379.

Hanson, B.A. and Lester, R.L. (1980) *Journal of Bacteriology* **142**, 79.

Harrison, J.S. (1971). *Journal of Applied Bacteriology* **34**, 173.

Hartelius, V. (1934). *Comptes Rendus des Travaux du Laboratoire Carlsberg* **20** (7), 1.

Hautera, P. and Lövgren, T. (1975). *Journal of the Institute of Brewing* **81**, 309.

Holliger, W. (1902). *Zentralblatt für Bakteriologie, Parasitenkunde, Infektionskrankheiten und Hygiene, Abteilung II* **9**, 305.

Hoogerheide, J.C. (1975). *Radiation and Environmental Biophysics* **11**, 295.

Hoseney, R.C., Finney, K.F., Shogren, M.D. and Pomeranz, Y. (1969). *Cereal Chemistry* **46**, 117.

Hsu, K.H., Hoseney, R.C. and Seib, P.A. (1979). *Cereal Chemistry* **56**, 419.

Huber, H. (1977). *Deutsche Müllerzeitung* **75**, 121.

Ingram, M. (1955). 'An Introduction to the Biology of Yeasts'. Sir Isaac Pitman and Sons, Ltd., London.

International Union of Pure and Applied Chemistry (1966). *Pure and Applied Chemistry* **13**, 403.

International Union of Pure and Applied Chemistry (1970). *Information Bulletin* No. 37.

International Union of Pure and Applied Chemistry (1971). *Information Bulletin, Technical Reports* No. 3.

Jäckl, H. (1978). *Deutsche Lebensmittel-Rundschau* **74** (1), 8.

Javes, R. (1971). *Baker's Digest* **45** (2), 56.

Johnson, J.A. and Sanchez, C.R.S. (1972). *Baker's Digest* **46** (4), 30.

Karwat, E. and Kruis, A. (1957). *In* 'Ullmans Encyklopädie der Technischen Chemie' (W. Foerst, ed.), vol. 9, 3rd edition, p. 748. Urban and Schwarzenberg, München-Berlin.

Kazanskaya, L.N., Gurina, O.F. and Afanas'eva, O.V. (1977). *Khlebopekarnaya i Konditerskaya Promyshlennost'*, No. 2, 25.

Kent-Jones, D.W. and Amos, A.J. (1967). 'Modern Cereal Chemistry', 6th edition. Food Trade Press Ltd., London.
Khoudokormoff, B. and Langejan, A. (1976). United States Patent 3,993,783.
Kiby, W. (1912). 'Handbuch der Presshefenfabrikation'. Friedrich Vieweg and Sohn, Braunschweig.
Kline, L. and Sugihara, T.F. (1971). Applied Microbiology 21, 459.
Kline, L., Sugihara, T.F. and McCready, L.B. (1970). Baker's Digest 44 (2), 48.
Koch, R.B., Geddes, W.F. and Smith, F. (1951). Cereal Chemistry 28, 424.
Koch, R.B., Smith, F. and Geddes, W.F. (1954). Cereal Chemistry 31, 55.
Koninklijke Nederlandsche Gist- en Spiritus-fabriek N. V. (1967). British Patent 1,152,286.
Kosmina, N.P. (1977). 'Biochemie der Brotherstellung'. VEB Fachbuchverlag, Leipzig.
Kretzschmar, H. (1955). 'Hefe und Alkohol'. Springer-Verlag, Berlin-Heidelberg.
Lange, N.A., ed. (1967). 'Handbook of Chemistry', 10th edition. McGraw-Hill Book Co., New York.
Langejan, A. (1974). United States Patent 3,843,800.
Larmour, R.K. and Bergsteinsson, H.N. (1936). Cereal Chemistry 13, 410.
Lawson, H.W. (1962). Proceedings of the American Society of Bakery Engineers, p. 251.
Lee, J.W., Cuendet, L.S. and Geddes, W.F. (1959). Cereal Chemistry 36, 522.
Lincoln, R.E. (1960). Journal of Biochemical and Microbiological Technology and Engineering 2, 481.
Linko, Y. and Johnson, J.A. (1963). Journal of Agricultural and Food Chemistry 11, 150.
Lorenz, K. (1974). Baker's Digest 48 (2), 14.
Lorenz, K. and Bechtel, W. G. (1964). Baker's Digest 38, (6), 59.
Lück, E., Oeser, H., Remmert, K.-H. and Sabel, J. (1975). Zeitschrift für Lebensmittel-Untersuchung und -Forschung 158, 27.
Luksas, A.J. (1971). United States Patent 3,615,695.
MacKenzie, R.M. (1958). Society of Chemical Industry Monograph 3, 127.
Makarenko, P.G., Krutova, E.A., Tsulina, E.P. and Anisimova, E.D. (1978). Kholodil'naya Tekhnika No. 6, 37.
Mandelstam, J. and McQuillen, K., eds. (1973). 'Biochemistry of Bacterial Growth', 2nd edition. Blackwell Scientific Publications, Oxford.
Marková, J. and Hampl, J. (1969). Brot und Gebäck 23, 74.
Marston, P.E. and Wannan, T.L. (1976). Baker's Digest 50 (4), 24, 49.
Maselli, J.A. (1959). Proceedings of the American Society of Bakery Engineers, p. 160.
Mathason, I.J. (1978). Baker's Digest 52 (3), 10.
Matz, S.A. (1972). 'Bakery Technology and Engineering', 2nd edition. The Avi Publishing Co., Westport, Connecticut.
Matz, S.A., Miller, J. and Davis, J. (1958). Food Technology (Chicago) 12, 625.
Mazur, P. (1961a). Biophysical Journal 1, 247.
Mazur, P. (1961b). Journal of Bacteriology 82, 662.
Mazur, P. and Schmidt, J.J. (1968). Cryobiology 5, 1.
Merritt, P.P. (1960). Baker's Digest 34 (4), 57.
Meyen, J. (1838). Archiv für Naturgeschichte 4, 1.
Meyer, B., Moore, R. and Buckley, R. (1956). Food Technology (Chicago) 10, 165.
Micka, J. (1955). Cereal Chemistry 32, 125.
Mitterhauzerova, L. and Sedlarova, L. (1966). Kvasny Prumysl 12, 222.
Mukherjee, S.K., Albury, M.N., Pederson, C.S., Van Veen, A.G. and Steinkraus, K.H. (1965). Applied Microbiology 13, 227.

Nagodawithana, T.W., Whitt, J.T. and Cutaia, A.J. (1977). *Proceedings of the American Society of Brewing Chemists* **35**, 179.

Napleton, L. (1976). 'A Guide to Microwave Catering". Northwood Publications Ltd., London.

Navarro, J.M. and Durand, G. (1978). *Annales de Microbiologie (Paris)* **129B**, 215.

Neish, A.C. and Blackwood, A.C. (1951). *Canadian Journal of Technology* **29**, 123.

Neuberg, C. and Hirsch, J. (1919). *Biochemische Zeitschrift* **96**, 175.

Neumann, M.P. and Pelshenke, P.F. (1954). 'Brotgetriede und Brot', Paul Parey Verlag, Berlin.

Ng, H. (1972). *Applied Microbiology* **23**, 1153.

Ng, H. (1976). *Applied and Environmental Microbiology* **31**, 395.

van Niel, C.B. and Anderson, E.H. (1941). *Journal of Cellular and Comparative Physiology* **17**, 49.

Olsen, A.J.C. (1960). *Chemistry and Industry* **17**, 416.

Olsen, A.J.C. (1961). *Society of Chemical Industry Monograph* **12**, 81.

Oszlanyl, A.G. (1980). *Baker's Digest* **54** (4), 16.

Oura, E. (1974) *Biotechnology and Bioengineering* **16**, 1197.

Oura, E. (1977a). *Process Biochemistry* **12** (3), 19.

Oura, E. (1977b). *EUCHEM Conference on Metabolic Reactions in the Yeast Cell in Anaerobic and Aerobic Conditions, Helsinki*, p. 23.

Oura, E. and Suomalainen, H. (1977). *In* 'CRC Handbook Series in Nutrition and Food' (M. Rechcigl, Jr., ed.), Section D: Nutritional Requirements, vol. 1, p. 171. CRC Press, Cleveland, Ohio.

Oura, E. and Suomalainen, H. (1978). *Journal of the Institute of Brewing* **84**, 283.

Oura, E., Suomalainen, H. and Collander, R. (1959). *Physiologia Plantarum* **12**, 534.

Parisi, F. (1970). 5. *Welt-Getreide- und Brotkongress, Dresden, DDR, Kongressbericht* **5**, 81.

Pelshenke, P.F. (1961). *Cereal Science Today* **6**, 325, 329.

Pena, A., Cinco, G., Garcia, A., Gomez Puyou, A. and Tuena, M. (1967). *Biochimica et Biophysica Acta* **148**, 673.

Peppler, H.J. (1960). *In* 'Bakery Technology and Engineering' (S.A. Matz, ed.), p. 35. The Avi Publishing Co., Westport, Connecticut.

Peppler, H.J., ed. (1967). 'Microbial Technology'. Reinhold Publishing Corporation, New York.

Piekarz, R.E. (1963). *Proceedings of the American Society of Bakery Engineers*, p. 118.

Pomeranz, Y. ed. (1971) 'Wheat Chemistry and Technology'. American Association of Cereal Chemists, Inc., St. Paul, Minnesota.

Pyler, E.J. (1973). 'Baking: Science and Technology', vol. 2. Siebel Publishing Co., Chicago, Illinois.

Rabe, E. (1977). *Getreide, Mehl und Brot* **31**, 230.

Racker, E. (1976). 'Mechanisms in Bioenergetics'. Academic Press, New York.

Radler, F. and Gerwarth, B. (1971). *Archiv für Mikrobiologie* **76**, 299.

Reed, G. (1972). *Baker's Digest* **46** (6), 16.

Reed, G. (1974). *Process Biochemistry* **9** (9), 11.

Reed, G. and Peppler, H.J. (1973). 'Yeast Technology'. The Avi Publishing Co., Westport, Connecticut.

Rehfeld, G. and Kraus, S. (1961). *Ernährungsforschung* **6**, 82.

Ringpfeil, M., Pochland, D., Schneider, J., Richter, K., Faulhaber, E., Liebetrau, L., Dickscheit, R. and Annemueller, W. (1974). East German Patent 105,001.

Robinson, R.J., Lord, T.H., Johnson, J.A. and Miller, B.S. (1958a). *Cereal Chemistry* **35**, 295.

Robinson, R.J., Lord, T.H., Johnson, J.A. and Miller, B.S. (1958b). *Cereal Chemistry* **35**, 306.
Rogosa, M. (1974). *In* 'Bergey's Manual of Determinative Bacteriology' (R.E. Buchanan and N.E. Gibbons, eds.), 8th edition, p. 576. The Williams and Wilkins Co., Baltimore, Maryland.
Rogosa, M. and Sharpe, M.E. (1959). *Journal of Applied Bacteriology* **22**, 329.
Rohrlich, M. (1953). *Zeitschrift für Lebensmittel-Untersuchung und -Forschung* **96**, 24.
Rohrlich, M. (1961). *Deutsche Lebensmittel-Rundschau* **57**, 83.
Rohrlich, M. and Essner, W. (1951). *Brot und Gebäck* **5**, 85.
Rohrlich, M. and Essner, W. (1960). *Die Mühle* **97**, 250.
Rohrlich, M. and Stegeman, J. (1958). *Brot und Gebäck* **12**, 41.
Rollag, N.L. and Enochian, R.V. (1967). *Baker's Digest* **41** (2), 54.
Rose, A.H. (1978). *In* 'Biochemistry and Genetics of Yeasts, Pure and Applied Aspects' (M. Bacila, B.L. Horecker and A.O.M. Stoppani, eds.), p. 197. Academic Press, New York and London.
Rothe, M. (1974). 'Aroma von Brot'. Akademie Verlag, Berlin.
Rothe, M. and Thomas, B. (1959). *Die Nahrung* **3**, 1.
Salovaara, H. (1979). 'Viljan tiet Kulutukseen'. Leipätoimikunta, Helsinki.
Samson, F.E., Katz, A.M. and Harris, D.L. (1955). *Archives of Biochemistry and Biophysics* **54**, 406.
Sandstedt, R.M. and Blish, M.J. (1934). *Cereal Chemistry* **11**, 368.
Saunders, R.M., Ng, H. and Kline, L. (1972). *Cereal Chemistry* **49**, 86.
Schatzmann, H. and Fiechter, A. (1974). *Chemie-Ingenieur-Technik* **46** (16), 1.
Schlegel, H.G. (1976). 'Allgemeine Mikrobiologie', 4th edition. Georg Thieme Verlag, Stuttgart.
Schultz, A.S., Fisher, R.A., Atkin, L. and Frey, C.N. (1943). *Industrial and Engineering Chemistry, Analytical Edition* **15**, 496.
Schulz, A. (1962). *In* 'Die Hefen' (R. Reiff, R. Kautzmann, H. Lüers and M. Lindemann, eds.), vol. 2, p. 695. Verlag Hans Carl, Nürnberg.
Schulz, A. (1965). *Brot und Gebäck* **19**, 61.
Schulz, A. (1966). *Baker's Digest* **40** (4), 77.
Schulz, A. (1972). *Getreide, Mehl und Brot* **26**, 129.
Selman, R.W. (1953). *Baker's Digest* **27** (4), 67.
Shcherbatenko, V.V., Nemtsova, E.S., Kuzminskij, R.V., Stoljarova, L.F., Elkin, S.I., Patt, V.A., Pashina, L.A., Kramynina, A.A. and Rzaev, R.I. (1972). Russian Patent 327,915.
Sher, H.N. (1969). *In* 'Continuous Cultivation of Microorganisms. Proceedings of the 4th Symposium' (J. Malek, ed.), p. 537. Academia, Praha.
Smith, D.E. and Coffman, J.R. (1960). *Analytical Chemistry* **32**, 1733.
Snyder, E. (1963). *Baker's Digest* **37** (4), 50.
Spicher, G. (1959). *Zentralblatt für Bacteriologie, Parasitenkunde, Infektionskrankheiten und Hygiene, Abteilung II* **113**, 80.
Spicher, G. (1966a). *Brot und Gebäck* **19**, 136.
Spicher, G. (1966b). *In* 'Brot in unserer Zeit' (W. Schäfer, ed.), p. 183. Verlag Moritz Schafer, Detmold.
Spicher, G. (1968a). *Brot und Gebäck* **22**, 61.
Spicher, G. (1968b). *Brot und Gebäck* **22**, 146.
Spicher, G. (1975). *Getreide, Mehl und Brot* **29**, 328.
Spicher, G. and Angermann, A. (1965). *Brot und Gebäck* **19**, 93.
Spicher, G. and Möllemann, A, (1975). *Getreide, Mehl und Brot* **29**, 206.

Spicher, G. and Rabe, E. (1979). Veröffentlichungen der Bundesforschungsanstalt für Getreide- und Kartoffelverarbeitung, No. 4644. Deltmold. **30.** *Tagung für Getreidechemie* 16.-18.5.1979.

Spicher, G. and Schöllhammer, K. (1977). *Getreide, Mehl und Brot* **31,** 215.

Spicher, G. and Schröder, R. (1978a). *Zeitschrift für Lebensmittel-Untersuchung und -Forschung* **167,** 342.

Spicher, G. and Schröder, R. (1978b). *Getreide, Mehl und Brot* **32,** 295.

Spicher, G. and Schröder, R. (1979a). *Zeitschrift für Lebensmittel-Untersuchung und -Forschung* **168,** 188.

Spicher, G. and Schröder, R. (1979b). *Zeitschrift für Lebensmittel-Untersuchung und -Forschung* **168,** 397.

Spicher, G. and Stephan, H. (1960). *Brot und Gebäck* **14,** 47.

Spicher, G. and Stephan, H. (1963). *Brot und Gebäck* **17,** 26.

Spicher, G. and Stephan, H. (1973). *Getreide, Mehl und Brot* **27,** 28.

Spicher, G., Schröder, R. and Schöllhammer, K. (1979). *Zeitschrift für Lebensmittel-Untersuchung und -Forschung* **169,** 77.

Sriranganathan, N., Seidler, R.J., Sandine, W.E. and Elliker, P.R. (1973). *Applied Microbiology* **25,** 461.

Stafford, H.R. (1967). *Process Biochemistry* **2** (4), 18.

Stegeman, J. and Rohrlich, M. (1958). *Brot und Gebäck* **12,** 65.

Stein, E., vom (1971). *Brot und Gebäck* **25,** 131.

Stephan, H. (1960). Merkblatt No. 41, Arbeitsgemeinschaft Getreideforschung, Detmold.

Stephan, H. (1970). Merkblatt No. 64, Arbeitsgemeinschaft Getreideforschung, Detmold.

Stokes, J.L. (1971). *In* 'The Yeasts' (A.H. Rose and J.S. Harrison, eds.), vol. 2, p. 119. Academic Press, London.

Sugihara, T.F. (1977). *Baker's Digest* **51** (5), 76.

Sugihara, T.F. (1978a). *Journal of Food Protection* **41,** 977.

Sugihara, T.F. (1978b). *Journal of Food Protection* **41,** 980.

Sugihara, T.F. and Kline, L. (1968). *Baker's Digest* **42** (5), 51.

Sugihara, T.F. and Kline, L. (1975). *Journal of Milk and Food Technology* **38,** 667.

Sugihara, T.F., Kline, L. and McCready, L.B. (1970). *Baker's Digest* **44** (2), 51.

Sugihara, T.F., Kline, L. and Miller, M.W. (1971). *Applied Microbiology* **21,** 456.

Suihko, M.-L. and Mäkinen, V. (1975). Valtion Teknillinen Tutkimuskeskus, Biotekniikan Laboratorio, Tiedonanto 11, Helsinki.

Suomalainen, H. and Nykänen, L. (1972). *Journal of the Institute of Brewing* **72,** 469.

Suomalainen, H., Oura, E. and Linnahalme, T. (1965). *Journal of the Institute of Brewing* **71,** 330.

Suomalainen, H., Dettwiler, J. and Sinda, E. (1972). *Process Biochemistry* **7** (5), 16.

Tanaka, Y. and Sato, T. (1969). *Journal of Fermentation Technology* **47,** 587.

Tanert, M.C. (1891a). *Bulletin de la Société Chimique de France* **5,** 724.

Tanert, M.C. (1891b). *Comptes Rendus Hebdomadaires des Séances de l'Academie des Sciences, Chimie Organique* **112,** 293.

Täufel, K., Rominger, K. and Hirschfeld, W. (1959). *Zeitschrift für Lebensmittel-Untersuchung und -Forschung* **109,** 1.

Thorn, A.J. and Reed, G. (1959). *Cereal Science Today* **4,** 198.

Umbreit, W.W., Burris, R.H. and Stauffer, J.F. (1972). 'Manometric and Biochemical Techniques', 5th edition. Burgess Publication Co., Minnesota.

Vaisey, M. and Unrau, A.M. (1964). *Agricultural and Food Chemistry* **12,** 84.

Varo, P., Westermark-Rosendahl, C., Hyvönen, L. and Koivistoinen, P. (1979). *Lebensmittel-Wissenschaft und Technologie* **12**, 153.

Vraná, D., Oura, E. and Suomalainen, H. (1968). *Suomen Kemistilehti* **B41**, 284.

Wagner, F. (1936). 'Presshefe and Gärungsalkohole'. E. Strache, Warnsdorf.

Walden, C.C. (1955). *Cereal Chemistry* **32**, 421.

Weldin, J.C. (1925). *Proceedings of the Iowa Academy of Science* **32**, 95.

Westermarck-Rosendahl, C. (1978). University of Helsinki, Department of Food Chemistry and Technology–series No. 464, Helsinki.

White, J. (1954). 'Yeast Technology'. Chapman and Hall, London.

Williams, R.A.D. (1975). *In* 'Lactic Acid Bacteria in Beverages and Food' (J.G. Carr, C.V. Cutting and G.C. Whiting, eds.), p. 351. Academic Press, London.

Williams, K.T. and Bevenue, A. (1951). *Cereal Chemistry* **28**, 416.

Windisch, S. and Steckowski, U. (1971a). West German Patent Application 1,927,402.

Windisch, S. and Steckowski, U. (1971b). *Branntweinwirtschaft* **111**, 197.

Windisch, S., Kowalski, S. and Zander, I. (1976). *European Journal of Applied Microbiology* **3**, 213.

Wolter, K. (1974). *Brotindustrie* **17**, 50.

Wood, P.S. (1980). *Process Biochemistry* **15**, (6), 12.

Wood, B.J., Cardenas, O.S., Yong, F.M. and McNulty, D.W. (1975). *In* 'Lactic Acid Bacteria in Beverages and Food' (J.G. Carr, C.V. Cutting and G.C. Whiting, eds.), p. 325. Academic Press, London.

Wood, T.H. (1956). *Advances in Biological and Medical Physics* **4**, 119.

Wutzel, H. (1968). United States Patent 3,410,692.

5. Cheeses

B. A. LAW

National Institute for Research in Dairying, Shinfield, Reading, England

I. INTRODUCTION

A. A Brief History

The manufacture of cultured dairy products represents the second most important fermentation industry (next to alcoholic beverages) accounting for approximately 20% of all fermented products produced World-wide in 1978. Cheese, in particular, is made in almost every country in one form or another, and in the developed World, where annual cheese consumption is highest (e.g. U.S.A., 2,200 million tonnes; E.E.C., 3,000 million tonnes in 1978), its manufacture involves high capital investment in complex automated equipment. However, all modern cheesemaking processes owe their historical origin to the exploitation, from as long as 9000 years ago, of the random and accidental infection and souring of milk by the then unrecognized lactic acid-producing bacteria. This process involves metabolic conversion of lactose into lactic acid and it largely prevents subsequent growth of spoilage organisms and of pathogens, so providing the basis for preserving the solids in surplus milk for later consumption. It was not until the beginning of the present century that cheesemakers developed the modern practice of using carefully selected pure strains of these bacteria, comprising the 'starter' cultures, which were deliberately added to cheese milk in standard amounts depending on the type of cheese required. The characteristics of a particular cheese variety are governed not only by the composition of the starter culture, but also by the temperature of manufacture, the coagulant used to gel the milk (chymosin or a microbial substitute; Green, 1977) and by the secondary microflora which may be present as chance contaminants (e.g. non-starter lactic-acid bacteria) or introduced into the cheesemaking process deliberately (e.g. spores of *Penicillium* sp).

B. Diversity Among Cheeses

An enormous range of cheese varieties exists but they can be grouped according to their moisture content and the complexity of their microfloras (Table 1). For example, soft cheeses have high (50–80%) water content and may be classed either as unripened (e.g. cottage cheese), with a simple mesophilic lactic acid bacterial flora (the

Table 1

Major cheese categories, their starter compositions and secondary microfloras

Cheese category	Example varieties	Moisture content (%)	Starter composition	Starter function	Secondary flora
Unripened, soft	Cottage	not > 80	*Streptococcus diacetylactis* *Leuconostoc* spp.	Acid and diacetyl production	none
	Mozzarella	> 50	*Strep. thermophilus* *Lactobacillus bulgaricus*	Acid production	none
Ripened, soft	Camembert	48 ⎱	*Strep. cremoris* and	Acid	*Penicillium*
	Brie	55 ⎰	*Strep. lactis*	production	*caseicolum*, yeasts
Semihard	Caerphilly	45	*Strep. diacetylactis* *Leuconostoc* spp. *Strep. cremoris* *Strep. lactis*	Acid and diacetyl production	Lactobacilli
	Limburg	45	*Strep. lactis* *Strep. cremoris*	Acid production	Yeasts, *Brevibacterium linens*
Hard	Cheddar	< 40	*Strep. cremoris* *Strep. lactis* *Strep. diacetylactis*[a] *Leuconostoc* spp.[a]	Acid production	Lactobacilli, pediococci
	Gouda	40	*Strep. cremoris* *Strep. lactis* *Strep. diacetylactis* *Leuconostoc* spp.	Acid and carbon dioxide production	Propionibacteria[b]
	Emmental	38	*Strep. thermophilus* *Lactobacillus helveticus* *L. lactis* *L. bulgaricus* *Propionibacterium shermanii*	Acid, carbon dioxide and propionic acid production	*Propionibacterium shermanii*, group D streptococci
	Gruyère	38–40	*Strep. thermophilus* *Lactobacillus helveticus* *L. lactis* *L. bulgaricus* *Propionibacterium shermanii*[c]	Acid, carbon dioxide and propionic acid production	*Propionibacterium shermanii*, group D streptococci plus yeasts and coryneforms including *B. linens*
Blue-vein	Roquefort Gorgonzola Stilton	40–45	*Strep. lactis* *Strep. diacetylactis* *Strep. cremoris* *Leuconostoc* spp.	Acid, carbon dioxide production	*Penicillium roqueforti*, yeasts, micrococci

[a] Not always included

[b] Sometimes present as adventitious bacteria but not vital to typical cheese characteristics (Kleter, 1976)

[c] Introduced with the starter but have no lactic acid-forming function. They grow as a secondary flora

acid-producing starter bacteria used to make the product) or as ripened, with a similar basic flora but with a surface mould growth which contributes flavour compounds characteristic of the mature cheese (e.g. Camembert after 6–8 weeks' storage). Semihard varieties are made with similar mesophilic starters but the manufacturing process is more complex and includes a short curd cooking stage to lower its moisture content to approximately 45% and render it firmer. Some varieties which are salted internally develop little flavour beyond that of fresh curd, though the texture softens due to protein breakdown (e.g. Caerphilly). Others are salted by immersion of the lightly pressed curd block in brine and subsequently by rubbing the surface with salt or a cloth soaked in brine. This treatment results in the development of a characteristic surface flora of yeasts and bacteria; Limburg is a classical example of this type. The hard cheeses (approximately 40% moisture or less) fall into three major categories; those with relatively simple microfloras resemble Cheddar and are made with the same mesophilic starters used in soft and semihard cheese. Subsequent bacterial growth in the cheese is limited by the acidity, salt concentration and redox potential so that a full, mature flavour may take up to 12 months to develop. The second group differs from the first by being inoculated with mould spores which germinate when air is admitted to the cheese by 'spiking'. The metabolism of the growing mycelium, and later of the spores, generates the flavour and aroma compounds characteristic of the cheese. Stilton, Danish Blue, Roquefort and Gorgonzola are examples of this type. Swiss cheese varieties (Emmental and Gruyère) form a third group and are distinguished both by the thermophilic starters used to make them, and by the subsequent growth in the cheese of propionic acid-producing bacteria which not only contribute to flavour development but also produce the gas-filled 'eyes' characteristic of this cheese. Swiss 'mountain cheese' varieties also have a surface flora of yeasts and bacteria which add to their basic flavour and aroma. Italian hard cheeses form a special group in which mammalian lipases are used to develop strong fatty-acid (rancid) flavours. Davis (1976) has reviewed the history of cheesemaking and the modern manufacturing methods used for different types. The present chapter describes the classification, composition and production of lactic acid-producing bacteria used as starters in the modern cheese industry, and considers their role, together with that of secondary microfloras of cheese, in the complete conversion of milk into mature cheese.

II. CHEESE STARTER BACTERIA

A. Classification

The lactic-acid bacteria used in the cheese industry belong to the genera *Streptococcus, Leuconostoc* and *Lactobacillus*. Streptococci are regarded as homofermentative, producing mainly the L(+) isomer of lactic acid (Fig. 1) from glucose or lactose. The mesophilic species were distinguished from *Strep. faecalis* (Group D streptococci; Lancefield, 1933) by Sherman (1938) and assigned to serological group N by Shattock and Mattick (1943; *Strep. lactis*), Briggs and Newland (1952; *Strep. cremoris*) and Briggs (1952; *Strep. diacetylactis*). The Eighth Edition of Bergey's Manual (Buchanan and Gibbons, 1974) designates *Strep. diacetylactis* as a subspecies (*Strept. lactis* subsp. *diacetylactis*, referred to hereafter as *Strep. diacetylactis*). The thermophilic lactic streptococci belong to the species *Strep. thermophilus*; they do not carry the group N antigen and are able to grow at 50°C. These and other distinguishing characteristics of the starter streptococci are shown in Table 2. Among the lactobacilli, only the homofermentative Thermobacterium group is important in providing starter cultures, and the useful strains are normally confined to *Lactobacillus helveticus, L. casei* and *L. bulgaricus*. However, species from the Streptobacterium (homofermentative) and Betabacterium (heterofermentative) groups may be significant in the secondary microflora of

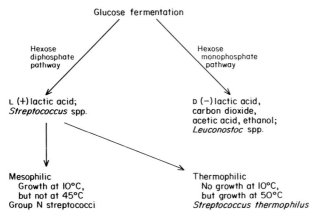

Fig. 1. Taxonomic relationships among lactic streptococci.

Table 2
Some differentiating characteristics of lactic streptococci

	Streptococcus lactis	Streptococcus diacetylactis	Streptococcus cremoris	Streptococcus thermophilus
Growth at 39.5°C	+	+	−	+
Growth at 50°C	−	−	−	+
Formation of ammonia from arginine	+	+	−	−
Formation of acid from maltose	+	+	−	−
Formation of carbon dioxide and diacetyl from citrate	−	+	−	−
Percentage of sodium chloride inhibiting growth	4.0–6.5	4.0–6.5	2.0–4.0	< 2.0
Presence of group-N antigen	+	+	+	−

+ indicates growth or positive reaction; −, no growth or a negative reaction

some hard-cheese varieties (Fryer, 1969). Some of the main differentiating characteristics of these groups are summarized in Table 3. Isolation and diagnostic media for starter streptococci and lactobacilli have been described and discussed in detail elsewhere (Shankar and Davies, 1977; Sharpe, 1978; Mullan and Walker, 1979; Kempler and McKay, 1980).

Table 3
Differentiation of lactobacilli

	Glucose fermentation products		
	Lactic acid		Lactic acid, carbon dioxide, acetic acid and ethanol
	Thermobacterium spp.	Streptobacterium spp.	Betabacterium spp.
Growth at 45°C	+	±	
Growth at 15°C	−	+	
Fermentation of ribose	−	+	
Formation of gas from gluconate	−	+	

B. Ecology

Lactic streptococci were originally isolated from souring milk. Lister (1878) described such an isolate which was later to be named *Strep. lactis* (Orla-Jensen, 1919) and its natural habitat seems to be plant material (Sandine *et al.*, 1972). *Streptococcus cremoris*, on the other hand, has never been isolated and unequivocally identified from any source other than milk, cream or related materials (Radich, 1968; King and Koburger, 1960; Cavett and Garvie, 1967). It may be present in the same environment as *Strep. lactis* in lower numbers (Sandine *et al.*, 1972), but present isolation methods cannot confirm this. The natural habitat of *Strep. thermophilus* is not known, but it can be isolated from heat-treated milk (Sherman, 1937). Its high growth temperature may reflect an intestinal origin, shared perhaps by the starter lactobacilli (e.g. Mitsuoka, 1968); they probably entered the milk habitat *via* the stomach preparations from sheep or calves used to aid milk fermentations in historical times. The general ecology of lactobacilli has been reviewed recently (Sharpe, 1980), and information on dairy lactobacilli is included.

C. Composition of Cheese Starter Cultures

The species of lactic-acid bacteria used as starters for the major cheese categories are listed as part of Table 1 (p. 149). They are used according to the manufacturing regime required for a particular cheese variety; those starters that produce only lactic acid make close-texture cheese, whereas the citrate-fermenting (carbon dioxide producing) species are used for cheeses that require either an open texture for aeration of germinating mould spores (see Section VII.E.1, p. 185) or formation of 'eyes' for typical appearance (Edam, Gouda). The ability of these starters to produce diacetyl from citrate also makes them suitable for manufacture of cottage cheese (see Section VI, p. 170). The mesophilic starters (group N streptococci) are used for cheeses whose manufacture requires no 'cooking' stage, or those whose 'cooking' stage is below 40°C (Davis, 1976), whereas the thermophilic bacteria (*Strep. thermophilus* and *Strep.* lactobacilli) acidify the high-temperature cooked-curd cheeses (50°C) such as the Swiss varieties and the Italian hard cheeses. In Swiss cheese, propionibacteria rather than *Strep. diacetylactis* produce the carbon dioxide for 'eye' formation (see Section VII.D, p. 182).

Commercial starter cultures fall into one of three main categories depending on the number of strains of streptococci and lactobacilli which they contain. Single-strain starters represent the simplest types but they now have little commercial application. Mutiple-strain starters are composed of several defined strains (usually six; Limsowtin *et al.*, 1977; Lawrence *et al.*, 1978), cultured separately until required for vat inoculation culture (see Section III.A, p. 156). Most single, paired single, and multiple starters employ strains of *Strep. cremoris*, though a limited number of *Strep. lactis* strains are sometimes included. Mixed-strain starters are the most common at present, and are composed of unknown mixtures of many different strains and species which are always propagated together. These starters may include strains of *Strep. diacetylactis* and *Leuconostoc* sp. as well as the two species already mentioned.

III. PRODUCTION AND USE OF STARTERS

A. Traditional Production Methods

Master cultures of cheese starters are normally held by the supply company or culture collection in the freeze-dried state. Before use in cheesemaking, they require several transfers in milk before they are completely revived as lactic acid producers (Cox *et al.*, 1978). Freshly revived starters are tested for their acid-producing activity in milk. The particular test depends on the type of cheese to be made; Cox *et al.* (1978) and Sandine (1979) have reviewed the alternatives in detail. Other tests may include those for yeasts and moulds, 'coli-aerogenes' strains and bacteriophage relationships (Cox *et al.*, 1978; Terzaghi and Sandine, 1975). Suitably screened starters must then be grown through at least two preparation stages of increasing scale before the culture is suitable for inoculation into the cheese vat. The inoculation level varies in order from approximately 10^6 colony-forming units ml^{-1} for low-acid cheese, or cheese in which acid can develop over long periods (e.g. Emmental), to 10^7 colony-forming units ml^{-1} for short-make acid cheese (e.g. Cheddar, Cheshire). These cell numbers are reached *via* a mother culture and the bulk starter culture, the latter being the culture that is actually added to the cheese vat (Fig. 2, p. 157). The detailed investigations which have been carried out to determine the appropriate temperature for growth of bulk starter cultures have been summarized by Cox (1977); for mixed-strain mesophilic Cheddar starters, the optimum is 22°C (for 17–18 hours), but thermophilic starters must be incubated at 37°C or higher. These conditions must be chosen carefully to avoid, on the one hand, a culture with too few cells (optimum range is $5 \cdot 10^8$ to $1 \cdot 10^9$ colony-forming units ml^{-1}) and, on the other, an over-acid culture in which the cells can be damaged by a low pH value so that they exhibit a long lag phase in the cheese vat. The bulk starter pH value should be 5.0 (Sozzi, 1972) but, if this falls to 4.2–4.4, glycolysis ceases (Lawrence *et al.*, 1976), the cells cannot maintain their relatively high internal pH value (Harold *et al.*, 1970), and intracellular proteins are damaged.

The quality-control methods and equipment used for starter production in modern creameries have been reviewed by Cox *et al.* (1978) and Jespersen (1979). In general, successful growth of bulk starter cultures is based on the use of antibiotic-free skim-milk powder, reconstituted at 12% (w/v) and heat-treated to inactivate natural inhibitors (see Sections IV.B, C and D, pp. 163–165). Both the starter streptococci and lactobacilli can be attacked and lysed by bacteriophages if they gain entry at this production stage, but the use of suitable aseptically inoculated tanks can avoid this problem (e.g. Jesperson, 1979; Cogan, 1980). Some factories grow their bulk starter in commercial phage-resistant media which are based on enzyme-digested skim milk, heavily buffered and treated with phosphates to lower the free concentration of calcium ions. This is claimed to prevent phage multiplication by interfering with the injection of phage DNA into the starter cells (Douth and Meanwell, 1953; Watanabe and Takasue, 1972). However, some phages can multiply in such media (Sandine, 1977) and their effectiveness is at present the subject of controversy (Gulstrom *et al.*, 1979; Erickson, 1980). In commercial practice, phage control is more difficult and complex during the cheese-making process, and interactions between starters and their phages will be discussed in detail when this stage is considered (see Section IV.A.3, p. 161).

During subculture and bulk starter growth, mixed starters may vary in composition because the constituent strains have different growth rates and survival rates at low pH values (Lode and Tufto, 1973). Although these factors do not normally cause starter failure, they can lead to unpredictable variations in acid-production rates in the cheese vat. Cox (1977) suggested that careful screening for acid-sensitive strains can reveal the most stable types thus allowing the balance of slow and fast strains to be monitored. Such problems do not arise with single or paired defined starters, but they are not sufficiently versatile for present-day commercial use. The multiple-starter concept (Limsowtin et al., 1977) is a development of the use of defined strains in which six selected strains are used to make cheese. As was the case with mixed starters, the culture as a whole can cope with changes in milk composition between 'makes', but the starter composition is kept constant because the six strains are not subcultured together; they are only mixed in the mother culture and co-exist for less than 20 generations before cheesemaking. This starter system also has advantages for phage control in commercial creameries (see Section IV.A.3, p. 162).

B. Concentrated Starters

Alternative methods for starter production and preparation have been developed recently, with the aim of eliminating either the master and mother culture stage or all preparation stages altogether, including the bulk starter (Fig. 2). These techniques are based on the practice of growing starters in clarified media from which they can be separated and concentrated by centrifugation. The concentrates are then either deep-frozen in liquid nitrogen or freeze-dried before distribution to creameries. Media formulations for starter concentrate production are often commercial secrets but are generally based either on proteolytic digests of whey, supplemented with peptone and/or yeast extract, or on digests of skim milk (Law et al., 1976b; Cox et al., 1978). Starter growth in milk or milk-based media is normally limited by the lactic acid they produce to populations of approximately 10^9 colony-forming units ml^{-1}. Consequently, fermentors designed for the production of starter concentrates are equipped with automatic titrators which maintain culture pH at 6.0–6.5 (streptococci) or 5.5 (lactobacilli) by the addition of sodium or ammonium hydroxide, and viable cell concentrations of 10^{10} ml^{-1} can be achieved at early stationary phase (Gilliland and Speck, 1974; Law et al., 1976b). Cells harvested at or near this point are most active after centrifugation and concentration (Bergère and Hermier, 1968; Stanley, 1977). Ammonium hydroxide is reported to give higher cell yields than

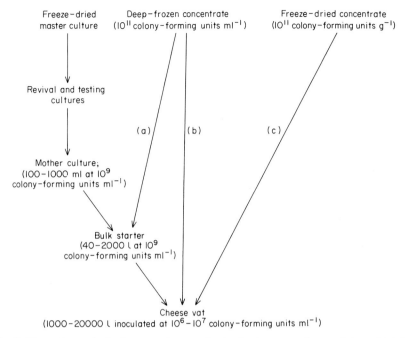

Fig. 2. Alternative methods of starter-culture production in cheese manufacture. Scheme (a) is for high-acid/short make time cheese, (b) for low-acid/long make time cheese, and (c) for all varieties.

sodium hydroxide as a neutralizing agent (Gilliland and Speck, 1974) and the ammonia-neutralized cells resist freeze-drying better (Efstathiou *et al.*, 1975). However, there is some evidence that acid-producing activity in these concentrates is less than that in concentrates grown with sodium hydroxide due to lower proteinase activity (Gilliland and Speck, 1974; Peebles *et al.*, 1969).

Continuous culture for starter concentrate production was investigated by Lloyd and Pont (1973) using single strains of mesophilic streptococci. They showed cultures could not be maintained free from phage, even with the most stringent precautions using filtered ultraviolet-treated air. Also, some cultures quickly became dominated by slow acid-producing variants. Maximum cell yields for lactic streptococci in continuous culture are approximately $3 \cdot 10^9$ colony-forming units ml^{-1} (Keen, 1972).

Osborne (1977) described a dialysis culture system for mesophilic starters which, although basically a batch fermentation, could be harvested 3–4 times from one inoculation. This was achieved by partially emptying the culture vessel when the cells had reached stationary phase, then refilling from a large medium reservoir *via* a cell containing a high surface area of cellulose acetate dialysis membrane. The most important advantage of this system is that the exponential phase of starter growth can be extended to give yields of 10^{11} cells ml^{-1} by pumping fresh medium through the dialysis cell during the fermentation in order to exchange continuously growth-limiting toxic waste products for fresh nutrients. The most significant such product appears to be ammonium lactate from the neutralizing agent (Osborne and Brown, 1980). Dialysis culture produces such high cell yields that centrifugal concentration can be eliminated, and the contents of the fermenter can be harvested and frozen directly.

C. Direct Vat Inoculation

Although deep-frozen starter concentrates can be used for inoculating the bulk starter for most cheese varieties, their application to direct cheese vat inoculation is limited to those types requiring a low level starter inoculum (0.1–0.2%, v/v) and slow acid production (e.g. Emmental, Camembert). Varieties such as Cheddar, required uneconomical amounts of concentrate (1.5–2.5%, equvalent to 150–250 litres for a 10,000 litre vat) unless a very long milk-ripening period is employed before renneting. This is unacceptable in factories in which vats are filled more than once a day and exposes the starter to an increased risk of phage infection. More recently, there has been renewed interest in freeze-dried starters for bulk-tank inoculation and even for direct vat inoculation (Sharpe, 1979). Careful selection of single *Strep. cremoris* and *Strep. lactis* strains, on the basis of resistance to freeze-drying, combined with the use of cryoprotective agents and easily available nutrients, have led to the development of paired and multiple starter systems containing about $5 \cdot 10^{11}$ colony-forming units g^{-1} which can be added directly to the cheese vat (Chapman, 1978). The lag phase before acid production is similar to that of traditional liquid bulk starter for Cheddar manufacture, though renneting and cooking temperatures require small adjustments to compensate for the absence of lactic acid in the starter inoculum and for the tendency towards too-rapid acid production while the vat temperature is raised to its maximum value. Camembert cheese has also been successfully made using this direct vat inoculation system

(Sharpe, 1979) and the Swiss varieties should present few problems since they only require a low inoculum. At present, however, there are few starters available which will recover quickly in milk after freeze-drying, but further screening and selection will undoubtedly increase the useable range in due course. Such development will be worthwhile because direct vat inoculation has several important advantages in addition to the obvious decrease in the degree of handling and preparation required. For example, no special low-temperature storage equipment is required; the starters are supplied in sachets which can be kept for short periods at room temperature, or for longer periods in ordinary refrigerators. Also, strain dominance ceases to be a problem since the defined strains are produced separately and only mixed in the cheese vat.

IV. GROWTH OF STARTERS IN THE CHEESE VAT

Starters must be able to multiply from 10^6–10^7 colony-forming units ml^{-1} to 10^8–10^9 colony-forming g^{-1} curd in a few hours in order to produce enough lactic acid to complete the conversion of milk into acid curd. Their ability to do this consistently is vital to the efficient and profitable operation of cheese factories, yet there are numerous and variable factors which cheesemakers have to try to control if failures or 'slow' vats are to be avoided. These factors range from bacteriophage infections to antibiotics derived from mastitis therapy, and are considered in detail below.

A. Bacteriophage

The association between the failure of mesophilic starter cultures to produce acid in cheese vats, and their infection and lysis by virulent (lytic) phages, was first identified by Whitehead and Cox (1935). Since then, the characteristics of phages which attack all types of starter lactic acid-producing bacteria have been studied intensively and numerous starter handling systems have been proposed to control their multiplication in the cheese vat.

1. *Morphology and Classification*

Phages for both streptococci and lactobacilli have been isolated and attempts to classify them are based on morphology revealed by electron microscopy, host range and serology. Lawrence *et al.* (1978) suggest that there are six morphological groups of mesophilic streptococcal phages distinguishable on the basis of head type (isometric or prolate) and size (small or large isometric), possession of a collar, and tail size. However, Lembke *et al.* (1980) found 14 distinct morphological types based on Bradley's group B and C (Bradley, 1967) and differing in tail, as well as head morphology. Similar morphological groups of phages show many different patterns of host range and some are not even species-specific (Chopin *et al.*, 1976; Lembke *et al.*, 1980). Serological typing can be used to subdivide morphological groups, and good correlations between serologically similar phages and host specificity have been shown for New Zealand starters and their phages (Jarvis, 1977).

Thermophilic starters may be less sensitive to phage than mesophilic starters (Lawrence and Thomas, 1979) and the phages themselves are very specific (Sozzi and Maret, 1975). Accolas and Spillman (1979a) described six morphologically similar phages from *Strep. thermophilus*, all belonging to group B and differing only in tail structure. Detailed studies of lactobacillus phage from yoghurt cultures indicate that those specific for *L. bulgaricus* belong to group B (Accolas and Spillman, 1979b), but group A phages attack *L. helveticus*, the cheese starter. Each group could be divided into two further groups on the basis of host-strain specificity.

2. Lytic Cycle and Lysogeny

Phage infection in lactic streptococci, as in other bacteria, involves adsorption, digestion of the cell wall and injection of DNA. Latent period and burst size vary according to temperature, e.g. 2–113 particles per cell in 23–56 minutes at 30°C or up to 139 particles per cell in 16–44 minutes at 37°C (Lawrence *et al.*, 1976). Phage multiplication during cheesemaking is limited by physical factors such as the size of the zone through which expelled whey diffuses from curd particles, carrying the phage with it to infect other cells. Not all phage infections of lactic streptococci result in lysis, and many examples of lysogenic strains (carrying temperate phages) have been observed after induction with ultraviolet radiation or mitomycin C (Reiter, 1949; Sandine *et al.*, 1962; Keogh and Shimmin, 1969; Lowrie, 1974; McKay and Baldwin, 1974; Park and McKay, 1975). Until recently, evidence for lysogeny has been based largely on the observation of phage particles released from induced cells, and indicator strains (those for which the temperate phage is lytic) have been difficult to find. However, Gasson and Davies (1980) have now completed the formal demonstration of lysogeny in *Strep. lactis* and *Strep. cremoris* by curing prophage-carrying strains and showing that the cured derivatives are indicators for their temperate phages. The cured strains were relysogenized after infection.

3. Sources of Phage and Control Methods

Phages are an ever-present hazard in commercial cheese factories because, unlike starter production, the cheese fermentation is not carried out under aseptic conditions; it would be impractical to do so on a large scale. Thus methods of phage control are based on an understanding of the sources of the phages, their multiplication factors, and the phage interrelatedness of the starter strains in use.

Raw milk is a possible source of phage since it usually contains between 10^2 and 10^3 colony-forming units of lactic acid-producing bacteria ml^{-1} which may be expected to induce spontaneously (Heap *et al.*, 1978). Many phages are resistant to the high-temperature short-time pasteurization (17 seconds at 72°C) given to cheese milk (Lawrence and Thomas, 1979; Chopin, 1980) and a few may enter the vat *via* this route. The starter cultures themselves contribute to the phage levels in factories because they contain lysogenic strains whose temperate phages, when released by induction, are usually lytic for other strains either in the same culture or in those used subsequently. Thorough disinfection of the vats, work surfaces, floors and air using sodium hypochlorite can minimize phage contamination of successive 'makes', but consistent cheesemaking in large Cheddar factories, whose high milk throughput (e.g. 500,000 litres per day) requires vats to be filled several times in one working day, is only possible with most starter systems if starter 'rotation' is used. These rotations are based on a knowledge of phage relationships between strains in the starter culture. The simplest system involves successive use of a series of phage-unrelated single-strain starters, and this was used successfully in New Zealand in factories where vats were filled once a day. However, the number of useful strains was limited by the tendency of some of them to produce characteristic flavour defects in the mature cheese (Lawrence and Pearce, 1972). Unfortunately, those starters that continued to multiply during the curd-cooking stage at above 35°C (the 'fast' starters), and thus achieved the desirable short manufacturing times, were the very starters that gave rise to most flavour defects. Also, their continued multiplication allowed their phages to proliferate to very high numbers. The 'slow' starters, which ceased multiplication as the cooking temperature rose above $35°C$, produced the best flavoured cheese and only low phage titres, but making times were too long (more than 6 hours from rennet to curd milling) to be economical in multifill factories. By using slow and fast starters in rotations of phage-unrelated pairs (2:1 ratio), the New Zealand workers were able to retain the advantages of short make times with those of low-phage titres and good flavour development (Lawrence and Pearce, 1972). However, even this development is limited by the relatively small number of suitable complementary starter pairs. Phage control with defined starter strains has proved most effective using the multiple system of Limsowtin et al. (1977) (also, see Section IIIA, p. 156). This involves six phage-unrelated strains isolated from commercial mixed cultures and selected for natural phage resistance using methods described by Heap and Lawrence (1976). One or two strains usually become susceptible to a lytic phage after the multiple starter has been in use for several months in the same factory. When this happens, and phage titres have become high (i.e. $10^8–10^9$ particles ml whey^{-1}), the strain or strains can be withdrawn and replaced with a new isolate. During the replacement period, even at such high phage titres, the starter culture as a whole continues to produce acid at almost the same rate as it did originally because the non-susceptible strains remain active. These multiple strain starters can be used in multifill factories without rotation. The first combinations consisted of three fast and three slow strains of *Strep. cremoris* but, more recently, *Strep. lactis* strains have been introduced to increase the uniformity of acid production in the presence of natural milk inhibitors to which they are less sensitive than *Strep. cremoris* (Lawrence et al., 1978).

Phage control with undefined mixed starters is more difficult to achieve by rotation, since phage relationships may vary as strain balance changes during subculture and bulk-starter production. Also, Lawrence and Thomas (1979) suggested that phage infections can occur without outside contamination, since spontaneous induction of lysogenic strains can frequently produce phage particles which are lytic for one or more of the other strains in the mixture. Ironically, this is more likely to lead to starter failure in the vat in those cheese factories which use aseptic bulk-starter production and phage-resistant medium, because there is no natural mechanism in operation to select out phage-sensitive strains. In Holland, Gouda starters are propagated without precautions against phage (Lawrence and Thomas, 1979) so that a constant natural contamination gives infection levels of up to 10^8 particles ml^{-1} and a favourable balance between phage-resistant and phage-susceptible strains is maintained. Transfer of such cultures in the protected environment of starter laboratories is kept to a minimum, and manufacturers have found that the frozen concentrated production system for these starters is ideal to achieve this end (Stadhouders, 1974). The same mixed starters from frozen stocks can be used continuously over long periods without rotation.

Phage resistance is conferred on starter cells by a number of mechanisms which may be exploited in the future to overcome the phage problem. For example, lysogeny protects a cell from infection by the same temperate phage, though not from other phages. If, however, a sufficient number of defective prophages could be introduced into a cured cell it is theoretically possible to give it immunity against all the phages likely to occur in a particular factory (Lawrence and Thomas, 1979). Phage-resistant mutants can be produced using mutagenic agents to alter phage-receptor sites, but additional cell-surface mutations produce other changes such as high-heat and salt resistance (Sinha, 1980) and may impair membrane-transport and proteinase systems sufficiently to render the starter incapable of growing rapidly enough in milk to produce acid for cheesemaking (Lawrence and Thomas, 1979). At present, it is preferable to select naturally phage-resistant mutants after exposure of cultures to phage-containing whey (e.g. Czulak *et al.*, 1979). Modification or restriction of phage DNA by the host can confer resistance, and these phenomena have been demonstrated in group N streptococci (Collins, 1956; Potter, 1970; Pearce and Lowrie, 1974; Sanders and Klaenhammer, 1980). However, more research is required into the nature of modification systems and the specificity of restriction endonucleases before these properties can be used rationally. Also, it may prove possible to engineer phage resistance by introducing modification/restriction enzymes from other bacteria *via* plasmid DNA.

B. Agglutinins

Agglutinating antibodies can cause some starter strains to sediment during cheesemaking, with detrimental effects on acid formation and curd texture. Since these agglutinins are inactivated by rennet (Stad-houders and Hassing, 1974), this problem is confined to soft cheeses whose manufacture involves little or no coagulant (e.g. cottage cheese). Emmons and Tuckey (1967) advocated the use of a simple screening test to identify agglutinin-sensitive strains and also suggested that aged milk, calcium-fortified milk and citrate-fermenting starters should be avoided. However, these precautions are based on general experience rather than detailed understanding since the mechanism of the agglutinating interaction with starter cells is not precisely known (Reiter, 1973).

C. Lactoperoxidase

The lactoperoxidase—thiocyanate–hydrogen peroxide system is another of the natural inhibitors of starters in milk (Jago and Morrison, 1962; Reiter *et al.*, 1964b) and is not completely inactivated by pasteurization used for cheese milk (Reiter, 1973). Only a few starters are sensitive to the system, but the use of severely heat-treated milk (in which the lactoperoxidase–thiocyanate–hydrogen peroxide system is inactivated) to grow bulk starter cultures allows such strains to survive to the cheesemaking stage where they may interfere with acid production in the pasteurized milk. Auclair and Vassal (1963) showed that mutants of non-sensitive strains, sensitive to the lactoperoxidase–thiocyanate–hydrogen peroxide system can arise and proliferate in starters grown in reconstituted skim milk. Thus, although screening for sensitive strains largely eliminates the problem, occasional slow starter cultures can be attributed to mutations.

D. Antibiotics

Antibiotics occur in milk if they have been used in bovine mastitis therapy (Mol. 1975). Many of the antibiotics used for this purpose are particularly active against streptococci (Wilson, 1964; Cox *et al.*, 1978) and, although withholding times for milk from treated cows are recommended by the antibiotic manufacturers, the antibiotics are often found in milk (Tramer, 1964) and account for some cases of starter failure in the cheese vat. Milk is therefore screened routinely for antibiotics ('Intertest'; Jacobs *et al.*, 1972) under a scheme established by the Joint Committee of the Milk Marketing Board and the Dairy Trades Federation. The presence of 0.02 international unit of penicillin equivalent ml^{-1} or above results in the rejection of milk for processing.

V. STARTER NUTRITION

Starter streptococci and lactobacilli are nutritionally fastidious bacteria which require complex media for cultivation. The fact that the dairy strains grow in milk indicates that it can supply them will all of their essential nutrients such as vitamins (biotin, niacin, pantothenic acid; Reiter and Oram, 1962), metals (Fe^{3+}, Mg^{2+}, Mo^{4+} and Se^{4+}; Olson and Qutub, 1970) and in some cases, purines and pyrimidines (Selby-Smith *et al.*, 1975; Taniguchi *et al.*, 1965). However, although an understanding of these requirements is helpful in explaining certain associative growth phenomena particularly in starters comprising streptococci and lactobacilli (Lawrence *et al.*, 1976), the utilization of milk proteins and lactose has the most direct relevance to their performance as cheese starters.

A. Utilization of Milk Proteins by Starters

Dairy lactic-acid bacteria require for growth (or are stimulated by) certain amino acids which they are unable to synthesize themselves (Reiter and Oram, 1962; Henderson and Snell, 1948; Shankar, 1977). The most detailed work has been carried out with single-strain mesophilic streptococci which must be supplied with leucine, isoleucine, valine, methionine, arginine, histidine, glutamic acid and, in some cases, proline and cystine (Law *et al.*, 1976b). Free amino-acid concentrations in milk are too low to support the starter growth required for cheese manufacture (Lawrence *et al.*, 1976) but the bacteria produce cell-bound extracellular proteinases which degrade whole casein to trans-portable peptides. Thomas *et al.* (1974) first reported the cell-wall location of a proteinase in *Strep. lactis* which was released from cells on formation of stable protoplasts by cell wall-degrading enzymes and, unlike the cell-free proteinase of *Strep. cremoris* (Williamson *et al.*, 1964), the cell-bound enzyme was EDTA-sensitive and unaffected by thiol-blocking agents. Exterkate (1975) described a casein-utilizing system in *Strep. cremoris* strain HP, which consisted of acid and neutral cell-wall proteinases, as well as membrane-bound endopeptidases (active against low molecular-weight synthetic peptide derivatives), two aminopeptidases and a pyrrolidone carboxylyl peptidase. Other strains of *Strep. cremoris* differ widely in the numbers and combinations of proteinases in their cell walls (Exterkate, 1976b) and one strain (AM1) produced a distinct cell-free proteinase (Exterkate, 1979b) whose stability was not Ca^{2+} dependent, unlike that of the cell-bound extracellular proteinases (Mills and Thomas, 1978; Exterkate, 1979b). The requirement for Ca^{2+} in the growth medium for accumulation of catalytically active proteinase in the cell wall of *Strep. cremoris*, and repression of enzyme synthesis by high concentrations of amino acids (Exterkate, 1979b), may explain the slow initial growth in milk of starters that have been propagated in non-milk media (low in Ca^{2+} compared with milk) supplemented with protein digests.

The possession of cell-bound extracellular proteinases by lactic streptococci has obvious advantages in that they produce peptides close to the cell and in proximity to their appropriate oligopeptide- and dipeptide-uptake systems whose exclusion and specificity properties appear to be similar to those of the *E. coli* systems (Law, 1977, 1978). Two of the inherently slow acid-producing strains of *Strep. cremoris* (AM2 and SK11) are deficient in dipeptide and/or oligopeptide uptake, though amino-acid uptake is normal. Intracellular hydrolysis of transported peptides to their constituent amino acids is mediated by a wide range of peptidases whose combined action ensures that all of the casein-derived peptide bonds can be cleaved (Law *et al.*, 1974; Mou *et al.*, 1975; Desmazeaud and Zevaco, 1977; Law, 1979a).

Novel cell-bound extracellular peptidases have recently been reported in *Strep. lactis* and *Strep. cremoris* (Law, 1979a). *Streptococcus lactis* appears to produce two peptidases which can be released from the cells by lysozyme. One has access to the cell exterior but the other is not detectable in whole-cell assays in the absence of peptide transport. This second enzyme is mercaptoethanol-sensitive and is also found in most strains of *Strep. cremoris*. Its absence from non-peptide-utilizing strains infers a role in peptide uptake. The interrelationship between the proteinases, peptidases and transport systems involved in protein utilization by the group N streptococci is shown in Figure 3.

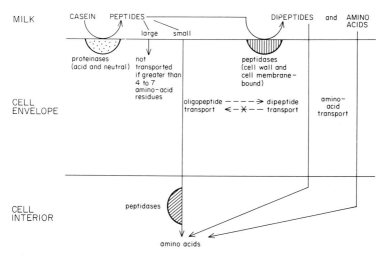

Fig. 3. Interrelationships among proteinases, peptidases and transport systems in group N streptococci.

Cultures of group N streptococci often contain small numbers of variants which grow only slowly in milk, yet grow normally in media containing amino acids or peptides. These are designated *prt*⁻ variants and one such strain has been characterized by the loss of an acid- and a neutral proteinase, present in the cell wall of the parent strain (Exterkate, 1976a). Efstathiou and McKay (1976) showed that loss of proteolytic activity from *Strep. lactis* was coincident with loss of plasmid DNA and there is now increasing evidence that all of the cell wall-bound extracellular proteinases are plasmid-coded in the starter streptococci.

Cell-bound extracellular proteinases and peptidases are not confined to group N streptococci. Argyle *et al.* (1976) showed that an EDTA-sensitive proteinase was released from *L. bulgaricus* by lysozyme treatment. Shankar (1977) showed that this organism had surface-bound proteinases which produced amino acids and peptides stimulatory for growth of *Strep. thermophilus*. The latter apparently used cell-bound and cell-free extracellular peptidases to hydrolyse lactobacillus-derived peptides. However, this work requires confirmation since evidence for the extracellular nature of the peptidases was not unequivocal. Intracellular peptidases in thermophilic starters have been described by Rabier and Desmazeaud (1973), Desmazeaud and Juge (1976) and El Soda *et al.* (1978); their function in protein and peptide utilization is presumably the same as that in mesophilic starters. Peptide utilization by lactobacilli is a well-established phenomenon and has been widely observed (e.g. Kihara and Snell, 1960; Leach and Snell, 1960), although the specificity of the uptake systems has not been the subject of definitive studies equivalent to those concerning the mesophilic starter streptococci.

B. Utilization of Lactose by Starters

The primary importance of fermentation of lactose to lactic acid by starters during cheesemaking has ensured that the process has been well studied and characterized in recent years, particularly in the case of the group N streptococci. The starters are facultative anaerobes, deriving their ATP from glycolysis, then regenerating reduced nicotinamide nucleotides (NADH) *via* (L +)-lactate dehydrogenase which operates in the direction of pyruvate reduction to lactic acid. Some of the properties of starter lactate dehydrogenase are used for taxonomic purposes (Garvie, 1980).

The first stage of lactose utilization involves its uptake. Earlier studies and comparisons with the phosphoenolpyruvate–phosphotransferase system in *Staphylococcus aureus* suggested that group N streptococci took up lactose as lactose phosphate, then hydrolysed it with phospho-β-D-galactosidase to glucose and galactose phosphate (McKay *et al.*, 1970; Molskness *et al.*, 1973; Postma and Roseman, 1976). However, the intermediate lactose 6-phosphate has only been found in *Staph. aureus*, and Thompson (1979) used studies on cells whose sugar transport was dissociated from glycolysis to show that both glucose and galactose moieties of lactose are phosphorylated by separate systems, both using phosphoenolpyruvate as the phosphoryl donor. After the disaccharide intermediate has been hydrolysed, glucose phosphate and galactose phosphate are metabolized to triose phosphate *via* the fructose phosphate and tagatose phosphate pathways, respectively (Thomas, 1979; Bissett and Anderson, 1974). Thereafter, the Embden–Meyerhof pathway operates to pyruvate. These stages of lactose utilization are summarized in Figure 4.

Some group N streptococci lack a functional lactose phosphotransferase or phospho-β-galactosidase (e.g. *Strep. lactis* 7962; Thompson and Thomas, 1977; Farrow and Garvie, 1979) and take up lactose as the free sugar. They possess only a β-galactosidase and use the Leloir pathway to

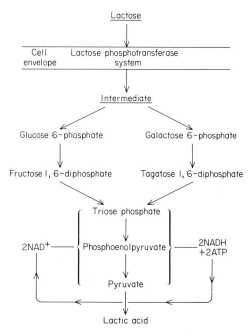

Fig. 4. Pathways for lactose utilization and lactic acid production in mesophilic starter streptococci.

convert glucose and galactose into glucose 6-phosphate, but their growth and acid production are insufficient to be of use in industrial fermentations. The phosphoenolpyruvate:lactose phosphotransferase and phospho-β-galactosidase in combination with fructose and tagatose-based pathways to triose phosphate are prerequisites for rapid homolactic lactose fermentation by mesophilic lactic streptococci. Although enzymes of the pentose phosphate pathway are present in these organisms (Oram and Reiter, 1966; Demko *et al.*, 1972), their specific activities are relatively low and they are thought to function in nucleotide synthesis rather than in energy metabolism (Law and Sharpe, 1978a).

Group N streptococci spontaneously produce mutants which are unable to utilize lactose, though they grow normally on glucose. They have been shown to lack two protein components of phosphoenolpyruvate:lactose phosphotransferase and phospho-β-galactosidase (McKay *et al.*, 1970; Cords and McKay, 1974) which may be coded on plasmid DNA (Efstathiou and McKay, 1976; Anderson and McKay, 1977).

Although lactose utilization by mesophilic starters has been studied intensively, much less is known about thermophilic species. Uptake by *Strep. thermophilus* appears to involve an energy-dependent inducible system (Reddy *et al.*, 1973; Somkuti and Steinburg, 1980), but there have been no reports of a high-affinity lactose phosphotransferase system similar to that of the mesophilic streptococci. Lactose is hydrolysed to galactose and glucose by β-galactosidase, but subsequent fermentation uses glucose preferentially, galactose being excreted (O'Leary and Woychick, 1976). Of the *Lactobacillus* species, *L. casei* only possesses phospho-β-galactosidase whereas *L. bulgaricus* and *L. lactis* have both β-galactosidase and phospho-β-galactosidase (Premi *et al.*, 1972). Part of the lactose-utilizing system of *L. casei* may be coded on plasmid DNA (Hofer, 1977).

VI. MANUFACTURE OF UNRIPENED CHEESE

Short shelf-life cheese is exemplified by cottage cheese whose properties are directly attributable to growth and metabolism of the starter culture introduced into the milk. Low-fat or skim milk is acidified with mixed-strain cultures of *Strep. cremoris, Strep. lactis* or *Strep. diacetylactis*. Details of manufacture are reviewed by Davis (1976) but some additional problems, relating to the bacteriological quality of milk for making cottage cheese, have been reported since then. For example, yields of curd from unstored milk are higher than from milk which has been stored raw for several days at 4°C before heat treatment and manufacture (Cousin and Marth, 1977a). Such storage methods are common in advanced dairy industries, and the lower yields are thought to be due to loss into the cheese whey of low molecular-weight casein-degradation products released by the action of heat-resistant extracellular proteinases of psychotrophic bacteria which dominate refrigerated milk microfloras (Law *et al.*, 1979a). Low molecular-weight nitrogenous compounds may also be the cause of starter stimulation which leads to unpredictable changes in cheesemaking times experienced with stored-market milks (Claydon and Koburger, 1961; Cousin and Marth, 1977a,b).

The basis of cottage cheese flavour is diacetyl, a byproduct of pyruvate metabolism, which should ideally be present at approximately 2 p.p.m. in the finished curd for manifestation of typical balanced flavour. Most of the diacetyl is produced from citrate in milk (Fig. 5). Citrate is an additional source of the pyruvate normally produced *via*

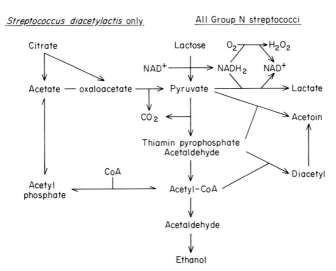

Fig. 5. Alternative pathways of carbohydrate metabolism in lactic streptococci.

glycolysis from lactose. In the course of its energy-yielding metabolism, the organism converts pyruvate into lactate *via* lactate dehydrogenase (see Section IIB, p. 168) in order to re-oxidize the NADH generated by glycolysis, although approximately 10% of this pyruvate is used for synthesis of new cell materials (Harvey and Collins, 1963). *Streptococcus lactis* and *Strep. cremoris*, which lack citrate permease, produce only enough pyruvate to meet these requirements and balance the need for NAD^+ regeneration; little is available for the expression of other pathways. In the *Strep. diacetylactis* strains used for manufacture of cottage cheese the pyruvate from citrate metabolism adds to the cellular pyruvate pool and allows formation of acetyl-CoA *via* pyruvate decarboxylase from the excess. Acetyl-CoA and the intermediate thiamin pyrophosphate–acetaldehyde complex react to form diacetyl and carbon dioxide (Cogan, 1976). In manufacture of cottage cheese, the carbon dioxide may cause floating curds if the gas becomes entrapped in the particles, but this problem can be avoided by making the cheese with *Strep. cremoris* or *Strep. lactis* as lactic-acid producers, then 'dressing' it with cream which has been cultured with *Strep. diacetylactis* as a separate source of diacetyl.

The most common flavour defect in cottage cheese is caused by reduction of diacetyl to flavourless acetoin by diacetyl reductase. Some starters produce this enzyme and must be avoided, but the water used to wash the curd is a possible source of diacetyl reductase-containing bacteria which is less easily controlled and may be especially troublesome since most of these contaminants are psychotrophs, able to grow and cause flavour loss even during refrigerated storage of the cheese. Some cottage cheese starters produce excessive amounts of acetaldehyde which can impart a harsh flavour to the product; ideally the ratio of diacetyl to acetaldehyde should be between 5:1 and 3:1 for balanced flavour (Lindsay et al., 1965). Factors governing the balance between pathways from pyruvate and acetyl-CoA to these flavour compounds are poorly understood but the problem of relatively high concentrations of acetaldehyde (approx. greater than 1 p.p.m.) is most easily overcome by including an acetaldehyde-utilizing strain of Leuconostoc in the starter culture (Keenan et al., 1966) to lower acetaldehyde concentrations.

VII. MANUFACTURE OF RIPENED CHEESE

A. Effect of Milk Microflora

The quality of the milk used to make cheeses is more critical when they are expected to remain in good condition during storage. The viable flora of raw milk is not directly important in most cheese since the milk is pasteurized before use. Thermoduric bacteria do not multiply in cheese (Law et al., 1973) (except heat-resistant strains of Lactobacillus), but the flora may influence the cheese indirectly through residual heat-resistant enzymes. The potential of raw milk bacteria to influence cheese quality has increased with the widespread adoption of bulk refrigerated storage and transport of raw milk in developed countries (e.g. Sinclair, 1979; Ellemor, 1979; Roberts, 1979). This system has eliminated obvious milk spoilage by mesophilic bacteria, but the microflora becomes dominated by Gram-negative psychotrophs, many of which produce heat-resistant lipases and proteinases. Characterization and enumeration of psychotrophs in raw milk have been reviewed recently (Cousins et al., 1977; Law et al., 1979a) and such detail is beyond the scope of this chapter. The effects of psychotroph enzymes on cheese maturation have been studied chiefly in Cheddar cheese in which excessive lipolysis due to lipases of Pseudomonas spp. produces early lipolytic rancidity if the bacterial numbers have risen above approximately 10^6 per ml in raw milk before pasteurization (Law et al., 1976d). Dumont et al. (1977) later showed that Camembert cheese made from stored milk could also become rancid if psychrotroph counts were increased to a value greater than 10^7 colony-forming units ml^{-1} by refrigerated storage. For a more detailed discussion of the relationship between the bacteriological quality of milk and maturation of cheese, the reader is referred to the review by Law (1979b).

B. Lipolysis and Proteolysis by Starters in Cheese

Starters are of fundamental importance to the manufacture of ripened, as well as unripened cheese but their influence on the textural and organoleptic characteristics of the final product varies considerably according to the type in question. Cheese maturation is a complex process involving breakdown of the curd by proteolysis, lipolysis and other enzyme-catalysed reactions to give rise to the flavour and texture changes typical of different varieties. The relationship between fat or protein breakdown and flavour development in some cheese types is not clearly understood but these processes are often used as indices of ripening and they probably contribute to development of background taste in all ripened cheeses by releasing free fatty acids and amino acids. Derivatives of these low molecular-weight degradation products may have definite and specific roles in some cheese types. Since the starter bacteria are common to all the varieties to be described in this section, a consideration of their lipolytic and proteolytic activities has general relevance. Specific data derived from studies of individual varieties will be discussed as they become relevant.

1. Lipolysis

Stadhouders and Veringer (1973) carried out the most definitive study of the lipolytic activity of mesophilic lactic streptococci and concluded that they hydrolysed mono- and diglycerides but that their activity against triglycerides was very weak. This explained discrepancies in previous results (Reiter et al., 1967, 1969) indicating that particular starters could cause different degrees of lipolysis in different cheeses. Presumably the extent of partial triglyceride breakdown by milk lipase (lipoprotein lipase) and heat-resistant psychotroph lipase governs the amount of substrate available to the starter lipase. Thus, the volatile fatty acids, thought to be important to the taste, rather than to the aroma of ripened cheese (Manning and Price, 1977), are likely to be produced in small amounts by the starter streptococci in most cheese varieties. Nakae and Elliot (1965) showed that volatile fatty acids are also produced by *Strep. diacetylactis* from amino acids *via* oxidative deamination, but it is not certain whether such pathways operate in cheese since Dulley and Grieve (1974) demonstrated that milk fat was necessary for production of volatile fatty acids other than acetate in Cheddar cheese.

Thermophilic starters are sufficiently lipolytic to contribute to fatty-acid formation in cheese. Paulsen *et al.* (1980) showed that, although volatile fatty acids in Swiss cheese were produced chiefly by carbohydrate metabolism of the lactic and propionic-acid bacteria, the same organisms also produced longer chain fatty acids during ripening by breaking down cheese fat.

2. Proteolysis

Much of the information about the proteolytic activity of starter streptococci has been derived from nutritional studies of the organisms (Thomas *et al.*, 1974; Exterkate, 1975; Law, 1979a), but the same proteinases and peptidases involved in releasing essential amino acids from casein during starter growth in milk probably continue to act in cheese after the starter bacteria have ceased growing (Law *et al.*, 1974). Little is known about the specificity of the proteinases, and their activity is masked initially in cheese by the high activity of the milk coagulant, chymosin. Studies with Cheddar and Gouda cheeses, made under bacteriologically controlled conditions with or without chymosin and/or starter bacteria, indicate that chymosin is responsible for proteolysis detectable by gel electrophoresis, gel filtration, and as pH 4.6-soluble nitrogenous compounds (i.e. large, medium and small peptides of molecular weights below 1400; O'Keefe *et al.*, 1976; Visser and De Groot-Mostert, 1977) (Fig. 6). Native milk proteinase appears to contribute only weakly, forming small peptides and amino acids. Proteinases produced by mesophilic starter bacteria may be significant in production of small, bitter-tasting (hydrophobic) peptides from casein, and from large, non-bitter peptides derived from rennet action on casein (Lowrie and Lawrence, 1972). These enzymes also have a secondary function in modifying the texture of ripening cheese by slowly degrading β-casein and α_{s1}1-casein which are responsible for forming the 'framework' of cheese structure (Zevaco and Desmazeaud, 1980).

Thermophilic starters are also proteolytic but their action in cheese has not been studied in great detail. Lactobacilli (particularly the thermobacteria) are more active than *Strep. thermophilus* (Tourneur, 1972).

Group N streptococci produce a range of peptidases which release free amino acids from the peptide products of starter proteinase and rennet. These peptidases have been described by several authors. Sorhaug and Sølberg (1973) detected five intracellular peptidase bands separable by starch-gel electrophoresis from cell-free extracts of *Strep. lactis*. Mou *et al.* (1975) showed that partially purified extracts of all group N streptococci contained aminopeptidases and a tripeptidase capable of complete hydrolysis of casein to amino acids (after prior proteolysis). Law (1979a) confirmed the presence of aminopeptidase, tripeptidase, and at least two different dipeptidase activities in intracellular extracts of *Strep. cremoris* and *lactis*, but also found cell wall-bound extracellular peptidase in *Strep.*

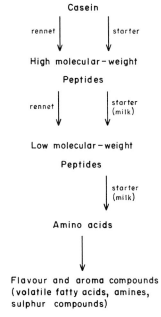

Fig. 6. Scheme for proteolysis of casein during cheese ripening, showing contributions made by rennet, milk proteinase and starter proteinase.

lactis. These enzymes were detectable in Cheddar-cheese extracts even after long periods of maturation (Law *et al.*, 1974; Cliffe and Law, 1979), and appeared to be released into the cheese matrix as soluble enzymes. Thermophilic starters also produce a range of peptidases which are thought to act in Swiss-type cheese during ripening, though the relative importance of these, and the peptidases of *Propionibacterium* spp., is not clearly understood (Section VIID, p. 183).

Lipolysis and proteolysis by starters therefore contribute to the supply of low molecular-weight compounds which are thought to provide the background to cheese flavour, and to provide flavour precursors, in the whole range of ripened cheese varieties in which they are used as acidifying agents. However, subsequent development of typical flavour in individual varieties follows distinct patterns according to the microflora which develops in or on the cheese during storage and ripening.

C. Cheddar Cheese

This variety is the most intensively studied, yet the least well understood of all cheeses. The enormous effort expended on determining which chemical compounds give its typical flavour is a reflection of its economic importance; it is manufactured on a large scale in more countries than is any other cheese variety.

The chemical basis of Cheddar cheese flavour and aroma is not known completely, though many workers have proposed roles for particular classes of compounds. For example, free fatty acids (Bills and Day, 1964; Ohren and Tuckey, 1969), volatile reduced sulphur compounds (Kristoffersen and Nelson, 1955; Lawrence, 1963; Manning and Robinson, 1973), methyl ketones (Harvey and Walker, 1960) and amines (Dahlberg and Kosikowski, 1948) have been suggested as specific flavour compounds, but convincing evidence of their involvement is lacking in most cases. The most complete evidence so far offered concerns the involvement of volatile sulphur-containing compounds, particularly methanethiol, in Cheddar flavour and aroma. For example, the concentration of this volatile in maturing cheeses and in experimental cheeses containing reducing agents correlates very closely with flavour intensity scores, irrespective of age (Manning *et al.*, 1976; Manning, 1979b). Also, the absence of methanethiol from cheese, and its removal from headspace volatiles, coincides with the absence of typical flavour and aroma, respectively (Manning, 1974; Manning and Price, 1977). Hydrogen sulphide appears to be a desirable component of aroma but its concentration is not critical for typical flavour (Manning, 1978), though very high concentrations give rise to 'sulphide' off-flavour (Manning, 1979a). However, despite increasing evidence for the involvement of sulphur-containing compounds, they cannot alone give balanced Cheddar flavour and aroma; many of the compounds cited by other authors could make as yet undefined contributions to the overall flavour profile. The following discussion will concern microbial processes that are most likely to generate flavour compounds during the manufacture and ripening of Cheddar cheese.

1. The Fermentation Stage

Production of volatile carbonyl compounds from the carbohydrate metabolism of starter streptococci in cottage cheese has already been considered, and the same pathways also contribute compounds such as diacetyl (thought to be important later in ripening; Manning and Robinson, 1973), acetate, ethanol and acetaldehyde to Cheddar curd. However, there is less emphasis on the use of *Strep. diacetylactis* strains since the high concentrations of diacetyl which they produce in cottage cheese are not necessary in Cheddar.

Alternative pathways of pyruvate metabolism (Fig. 5, p. 171) operate in other starters because, although they are regarded as homofermentative when lactose is used as an energy source, many strains have an inducible (by aeration) NADH oxidase (Bruhn and Collins, 1970; Anders *et al.*, 1970) so that they have less need to produce lactate from pyruvate as a means of re-oxidizing NADH. Not only can some pyruvate be diverted to other pathways leading to volatile carbonyls, but NAD^+ also has a direct function in re-oxidizing cofactors in oxidative decarboxylation of pyruvate to the acetyl-CoA required in these alternative pathways. The importance of these pathways was demonstrated by Czulak *et al.* (1974) who showed that pyruvate dehydrogenase in group N streptococci was inhibited in cheese made with milk containing high concentrations of polyunsaturated fat. The cheeses developed a bland flavour due to low concentrations of diacetyl and other carbonyls.

In addition to aeration, restricted growth may also cause normally homofermentative starters to produce alternatives to lactate from pyruvate. Thomas (1979) showed that lactose limitation in a chemostat causes *Strep. cremoris* and *Strep. lactis* to 'switch' fermentations and produce formate, acetate and ethanol, but not lactate. Similar results were obtained with *L. casei* (de Vries *et al.*, 1970), although in neither case was it clear whether these observations could be extrapolated to conditions in cheese vats. However, it is clear that during cheesemaking metabolizing starters produce a variety of volatile aroma compounds which, although they cannot combine at this early stage to give the typical aroma of mature cheese, may contribute, with other compounds, later in ripening. Starter metabolism also has important physicochemical consequences which affect the course of the next stage in production of mature cheese.

2. The Maturation Stage

Lactic acid produced by reduction of pyruvate lowers the pH value in the cheese to 5.2, inhibiting growth of many potential pathogens which may otherwise have spoiled the cheese during ripening (e.g. staphylococci, clostridia, coliforms). Also, lactic acid itself may be inhibitory and many strains produce antibiotics (Babel, 1977). The fermentation stage of cheesemaking lowers the redox potential in cheese to about -150 to -200 mV (Galesloot, 1960; Law *et al.*, 1976c). This has the effect of stabilizing reduced sulphur compounds and may explain in part their absence (Manning, 1974) from stored experimental cheese made without starter bacteria (acidified with δ-gluconic acid lactone) in which the redox potential remains at about $+300$ mV (Law *et al.*, 1976c). The low redox potential in cheese made with starter may actually be instrumental in production of hydrogen sulphide and methanethiol; Manning (1979b) demonstrated their non-enzymic formation in cheese made with δ-gluconic acid lactone, and artificially reduced with dithiothreitol or glutathione. He proposed a scheme in which hydrogen sulphide is produced from cystine/cysteine by an unknown mechanism under reducing conditions; the sulphide then reacts with free methionine (produced by starter peptidases) or with casein, to split C–S bonds and release methanethiol. A mechanism for the final reaction was not proposed but a study of products in the experimental system should allow a distinction to be made between reduction and substitution. Observations made in the course of normal cheese ripening show that hydrogen sulphide is produced before methanethiol, supporting the hypothetical sequence, and Law *et al.* (1976a) noted that hydrogen sulphide was even present in fresh cheese curd, presumably formed by heat treatment of the milk and during the curd-making stage. Additional hydrogen sulphide may be formed during maturation by some species of lactobacilli (Sharpe and Franklin, 1962). Although reduction of cheese was achieved artificially by Manning (1979b), the reducing conditions created by the starter bacteria in normal cheese may be sufficient to initiate methanethiol production in the same way, and such conditions would certainly maintain the compound in its reduced state. This would explain why only cheeses made with starters contain methanethiol, yet the starters themselves are unable to produce it enzymically from methionine (Law and Sharpe, 1978b).

Thus, although starter bacteria are essential for typical flavour formation in Cheddar cheese (Reiter *et al.*, 1967), their role may be largely indirect in that some of the key flavour compounds are produced non-enzymically, as a result of conditions created in the cheese by starters rather than by their cellular metabolism or their residual enzymes. This concept of the indirect contribution of starters to production of flavour compounds was suggested by Law *et al.* (1976c). It was consistent with the observation that there did not appear to be a relationship between concentrations of released intracellular starter enzymes and the flavour intensity of Cheddar cheese. If the enzymes themselves catalysed reactions leading directly to formation of flavour compounds, such a relationship would exist.

3. Off-Flavour Production by Starters

Lowrie and Lawrence (1972) and Law *et al.* (1979b) noted that cheeses in which starters had multiplied to relatively low numbers (about 10^8 colony-forming units g^{-1}; the 'slow' starters) during manufacture subsequently developed the best flavour, provided that acid production was sufficient to give the required moisture expulsion and final pH value. On the other hand, the 'fast' starters (reaching more than 10^9 colony-forming units g^{-1} curd) imparted bitter off-flavours to cheese and intensity scores for typical flavour were lower. Lowrie and Lawrence (1972) suggested that the cheeses made with 'fast' starters became bitter because the large numbers of starter cells contributed excessive amounts of proteinases which released bitter-tasting peptides from large, rennet-derived peptides. These workers (Lowrie and Lawrence, 1972; Lowrie *et al.*, 1974) were able to make non-bitter cheese with intense typical flavour from the same starters by restricting their multiplication (using higher curd scalding or by controlled bacterio-phage infection) so that viable counts in curd were similar to those in cheeses made with slow, non-bitter starters. Direct evidence for the involvement of cell-wall proteinases from starters had come from the recent observation that proteinase-negative variants of fast starters produced less bitterness in cheese than the parent strains, even when total starter-cell populations were similar (Mills and Thomas, 1980).

Bitter taste itself is caused by peptides with molecular weights ranging from around 1,000 to 12,000 which characteristically contain a high proportion of hydrophobic amino-acid residues (e.g. leucyl, prolyl, phenylalanyl; Harwalkar, 1972; Richardson and Creamer, 1973). Richardson and Creamer (1973) showed that bitter peptide in New Zealand Cheddar cheese originated from near the chymosin-sensitive bond of α_{s1}-casein, supporting the idea that peptides released by rennet were the substrate for starter endopeptidases. These hydrophobic bitter peptides tend to accumulate in cheese, probably because they are degraded only slowly by the peptidases of starter bacteria. Their rates of degradation are likely to be influenced by the peptidase activity in the starter cells, which is strain dependent (Sullivan *et al.*, 1973) and by the rate of release of peptidases from lysing cells.

Stadhouders and Hup (1975) considered that the explanation of bitterness formation in Cheddar cheese is not wide enough in scope to extend to the situation in Gouda cheese. They suggested that, at the relatively low cooking temperatures used for this cheese, the amount of rennet retained in the curd was of increased importance since it can produce bitter peptides. The starters not only influence the extent of bitterness through their numbers in curds, but also by the amounts and types of proteinases in their cell walls (Exterkate, 1975, 1976a,b), as well as their different peptidase specificities for bitter peptides. These factors were discussed in a brief review by Crawford (1977).

In addition to bitterness, other off-flavours are associated with the metabolism or enzymes of starters. For example, starters that utilize citrate or have induced NADH oxidase, and *Leuconostoc cremoris* strains with high alcohol dehydrogenase activity, are more likely to produce ethanol (among other alternatives to lactate) which, over long storage periods in cheese, reacts with butyric or hexanoic acids (lipolysis products) to form fruity-flavoured esters (Bills *et al.*, 1965). Other starter strains (*Strep. lactis* var *maltigenes*) can transaminate and decarboxylate leucine to 3-methylbutanal which is the basis of an unpleasant malt-like flavour defect in cheese (Sheldon *et al.*, 1971).

4. *Influence of Non-Starter Flora*

Cheddar cheeses made under controlled bacteriological conditions and containing only starter streptococci develop balanced, typical flavour, but Reiter *et al.* (1967) suggested that the non-starter flora (heat-resistant bacteria from raw milk and post-pasteurization contaminants from the creamery environment) also contribute to ripening. These workers isolated, from commercial creameries, groups of these adventitious bacteria consisting of group D streptococci, Gram-negative rods, staphylococci, micrococci, lactobacilli and pediococci. When these organisms (called 'reference floras') were added to experimental cheese, they developed flavours similar to those of the original commercial cheese from which the floras were derived. Despite the development of flavour defects (yeasty, fruity and sour, thought to be due to heterofermentative lactobacilli), intense typical Cheddar flavour developed more rapidly than in starter-only cheese. However, subsequent attempts to define the most important components of the reference floras have failed (Law *et al.*, 1976a) even though the inclusion of different reference-flora groups affected formation of several potential flavour compounds in different ways; concentrations of volatile fatty acids, higher fatty acids and hydrogen sulphide were higher in reference-flora cheeses than in starter-only cheese. The principal sources of volatile fatty acids were non-starter lactic-acid bacteria (acetate) and the lactobacilli and Gram-negative rods (butyrate), but concentrations of volatile fatty acid in cheese did not appear to be critical in relation to flavour. Manning and Price (1977) reported that volatile fatty acids did not influence Cheddar aroma at low pH values, but they did not rule out a role in the general background taste of cheese. Higher free fatty acids do not themselves appear to be directly beneficial to Cheddar flavour; indeed, at the high concentrations produced by heat-resistant psychotroph lipases, they can be detected organoleptically as rancid, soapy flavour defects (Law *et al.*, 1976d). Also, high concentrations of butyrate and hexanoate, combined with high concentrations of ethanol from some starters, lead to fruity flavours often associated with lipolytic rancidity.

Levels of hydrogen sulphide in reference-flora cheeses are highest in the presence of lactobacilli, some strains being able to desulphurylate cysteine (Sharpe and Franklin, 1962). Hydrogen sulphide is a desirable component of Cheddar aroma (Manning and Price, 1977) and it may be involved in the generation of methanethiol (see p. 178) but, unlike methanethiol, its concentration in cheese does not appear to be critical. In a later study, Law and Sharpe (1978b) showed that only Gram-negative rods and coryneform bacteria produced methanethiol from methionine, but the enzymes involved in the reaction were too unstable, and unsuited to conditions in cheese, to influence thiol concentrations during ripening. These data support the conclusions above that this important compound is produced by non-enzymic reactions, independently of the cheese microflora.

Observed variations in ketone levels in Cheddar cheese (Harvey and Walker, 1960; Law et al., 1976a) are difficult to explain, since the accepted mechanism for their production in other varieties involves β-oxidation of fatty acids by moulds. However, milk glycerides contain a small proportion of esterified β-keto acids which can be released by lipolysis (Kinsella, 1969). The free acids can undergo spontaneous decarboxylation to the corresponding ketones in stored dairy products because the activation energy of the reaction is low (Schwartz et al., 1966). Although it is doubtful whether ketones are important in Cheddar-cheese flavour, Manning (1979a) suggested that the concentration of pentanone in maturing cheese was a good index of its state of maturation.

D. Swiss Cheese

Emmental and Gruyère, the Swiss varieties, differ from Cheddar and most other hard cheeses in that their manufacture involves a high-temperature cooking stage (50°C) which only thermophilic starters can survive (Davis, 1976). Mixed strains of Strep. thermophilus and L. helveticus or L. lactis are used to produce the necessary lactic acid. Propionibacteria are also added to the cheesemilk but they do not contribute to the cheesemaking at this early stage. Lactic-acid production is greatest in newly pressed cheese, first by streptococci and later by lactobacilli

(Accolas *et al.*, 1978a). Homofermentative pathways to lactate predominate, though some acetate may be produced at this stage (Krett *et al.*, 1952). Most of the sugar in the cheese should have been fermented to lactate within 24 hours of pressing (Moquot, 1979). Unlike Cheddar manufacture, the Swiss-cheese process involves curd syneresis (water expulsion in response to acid formation) during the pressing stage rather than the vat stage, and development of the lactic microflora is very critical during this period. The cheese must be cooled carefully in the press so that the internal temperature gradient allows optimum lactose utilization by a stable *Strep. thermophilus* population and a growing *L. helveticus* population. It is most important that there should be a uniform lactate concentration (with no residual lactose) to allow the secondary propionic acid fermentation to produce the correct size range of carbon dioxide holes ('eyes'). The market value of the cheese is governed by eye formation as much as by flavour.

Fermentation of lactate and residual sugars by propionic acid bacteria is thus a vital stage in Swiss-cheese ripening, and follows initial lactic fermentation by the starters. Hettinga and Reinbold (1972) have reviewed the extensive literature on the propionate fermentation, which involves a complex double cycle *via* pyruvate, catalytic amounts of methylmalonyl-CoA, and an apparently unique biotin-dependent transcarboxylation reaction (Fig. 7). The products are carbon dioxide (which forms the 'eyes'), and propionic and acetic acids which, with lactic acid, dominate the flavour of Emmental cheese in particular. Production of volatile fatty acids by lipolysis is not significant in desirable Swiss-cheese flavour, but higher fatty acids, produced by the starter and propionic-acid bacteria, may contribute to typical flavour. Lipolytic rancidity develops if *Clostridium tyrobutyricum* (derived from feed silage) is present in the cheese (Langsrud and Reinbold, 1974). Protein breakdown is important in Swiss cheese not only for texture development but also for the production of small peptides and amino acids which apparently contribute 'nutty', 'brothy' and sweet tastes (Biede and Hammond, 1979). Chymosin contributes to gross proteolysis as it does in other cheese varieties. Peptides and amino acids are formed through the further action of peptidases released from cells of the starter streptococci and lactobacilli. El Soda *et al.* (1978) separated and identified an endopeptidase, dipeptidase, aminopeptidase and carboxypeptidase from *L. casei*, and Desmazeaud and Juge (1976) detected dipeptidase, aminopeptidase and proteinase activities in *Strep. thermo-*

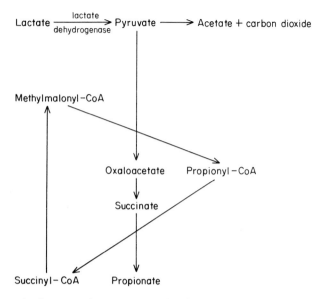

Fig. 7. Pathways leading to propionate, acetate and carbon dioxide production in *Propionibacterium shermanii.*

philus. Propionibacteria also produce intracellular peptidases which are released by autolysis into the cheese (Langsrud *et al.*, 1978). Extracts of *Proprionibacterium shermanii* contained 12 proteinase and 7 peptidase components, separable by electrophoresis, and earlier observations by Langsrud *et al.* (1977) had suggested that proline-releasing peptidases predominate in this organism. Proline is reported to have a sweet taste and its presence in particularly high concentrations in Emmental cheese (Virtanen and Kreula, 1948) led Langsrud and Reinbold (1973) to suggest that it was the basis of this aspect of Swiss-cheese flavour. *Propionibacterium shermanii* may also contribute to sweet flavour by producing dimethyl sulphide (Keenan and Bills, 1968; Dykstra *et al.*, 1971).

    Although the characteristic sweet, nutty flavour of Emmental cheese is regarded as manifesting a complete, balanced state of maturation, mountain varieties of Gruyère cheese develop additional flavour due to growth of a surface flora. This consists initially of lactate-utilizing yeasts whose growth raises the surface pH value towards 6 so that *Brevibacterium linens* can also grow as a secondary flora (Accolas *et al.*, 1978b). These orange-pigmented coryneform bacteria are probably stimulated by vitamins and amino acids produced by the yeasts (cf. Limburg cheese flora; Purko *et al.*, 1951a,b). The role of the surface microflora in flavour development is not clearly understood, but both yeasts and *B. linens* are proteolytic and the latter produces a range of peptidases (Torgersen and Sorhaug, 1978) of which the aminopeptidase is the most thoroughly studied (Foissy, 1978). The aroma and taste near the cheese surface is mildly putrid, partly due to formation of methanethiol by *B. linens* (Sharpe *et al.*, 1977; Dumont *et al.*, 1976a; Dumont and Adda, 1978). Esterification by the concerted action of *B. linens* and micrococci (Cuer *et al.*, 1979) of methanethiol and acetic or propionic acids produces thio-esters which have a cheesy aroma. However, the same authors found many more volatile compounds in Gruyère cheese (esters, alcohols, carbonyls and hydrocarbons) whose origins and contributions to flavour are impossible to assess at present. The surface flora probably adds to the antimicrobial action of the lactic flora (e.g. *Strep. thermophilus*; Pulsani *et al.*, 1979) by producing fatty acids which are inhibitory to *Clostridium botulinum* (Grecz *et al.*, 1959). This is a common surface contaminant and would normally be able to germinate and grow because the surface pH value is raised by yeast growth.

E. Mould-Ripened Cheese

Cheese varieties whose major flavour characteristics are attributable to products of mould growth can be divided in two broad categories, depending on whether the organisms grow inside, or on the surface of, the cheese.

1. Blue-Vein Cheese

The curds of these varieties are inoculated with spores of blue–green *Penicillium* moulds (*P. roqueforti*). When the freshly made cheeses are 'spiked' to admit air, the spores germinate and the resulting mould growth spreads throughout the inside of the cheese, aided by an open texture conferred by inclusion of gas-forming leuconostocs in the starter (Moreau, 1980). Mesophilic starters are used to form the curds so that the general considerations of proteolysis and lipolysis have already been outlined. The spores of *Penicillium* spp. germinate early in maturation, and mycelia are visible after 8–10 days. Maximum growth is reached by one to three months. The mycelia and spores are strongly lipolytic and proteolytic and their action in the cheese is superimposed on that of the starters. The extracellular acid and neutral proteinases of *P. roqueforti* have been shown to act synergistically with the starter proteinases and peptidases (Desmazeaud *et al.*, 1976; Gripon *et al.*, 1977). The mould enzymes produce peptides of molecular weights greater than 1000 from which the starters produce small peptides and amino acids. According to Kinsella and Hwang (1976), production of high concentrations of free amino acids buffers the cheese at pH 6.5, aiding other enzyme activities and providing background flavour.

Blue-vein cheeses normally have a surface slime containing moulds, yeasts and micrococci. Devoyod and Bret (1966) showed that these three groups became dominant, at the expense of coliforms and staphylococci, as the cheese was acidified by the lactic flora early in maturation. Changes occur in the composition of the yeast flora during ripening in response to surface salting (Galzin *et al.*, 1970). *Candida lipolytica* is a common species which is thought to contribute directly to flavour development by its lipolytic activity (Kinsella and Hwang, 1976). Salt is also assumed to be the source of micrococci in the surface slime and, although their role in ripening is not known, they tend to be more lipolytic than starters and may contribute through such action (Devoyod, 1969).

The overriding flavour notes in blue-vein cheese are those of fatty acids and methyl ketones, both classes of compounds being produced by *P. roqueforti*. The rate of release of fatty acids by the lipases of *Penicillium* sp. governs the rate of ketone formation (Kinsella and Hwang, 1976). Methyl ketones are formed from fatty acids, by partial oxidation *via* the fatty acid β-oxidation pathway (Fig. 8), which occurs in both spores and mycelium. The β-oxo-acyl-CoA, formed by enzymic dehydrogenation of fatty-acyl β-hydroxyacyl-CoA derivatives, is deacylated by thiohydrolase to form free β-keto acid. The decarboxylase which then catalyses formation of corresponding methyl ketones is most active in mycelia and β-oxolauric acid (yielding 2-undecanone) is its preferred substrate, but the preponderance of heptan-2-one in cheese is due to the preference of the preceding thiohydrolase for β-oxo-octanyl-CoA.

Flavour-simulation studies suggest that δ-lactones contribute to cheese flavour (Wong *et al.*, 1973), and Jolly and Kosikowski (1975) increased the quality of blue-cheese flavour by increasing the concentrations of δ-tetradecalactone and δ-dodecalactone. Two mechanisms for δ-lactone formation are possible. Traces of δ-hydroxy acid may be present in milk glycerides, and they can be released by lipases during

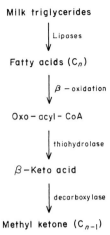

Fig. 8. Metabolic pathway for formation of methyl ketones by mycelium and spores of *Penicillium roqueforti; n* has a value of 6,8,10 or 12.

cheese ripening. The free acids may spontaneously undergo ring closure to form lactones, or they may be converted enzymically (Boldingh and Taylor, 1962). Alternatively, esterified δ-keto acids in milk glycerides may be released by lipases, reduced to hydroxy acids and then converted into lactones (Fig. 9). Boldingh and Taylor (1962) cited examples of yeasts and moulds which could produce lactones by this pathway.

In addition to flavour compounds, *P. roqueforti* can produce myco-

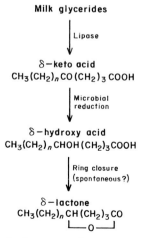

Fig. 9. Metabolic pathway for formation of δ-lactones in cheese; *n* has a value of 4 or 6.

toxins (e.g. PR toxin) and alkaloids (e.g. roquefortine). However, the PR toxin has never been found in Roquefort cheese and is only produced by specific strains under narrowly defined conditions (Moreau, 1980). Alkaloids, on the other hand, have been isolated from Stilton, Gorgonzola and Roquefort cheeses (Scott and Kennedy, 1976; Polonsky *et al.*, 1977; Ware *et al.*, 1980) and, although Moreau (1980) considered that the concentrations found in cheese were too small to be of concern to toxicologists, further studies on their action would be prudent.

2. Surface-Mould Cheese

These are soft-cheese varieties such as Camembert and Brie, prepared with mesophilic starters, but uncooked so that curd moisture is high. Flavour compounds are produced by proteolytic and lipolytic white moulds (*P. candidum*, *P. caseicolum* or *P. camemberti*) which grow on the cheese surface after it has been dry-salted and seeded with spores. Yeasts are also present in the surface flora, presumably as contaminants, and their presence at up to 10^9 colony-forming units g^{-1} suggests that they may contribute to the formation of flavour compounds in the cheese, though the relevant metabolic pathways are not known at present, beyond the familiar lipolytic and proteolytic routes. The yeasts of Camembert are dominated by species of *Saccharomyces*, *Kluyveromyces*, *Torulopsis* and *Candida* (Schmidt and Lenoir, 1980a,b); all known strains ferment lactose. Complex yeast microfloras are a common feature of many French cheese varieties and, although they have been well characterized (Schmidt and Lenoir, 1978), their role in maturation remains obscure.

Brie is characterized by a nutty, faintly ammoniacal flavour, reflecting breakdown of proteins, peptides and amino acids by the mould. No specific references exist to deamination of amino acids during maturation of surface-mould cheese, but it is generally assumed to be the source of ammonia.

Camembert is usually ripened longer than Brie and it sometimes develops a secondary surface flora of *B. linens*. This organism is probably the source of the putrid methanethiol aroma of well-ripened Camembert (the thiol is present at p.p.m. concentrations, rather than p.p.b. as in Cheddar cheese; Manning, 1979c). Condensation of methanethiol with formaldehyde yields bis(methylthio)methane, which is thought to be an important component of Camembert flavour (Dumont *et al.*,

1976b). Aromatic hydrocarbons of the alkyl and alkenyl benzene types have been found in Camembert and blue cheese. They are thought to give the cheeses a musty, earthy aroma and are produced by species of *Penicillium* by unknown pathways.

VIII. OCCURRENCE AND BEHAVIOUR OF PATHOGENS IN CHEESE

Both mesophilic and thermophilic starter strains normally produce enough lactic acid to suppress growth of the pathogens most commonly found in milk and curds (Reiter *et al.*, 1964a; Gilliland and Speck, 1969, 1972; Haines and Harmon, 1973; Dahiya and Speck, 1968; Pulusani *et al.*, 1979). Pathogens isolated from cheese include species of *Staphylococcus*, *Salmonella*, enteropathogenic strains of *E. coli*, species of *Shigella*, *Brucella*, group D streptococci, species of *Clostridium* and *Mycobacterium tuberculosis*. Most of these organisms can only multiply if starter failure or slowness results in production of a high pH value, low-acid cheese, though *Mycobacterium tuberculosis* can survive in normal cheese and is best avoided by primary eradication. Viral infections from cheese are rare but not unknown (Gresikova, 1974; encephalitis from sheep-milk cheese) and several types of virus have been shown to survive in cheese (coxsackie, echovirus, and foot-and-mouth disease virus; Blackwell, 1975; Barzu, 1970; Potter, 1970, 1973; Cliver, 1973). A comprehensive literature review of pathogens in milk and cheese was prepared recently under the auspices of the International Dairy Federation (Moquot *et al.*, 1980) to which the interested reader is referred for more details.

IX. ACCELERATED CHEESE RIPENING

Resistance to rapid decomposition, which is an inherent characteristic of the hard and semi-hard cheese varieties, represents their major attribute as basic foods. However, there is increasing interest in methods which may enable cheese manufacturers to produce mature flavoured products in a short time, thus yielding savings both of capital cost and of running costs for storage facilities. There is also a need for flavour intensification of bland low-fat variants of traditional cheeses. A rational, scientific approach to devising methods for accelerated cheese ripening is impeded by the lack of information available on the mechanisms involved in the formation of flavour compounds in maturing cheese. However, it is generally recognized that breakdown of protein and, in some cases, fat are good indices of the progress of maturation. The products of these processes may contribute to background taste as well as providing a pool of flavour precursors. This has led many investigators to attempt rapid ripening of hard cheese by addition of microbial proteinases and lipases of commercial origin (Kosikowski and Iwasaki, 1975; Nakanishi and Itoh, 1973, 1974; Malkki et al., 1977). In cheeses whose characteristic flavour depends on high concentrations of free fatty acids (e.g. Italian cheese), the use of added lipase is almost guaranteed a measure of success (Huang and Dooley, 1976). Acceleration of more subtle flavours, like that of Cheddar, requires careful selection of enzymes in order to avoid flavour imbalance from overproduction of bitter peptides by proteinases, and of rancidity inducing fatty acids by lipases. Recent results of Sood and Kosikowski (1979) have indicated that suitably screened food-grade microbial proteinases and lipases, added to Cheddar curd at very low concentrations (0.005 and 0.00005%, w/v, respectively), can halve ripening times normally required to produce balanced mature flavour, without off-flavour production. Other examples of accelerated cheese ripening with microbial enzymes have been reviewed by Law (1980).

As an alternative to direct addition of enzymes, accelerated ripening has been attempted by increasing the numbers of lactic-acid bacteria in cheese, since these bacteria are a source of proteinases and peptidases with a natural balance of specificities. However, methods must be found to prevent the additional cells from contributing to acid production of the normal starter inoculum. Some workers have used mutants which overproduce proteinases (Dilanyan et al., 1976) or do not ferment lactose, yet have normal proteinase-peptidase activity (lac⁻ mutants; Dulley et al., 1978). Starter cells can be solvent-treated to enhance the activity of membrane-bound proteinases by up to ten-fold (Exterkate, 1979a) and should be beneficial in accelerating ripening of Gouda cheese. Sublethal heat treatment at 59°C or 69°C substantially delays acid production by mixed-strain mesophilic and thermophilic starters, but only lowers their proteolytic activities by 10–30%. Petersson and Sjöstrom (1975) have used these 'heat-shocked' starters to speed up ripening of Swedish cheeses, and Thompson (1980) reported preliminary results of flavour-enhancement studies in low-fat Cheddar cheese. Lysozyme sensitization was used by Law *et al.* (1976c) as an experimental tool to study accelerated formation of amino acids in Cheddar but the method has no commercial application. Other acceleration methods involve artificial cheese systems and innovations in technology. As such, they are beyond the scope of this chapter and the interested reader is referred to the review by Law (1980).

REFERENCES

Accolas, J.-P. and Spillman, H. (1979a). *Journal of Applied Bacteriology* **47**, 135.

Accolas, J.-P. and Spillman, H. (1979b). *Journal of General Microbiology* **47**, 309.

Accolas, J.-P., Melcion, D. and Vassal, L. (1978a). *Proceedings of the 20th International Dairy Congress, Paris*, **E**, 762.

Accolas, J.-P., Veaux, M., Vassal, L. and Moquot, G. (1978b). *Le Lait* **58**, 118.

Anders, R.F., Hogg, D.M. and Jago, G.R. (1970). *Applied Microbiology* **19**, 608.

Anderson, D.G. and McKay, L.L. (1977). *Journal of Bacteriology* **129**, 367.

Argyle, P.J., Mathison, G.E. and Chandan, R.C. (1976). *Journal of Applied Bacteriology* **41**, 175.

Auclair, J. and Vassal, Y. (1963). *Journal of Dairy Research* **30**, 345.

Babel, F.J. (1977). *Journal of Dairy Science* **60**, 815.

Barzu, A. (1970). *Igiena* **19**, 345.

Bergère, J.L. and Hermier, J. (1968). *Le Lait* **48**, 13.

Biede, S.L. and Hammond, E.G. (1979). *Journal of Dairy Science* **62**, 227.

Bills, D.D. and Day, E.A. (1964). *Journal of Dairy Science* **47**, 733.

Bills, D.D., Morgan, M.E., Libbey, L.M. and Day, E.A. (1965). *Journal of Dairy Science* **48**, 1168.

Bissett, D.L. and Anderson, R.L. (1974). *Journal of Bacteriology* **117**, 318.

Blackwell, J.H. (1975). *Journal of Dairy Science* **58**, 784.

Boldingh, J. and Taylor, R.J. (1962). *Nature, London* **194**, 909.

Bradley, D.E. (1967). *Bacteriological Reviews* **31**, 230.

Briggs, C.A.E. (1952). *Journal of Dairy Research* **19**, 167.

Briggs, C.A.E. and Newland, L.G.M. (1952). *Journal of Dairy Research* **19**, 160.

Bruhn, J.C. and Collins, E.B. (1970). *Journal of Dairy Science* **53**, 867.

Buchanan, R.E. and Gibbons, N.E. eds. (1974). *Bergey's Manual of Determinative Bacteriology*, 8th edn. Williams and Wilkins, Baltimore.

Cavett, J.J. and Garvie, E.I. (1967). *Journal of Applied Bacteriology* **30**, 377.

Chapman, H.R. (1978). *Journal of the Society of Dairy Technology* **31**, 99.

Chopin, M.-C. (1980). *Journal of Dairy Research* **47**, 131.

Chopin, M.-C., Chopin, A. and Roux, C. (1976). *Applied and Environmental Microbiology* **32**, 741.

Claydon, T.J. and Koburger, J.A. (1961). *Applied Microbiology* **9**, 117.

Cliffe, A.J. and Law, B.A. (1979). *Journal of Applied Bacteriology* **47**, 65.

Cliver, D.O. (1973). *Journal of Dairy Science* **56**, 1329.

Cogan, T.M. (1976). *Dairy Industries International* **41**, 12.

Cogan, T.M. (1980). *Le Lait* **60**, 397.

Collins, E.B. (1956). *Virology* **2**, 261.

Cords, B.R. and McKay, L.L. (1974). *Journal of Bacteriology* **119**, 830.

Cousin, M.A. and Marth, E.H. (1977a). *Cultured Dairy Products Journal* **12**, 15.

Cousin, M.A. and Marth, E.H. (1977b). *Journal of Dairy Science* **60**, 1048.

Cousins, C.M., Sharpe, M.E. and Law, B.A. (1977). *Dairy Industries International* **42**, 12.

Cox, W.A. (1977). *Journal of the Society of Dairy Technology* **30**, 5.

Cox, W.A., Stanley, G. and Lewis, J.E. (1978). *In* 'Streptococci' (F.A. Skinner and L.B. Quesnel, eds.), Society of Applied Bacteriology Symposium Series, vol. 7, pp. 279–296. Academic Press, London.

Crawford, R.J.M. (1977). *International Dairy Federation Bulletin* **97**, 1.

Cuer, A., Dauphin, G., Kergomard, A., Dumont, J.-P. and Adda, J. (1979). *Agricultural and Biological Chemistry* **43**, 1783.

Czulak, J., Hammond, L.A. and Horwood, J.F. (1974). *Australian Journal of Dairy Technology* **29**, 124.

Czulak, J., Bant, D.J., Blyth, S.C. and Crace, J.B. (1979). *Dairy Industries International* **44**, 17.

Dahiya, R.S. and Speck, M.L. (1968). *Journal of Dairy Science* **51**, 1568.

Dahlberg, A.C. and Kosikowski, F.V. (1948). *Journal of Dairy Science* **31**, 305.

Davis, J.G. (1976). 'Cheese', vol. 3, pp. 487–920. Churchill Livingstone, London.

Demko, G.M., Blanton, S.J.B. and Benoit, R.E. (1972). *Journal of Bacteriology* **112**, 1335.

Desmazeaud, M.J. and Juge, M. (1976). *Le Lait* **56**, 241.

Desmazeaud, M.J. and Zevaco, C. (1977). *Annales de Biologie Animale Biochimie et Biophysique* **17**, 723.

Desmazeaud, M.J., Gripon, J.-C., Le Bars, D. and Bergere, J.L. (1976). *Le Lait* **56**, 379.

Devoyod, J.J. (1969). *Le Lait* **49**, 20.

Devoyod, J.J. and Bret, G. (1966). *Proceedings of the 17th International Dairy Congress*, **D**, 585.

De Vries, W., Kapteijn, W.M.C., Van Der Beck, E.G. and Stouthamer, A.H. (1970). *Journal of General Microbiology* **63**, 333.

Dilanyan, Z., Makarian, K. and Chuprina, D. (1976). *Milchwissenschaft* **31**, 217,

Douth, D.P. and Meanwell, L.J. (1953). *Proceedings of the 13th International Dairy Congress* **3**, 1114.

Dulley, J.R. and Grieve, P.A. (1974). *Australian Journal of Dairy Technology* **29**, 120.

Dulley, J.R., Brooks, D.E.J. and Grieve, P. (1978). *Proceedings of the 20th International Dairy Congress. Paris* **E**, 485.

Dumont, J.-P. and Adda, J. (1978). *Journal of Agricultural and Food Chemistry* **26**, 364.

Dumont, J.-P., Pradel, G., Roger, S. and Adda, J. (1976a). *Le Lait* **56**, 18.

Dumont, J.-P., Roger, S. and Adda, J. (1976b). *Le Lait* **56**, 595.

Dumont, J.P., Delespaul, G., Miguot, B. and Adda, J. (1977). *Le Lait* **57**, 619.

Dykstra, G.J., Drerup, D.L., Branen, A.L. and Keenan, T.W. (1971). *Journal of Dairy Science* **54**, 168.

Efstathiou, J.D. and McKay, L.L. (1976). *Applied and Environmental Microbiology* **32**, 38.

Efstathiou, J.D., McKay, L.L., Morris, H.A. and Zottola, E.A. (1975). *Journal of Milk and Food Technology* **38**, 44.

Ellemor, R. (1979). *Journal of the Society of Dairy Technology* **32**, 67.

El Soda, M., Desmazeaud, M.J. and Bergère, J.-L. (1978). *Journal of Dairy Research* **45**, 445.

Emmons, D.B. and Tuckey, S.L. (1967). 'Cottage Cheese and Other Cultured Milk Products'. Pfizer and Co. Inc., New York.

Erickson, R.J. (1980). *Dairy Industries International* **45**, 37.

Exterkate, F.A. (1975). *Netherlands Milk and Dairy Journal* **29**, 303.

Exterkate, F.A. (1976a). *Netherlands Milk and Dairy Journal* **30**, 3.

Exterkate, F.A. (1976b). *Netherlands Milk and Dairy Journal* **30**, 95.

Exterkate, F.A. (1979a). *Journal of Dairy Research* **46**, 473.

Exterkate, F.A. (1979b). *Archives of Microbiology* **120**, 247.

Farrow, J. and Garvie, E.I. (1979). *Journal of Dairy Research* **46**, 121.

Foissy, H. (1978). *Milchwissenschaft* **33**, 221.

Fryer, T.F. (1969). *Dairy Science Abstracts* **31**, 471.

Galesloot, Th.E. (1960). *Netherlands Milk and Dairy Journal* **14**, 176.

Galzin, M., Galzy, P. and Bret, G. (1970). *Le Lait* **50**, 1.

Garvie, E.I. (1980). *Microbiological Reviews* **44**, 106.

Gasson, M.J. and Davies, F.L. (1980). *Applied and Environmental Microbiology* **40**, 964.

Gilliland, S.E. and Speck, M.L. (1969). *Applied Microbiology* **17**, 797.

Gilliland, S.E. and Speck, M.L. (1972). *Journal of Milk and Food Technology* **35**, 307.

Gilliland, S.E. and Speck, M.L. (1974). *Journal of Milk and Food Technology* **37**, 107.

Grecz, N., Wagenaar, R.O. and Dack, G.M. (1959). *Applied Microbiology* **7**, 228.

Green, M.L. (1977). *Journal of Dairy Research* **44**, 159.

Gresikova, M. (1974). *Dairy Sciences Abstracts* **38**, 379.

Gripon, J.-C., Desmazeaud, M.J., Le Bars, D. and Bergère, J.-L. (1977). *Journal of Dairy Science* **60**, 1532.

Gulstrom, T.J., Pearce, L.E., Sandine, W.E. and Elliker, P.R. (1979). *Journal of Dairy Science* **62**, 208.

Haines, W.C. and Harmon, L.G. (1973). *Applied Microbiology* **25**, 436.

Harold, F.M., Pavlasova, E. and Baarda, J.R. (1970). *Biochimica et Biophysica Acta* **196**, 235.

Harvey, R.J. and Collins, E.B. (1963). *Journal of Bacteriology* **86**, 1301.

Harvey, R.J. and Walker, J.R.L. (1960). *Journal of Dairy Research* **27**, 335.

Harwalkar, V.R. (1972). *Journal of Dairy Science* **55**, 735.

Heap, H.A. and Lawrence, R.C. (1976). *New Zealand Journal of Dairy Science and Technology* **11**, 16.

Heap, H.A., Limsowtin, G.K.Y. and Lawrence, R.C. (1978). *New Zealand Journal of Dairy Science and Technology* **13**, 16.

Henderson, L.M. and Snell, E.E. (1948). *Journal of Biological Chemistry* **172**, 15.

Hettinga, D.H. and Reinbold, G.W. (1972). *Journal of Milk and Food Technology* **35**, 358.

Hofer, F. (1977). *Federation of European Microbiological Societies Microbiology Letters* **1**, 167.

Huang, H.T. and Dooley, J.G. (1976). *Biotechnology and Bioengineering* **18**, 909.

Jacobs, J., Klusens, M. and Pennings, A. (1972). *Tijdschrift voor Diergneeskunde* **97**, 548.

Jago, G.R. and Morrison, M. (1962). *Proceedings of the Society of Experimental Biology, New York* **111**, 585.

Jarvis, A.W. (1977). *New Zealand Journal of Dairy Science and Technology* **12**, 176.

Jespersen, N.J.T. (1979). *Journal of the Society of Dairy Technology* **34**, 190.

Jolly, R.C. and Kosikowski, F.V. (1975). *Journal of Agricultural and Food Chemistry* **23**, 1175.

Keen, A.R. (1972). *Journal of Dairy Research* **39**, 141.

Keenan, T.W. and Bills, D.D. (1968). *Journal of Dairy Science* **51**, 797.

Keenan, T.W., Lindsay, R.C. and Day, E.A. (1966). *Applied Microbiology* **14**, 802.

Kempler, G.M. and McKay, L.L. (1980). *Applied and Environmental Microbiology* **39**, 926.

Keogh, B.P. and Shimmin, P.D. (1969). *Journal of Dairy Research* **36**, 87.

Kihara, H. and Snell, E.E. (1960). *Journal of Biological Chemistry* **235**, 1409.

King, N.S. and Koburger, J.A. (1960). *Journal of Dairy Science* **53**, 403.

Kinsella, J.E. (1969). *Chemistry and Industry* **2**, 36.

Kinsella, J.E. and Hwang, D. (1976). *Biotechnology and Bioengineering* **18**, 927.

Kleter, G. (1976). *Netherlands Milk and Dairy Journal* **30**, 254.

Kosikowski, F.V. and Iwasaki, T. (1975). *Journal of Dairy Science* **58**, 963.

Krett, O.J., Le Boyer, A., Daume, H.E. and Stine, J.B. (1952). *Journal of Dairy Science* **35**, 480.

Kristoffersen, T. and Nelson, F.E. (1955). *Journal of Dairy Science* **38**, 1319.

Lancefield, R.C. (1933). *Journal of Experimental Medicine* **57**, 571.

Langsrud, T. and Reinbold, G.W. (1973). *Journal of Milk and Food Technology* **36**, 593.

Langsrud, T. and Reinbold, G.W. (1974). *Journal of Milk and Food Technology* **37**, 26.

Langsrud, T., Reinbold, G.W. and Hammond, E.G. (1977). *Journal of Dairy Science* **60**, 16.

Langsrud, T., Reinbold, G.W. and Hammond, E.G. (1978). *Journal of Dairy Science* **61**, 303.

Law, B.A. (1977). *Journal of Dairy Research* **44**, 309.

Law, B.A. (1978). *Journal of General Microbiology* **105**, 113.

Law, B.A. (1979a). *Journal of Applied Bacteriology* **46**, 455.

Law, B.A. (1979b). *Journal of Dairy Research* **46**, 573.

Law, B.A. (1980). *Dairy Industries International* **45**, 15.

Law, B.A. and Sharpe, M.E. (1978a). *In* 'Streptococci' (F.A. Skinner and L.B. Quesnel, eds.), Society of Applied Bacteriology Symposium Series, vol. 7, pp. 263–278. Academic Press, London.

Law, B.A. and Sharpe, M.E. (1978b). *Journal of Dairy Research* **45**, 267.

Law, B.A., Sharpe, M.E., Mabbitt, L.A. and Cole, C.B. (1973). *In* 'Sampling—Microbiological Monitoring of Environments' (R.G. Board and D.W. Lovelock, eds.), Society of Applied Bacteriology Technical Series vol. 7, pp. 6–7. Academic Press, London.

Law, B.A., Sharpe, M.E. and Reiter, B. (1974). *Journal of Dairy Research* **41**, 137.

Law, B.A., Castanon, M.J. and Sharpe, M.E. (1976a). *Journal of Dairy Research* **43**, 117.

Law, B.A., Sezgin, E. and Sharpe, M.E. (1976b). *Journal of Dairy Research* **43**, 291.

Law, B.A., Castanon, M.J. and Sharpe, M.E. (1976c). *Journal of Dairy Research* **43**, 301.

Law, B.A., Sharpe, M.E. and Chapman, H.R. (1976d). *Journal of Dairy Research* **43**, 459.

Law, B.A., Cousins, C.M., Sharpe. M.E. and Davies, F.L. (1979a). *In* 'Cold Tolerant Microbes in Spoilage and the Environment' (A.D. Russell and R. Fuller, eds.), Society of Applied Bacteriology Technical Series, vol. 13, pp. 137–151. Academic Press, London.

Law, B.A., Hosking, Z.D. and Chapman, H.R. (1979b). *Journal of the Society of Dairy Technology* **32**, 87.

Lawrence, R.C. (1963). *Journal of Dairy Research* **30**, 235.

Lawrence, R.C. and Pearce, L.E. (1972). *Dairy Industries* **37**, 73.

Lawrence, R.C. and Thomas, T.D. (1979). *In* '29th Symposium of the Society for General Microbiology' (A.T. Bull, D.C. Ellwood and C. Ratledge, eds.), pp. 187–219. Cambridge University Press, Cambridge.

Lawrence, R.C., Thomas, T.D. and Terzaghi, B.E. (1976). *Journal of Dairy Research* **43**, 141.

Lawrence, R.C., Heap, H.A., Limsowtin, G.K.Y. and Jarvis, A.W. (1978). *Journal of Dairy Science* **61**, 1181.

Leach, F.R. and Snell, E.E. (1960). *Journal of Biological Chemistry* **235**, 3523.

Lembke, J., Krusch, U., Lompe, A. and Teuber, M. (1980). *Zentralblatt für Bakteriologie, Originale C* **1**, 79.

Limsowtin, G.K.Y., Heap, H.A. and Lawrence, R.C. (1977). *New Zealand Journal of Dairy Science and Technology* **12**, 101.

Lindsay, R.C., Day, E.A. and Sandine, W.E. (1965). *Journal of Dairy Science* **48**, 863.

Lister, J. (1878). *Transactions of the Pathology Society* **29**, 425.

Lloyd, G.T. and Pont, E.G. (1973). *Journal of Dairy Research* **40**, 149.

Lode, A. and Tufto, G.N. (1973). *Meieriposten* **62**, 687.

Lowrie, R.J. (1974). *Applied Microbiology* **27**, 210.

Lowrie, R.J. and Lawrence, R.C. (1972). *New Zealand Journal of Dairy Science and Technology* **7**, 51.

Lowrie, R.J., Lawrence, R.C. and Peberdy, H.F. (1974). *New Zealand Journal of Dairy Science and Technology* **9**, 116.

Malkki, Y., Rouhianen, L., Mattsson, R. and Markkonen, P. (1977). United States Patent 4, 062, 730.

Manning, D.J. (1974). *Journal of Dairy Research* **41**, 81.

Manning, D.J. (1978). *Journal of Dairy Research* **45**, 479.

Manning, D.J. (1979a). *Journal of Dairy Research* **46**, 523.

Manning, D.J. (1979b). *Journal of Dairy Research* **46**, 531.

Manning, D.J. (1979c). *Journal of Dairy Research* **46**, 539.

Manning, D.J. and Price, J.C. (1977). *Journal of Dairy Research* **44**, 357.

Manning, D.J. and Robinson, H.M. (1973). *Journal of Dairy Research* **40**, 63.

Manning, D.J., Chapman, H.R. and Hosking, Z.D. (1976). *Journal of Dairy Research* **43**, 313.

McKay, L.L. and Baldwin, L.A. (1974). *Applied Microbiology* **28**, 342.

McKay, L.L., Miller, A., Sandine, W.E. and Elliker, P.R. (1970). *Journal of Bacteriology* **102**, 804.

Mills, O.E. and Thomas, T.D. (1978). *New Zealand Journal of Dairy Science and Technology* **13**, 209.

Mills, D.E. and Thomas, T.D. (1980). *New Zealand Journal of Dairy Science and Technology* **15**, 131.

Mitsuoka, T. (1968). *Proceedings of the 10th International Association of Microbiological Societies Symposium*, p. 89.

Mol, H. (1975). 'Antibiotics in Milk'. Balkema, Rotterdam.

Molskness, T.A., Lea, D.R., Sandine, W.E. and Elliker, P.R. (1973). *Applied Microbiology* **25**, 373.

Moquot, G. (1979). *Journal of Dairy Research* **46**, 133.

Moquot, G., Keogh, B.P., Waes, I., Elliot, J.A., Stadhouders, J., Schwab, H., Sharpe, M.E. and Terplan, G. (1980). *In* 'Behaviour of Pathogens in Cheese' *International Dairy Federation, Document 122*, pp. 4–23.

Moreau, C. (1980). *Le Lait* **60**, 254.

Mou, L., Sullivan, J.J. and Jago, G.R. (1975). *Journal of Dairy Research* **42**, 147.

Mullan, M.A. and Walker, A.L. (1979). *Dairy Industries International* **44**, 13.

Nakae, T. and Elliot, J.A. (1965). *Journal of Dairy Science* **48**, 287.

Nakanishi, T. and Itoh, M. (1973). *Japanese Journal of Dairy Science* **22**, A110.

Nakanishi, T. and Itoh, M. (1974). *Japanese Journal of Dairy Science* **23**, A121.

Ohren, J.A. and Tuckey, S.L. (1969). *Journal of Dairy Science* **52**, 598.

O'Keefe, R.B., Fox, P.F. and Daly, C. (1976). *Journal of Dairy Research* **43**, 97.

O'Leary, V.S. and Woychik, J.H. (1976). *Applied and Environmental Microbiology* **32**, 89.

Olson, H.C. and Qutub, A.H. (1970). *Cultured Dairy Products Journal* **5**, 12.

Oram, J.D. and Reiter, B. (1966). *Biochemical Journal* **100**, 373.

Orla-Jensen, S. (1919). 'The Lactic Acid Bacteria', 2nd edn., 1942. Copenhagen.

Osborne, R.J.W. (1977). *Journal of the Society of Dairy Technology* **30**, 40.

Osborne, R.J.W. and Brown, J.V. (1980). *Journal of Dairy Research* **47**, 141.

Park, C. and McKay, L.L. (1975). *Journal of Milk and Food Technology* **38**, 594.

Paulsen, P.V., Kowalewska, J., Hammond, E.G. and Glatz, B.A. (1980). *Journal of Dairy Science* **63**, 912.

Pearce, L.E. and Lowrie, R.J. (1974). *Proceedings of the 19th International Dairy Congress, New Delhi* **1E**, 410.

Peebles, M.M., Gilliland, S.E. and Speck, M.L. (1969). *Applied Microbiology* **17**, 805.

Petersson, H.E. and Sjöstrom, G. (1975). *Journal of Dairy Research* **42**, 313.

Polonsky, J., Merrion, M.A. and Scott, P.M. (1977). *Annales de Nutrition et Alimentation* **31**, 693.

Postma, P.W. and Roseman, S. (1976). *Biochimica et Biophysica Acta* **457**, 213.

Potter, N.N. (1970). *Journal of Dairy Science* **53**, 1358.

Potter, N.N. (1973). *Journal of Milk and Food Technology* **36**, 307.

Premi, L., Sandine, W.E. and Elliker, P.R. (1972). *Applied Microbiology* **24**, 51.

Pulusani, S.R., Rao, D.R. and Sunki, G.R. (1979). *Journal of Food Science* **44**, 575.

Purko, M., Nelson, W.O. and Wood, W.A. (1951a). *Journal of Dairy Science* **34**, 699.

Purko, M., Nelson, W.O. and Wood, W.A. (1951b). *Journal of Dairy Science* **34**, 874.

Rabier, D. and Desmazeaud, M.J. (1973). *Biochimie* **55**, 389.

Radich, P.C. (1968). Ph.D. Thesis: Oregon State University.

Reddy, M.S., Williams, F.D. and Reinbold, G.W. (1973). *Journal of Dairy Science* **56**, 634.

Reiter, B. (1949). *Nature, London* **164**, 667.

Reiter, B. (1973). *Journal of the Society of Dairy Technology* **26**, 3.

Reiter, B. and Oram, J.D. (1962). *Journal of Dairy Research* **29**, 63.

Reiter, B., Fewins, B.G., Fryer, T.F. and Sharpe, M.E. (1964a). *Journal of Dairy Research* **31**, 261.

Reiter, B., Pickering, A. and Oram, J.D. (1964b). *In* 'Microbial Inhibitors in Food' (N. Molin, ed.), p. 297. Almqvist and Wiksell, Stockholm.

Reiter, B., Fryer, T.F., Pickering, A., Chapman, H.R., Lawrence, R.C. and Sharpe, M.E. (1967). *Journal of Dairy Research* **34**, 257.

Reiter, B., Sorokin, Y., Pickering, A. and Hall, A.J. (1969). *Journal of Dairy Research* **36**, 65.

Richardson, B.C. and Creamer, L.K. (1973). *New Zealand Journal of Dairy Science and Technology* **8**, 46.

Roberts, A.W. (1979). *Journal of the Society of Dairy Technology* **32**, 24.

Sanders, M.E. and Klaenhammer, T.R. (1980). *Applied and Environmental Microbiology* **40**, 500.

Sandine, W.E. (1977). *Journal of Dairy Science* **60**, 822.

Sandine, W.E. (1979). *In* 'Lactic Starter Culture Technology', Pfizer Cheese Monographs, vol. 6, pp. 42–50. Pfizer, New York.

Sandine, W.E., Elliker, P.R. and Hays, H. (1962). *Canadian Journal of Microbiology* **8**, 161.

Sandine, W.E., Radich, P.C. and Elliker, P.R. (1972). *Journal of Milk and Food Technology* **35**, 176.

Schmidt, J.L. and Lenoir, J. (1978). *Le Lait* **58**, 355.

Schmidt, J.L. and Lenoir, J. (1980a). *Le Lait* **60**, 272.

Schmidt, J.L. and Lenoir, J. (1980b). *Le Lait* **60**, 343.

Schwartz, D.P., Parks, O.W. and Yoncoskie, R.A. (1966). *Journal of the American Oil Chemists' Society* **43**, 128.

Scott, P.M. and Kennedy, P.B. (1976). *Journal of Agricultural and Food Chemistry* **24**, 865.

Selby-Smith, J., Hillier, A.J., Lees, G.J. and Jago, G.R. (1975). *Journal of Dairy Research* **42**, 123.

Shankar, P.A. (1977). Ph.D. Thesis: University of Reading.

Shankar, P.A. and Davies, F.L. (1977). *Journal of the Society of Dairy Technology* **30**, 28.

Sharpe, M.E. (1978). *In* 'Streptococci' (F.A. Skinner and L.B. Quesnel, eds.), Society of Applied Bacteriology Symposium Series, vol. 7, pp. 386–391. Academic Press, London.

Sharpe, M.E. (1979). *Journal of the Society of Dairy Technology* **32**, 9.

Sharpe, M.E. (1980). *In* 'Prokaryotes' (M.P. Starr, H. Sholp, H.G. Truper, A. Balows and H.G. Schlegel, eds.), p. 2674. Springer Verlag, Berlin, Heidelberg, New York.

Sharpe, M.E. and Franklin, J.G. (1962). *Proceedings of the 8th International Congress of Microbiology* **B.11.3.**, 46.

Sharpe, M.E., Law, B.A., Phillips, B.A. and Pitcher, D.G. (1977). *Journal of General Microbiology* **101**, 345.

Shattock, P.M.F. and Mattick, A.T.R. (1943). *Journal of Hygiene* **43**, 173.

Sheldon, R.M., Lindsay, R.C., Libbey, L.M. and Morgan, M.E. (1971). *Applied Microbiology* **22**, 263.

Sherman, J.M. (1937). *Bacteriological Reviews* **1**, 3.

Sherman, J.M. (1938). *Journal of Bacteriology* **35**, 81.

Sinclair, G. (1979). *Journal of the Society of Dairy Technology* **32**, 63.

Sinha, R.P. (1980). *Applied and Environmental Microbiology* **40**, 326.

Somkuti, G.A. and Steinberg, H.D. (1980). *Journal of Dairy Science* **63** (Suppl. 1), 53.

Sood, V.K. and Kosikowski, F.V. (1979). *Journal of Dairy Science* **62**, 1865.

Sorhaug, T. and Sølberg, P. (1973). *Applied Microbiology* **25**, 388.

Sozzi, T. (1972). *Milchwissenschaft* **27**, 503.

Sozzi, T. and Maret, R. (1975). *Le Lait* **55**, 269.

Stadhouders, J. (1974). *Milchwissenschaft* **29**, 329.

Stadhouders, J. and Hassing, F. (1974). *Proceedings of the 19th International Dairy Congress*, New Delhi **1E**, 369.

Stadhouders, J. and Hup, G. (1975). *Netherlands Milk and Dairy Journal* **29**, 335.

Stadhouders, J. and Veringer, H.A. (1973). *Netherlands Milk and Dairy Journal* **27**, 77.

Stanley, G. (1977). *Journal of the Society of Dairy Technology* **30**, 36.

Sullivan, J.J., Mou, L., Rood, J.I. and Jago, G.R. (1973). *Australian Journal of Dairy Technology* **28**, 20.

Taniguchi, K., Nagao, A. and Tsugo, T. (1965). *Japanese Journal of Zootechnical Science* **36**, 376.

Terzaghi, B.E. and Sandine, W.E. (1975). *Applied Microbiology* **29**, 807.

Thomas, T.D. (1979). *New Zealand Journal of Dairy Science and Technology* **14**, 12.

Thomas, T.D., Jarvis, B.D.W. and Skipper, N.A. (1974). *Journal of Bacteriology* **118**, 329.

Thompson, J. (1979). *Journal of Bacteriology* **140**, 774.

Thompson, J. and Thomas, T.D. (1977). *Journal of Bacteriology* **130**, 583.

Thompson, M.P. (1980). *Dairy and Ice Cream Field* **163**, 74F.

Torgersen, H. and Sorhaug, T. (1978). *Proceedings of the 20th International Dairy Congress*, Paris **E**, 478.

Tourneur, C. (1972). *Le Lait* **52**, 149.

Tramer, J. (1964). *Journal of the Society of Dairy Technology* **17**, 95.

Virtanen, A.I. and Kreula, M. (1948). *Meijeriten Aikakauska* **10**, 13.

Visser, F.M.W. and De Groot-Mostert, H.E.A. (1977). *Netherlands Milk and Dairy Journal* **31**, 247.

Ware, G.W., Thorpe, C.W. and Pohland, A.E. (1980). *Journal of the Association of Official Analytical Chemists* **63**, 637.

Watanabe, K. and Takasue, S. (1972). *Journal of General Virology* **17**, 19.

Whitehead, H.R. and Cox, G.A. (1935). *New Zealand Journal of Science and Technology* **16**, 319.

Williamson, W.T., Tove, S.B. and Speck, M.L. (1964). *Journal of Bacteriology* **87,** 49.
Wilson, C.D. (1964). *Journal of the Society of Dairy Technology* **17,** 142.
Wong, N.P., Ellis, R., La Croix, D.E. and Alford, J.A. (1973). *Journal of Dairy Science* **56,** 636.
Zevaco, C. and Desmazeaud, M.J. (1980). *Journal of Dairy Science* **63,** 15.

6. Fermented Milks

EBENEZER R. VEDAMUTHU

Microlife Technics, Sarasota, Florida 33578, U.S.A.

I. Introduction

The origins of cultured dairy products probably date back to the dawn of civilization. Their history could likely be traced to domestication of cattle by man. Mention of cultured dairy products is found in some of the earliest writings of civilized man, for example, in the Bible and the *Vedas*, the sacred books of Hinduism. It is possible that serendipity, chance contamination, favourable environmental and climatic conditions together played a role in the development of many of the cultured dairy products we know of today. In all these instances, the chance circumstances were favourable for producing the right kind of changes in milk that resulted in a product that caught the fancy of the first man who tasted it! But these right combinations of factors were not always present when man tried to repeat these events. Probably this gave rise to 'starters'—more appropriately 'back-slops', that is, using a small portion of the previous lot to start a new batch—and rudiments of technology. Over a period of time, man learned to repeat the events and refine the steps to arrive consistently at an edible product (Vedamuthu, 1979a).

From the current scientific understanding of food preparation and preservation, we can ascribe two basic reasons for converting milk into cultured dairy products. The first and foremost reason is to preserve the high-quality nutrients present in the readily perishable fluid milk in the form of relatively more stable products. This is accomplished by accumulation of lactic acid (or the lowering of the pH value) during fermentation, partial depletion from fluid milk of an energy and carbon source for micro-organism namely lactose, removal of a portion of the moisture in milk in the form of whey which results in the lowering of the a_w value (water activity), and production of metabolites (lactic acid, acetic acid and hydrogen peroxide) and antibiotics (nisin, diplococcin, acidophilin, acidolin, bulgarican, other peptide factors and undefined low molecular-weight compounds) inhibitory to certain pathogenic and spoilage microflora. The second major reason for production of cultured dairy products is to provide variety in foods. Fermentation of milk by specific microflora and the accompanying manufacturing steps induce changes in body, texture, colour, flavour, and nutritive properties to create a wide variety of delectable foods. The aforementioned changes involve microbial modification of milk constituents especially lactose, citrate, milk proteins and lipids (Kilara and Shahani, 1978).

In Western Countries, the fermented milk products of major importance are ripened cream butter, cultured buttermilk, yogurt and sour cream. In Slavic countries, kefir, koumiss and bulgarian milk are popular. In Finland, vilia is consumed by large sections of the population. Over the years, there has been interest in Europe and North America in acidophilus milk and yogurt containing bifidobacteria. Leben and dahi consumed in the Middle East and India, respectively, are variations of yogurt. In Japan, interest in 'health foods' has resulted in the development of various fermented milks of which yakult is well known. These products will be discussed separately later in this review.

II. MICROBIAL MODIFICATION OF MILK CONSTITUENTS

A. Lactose Metabolism.

The single most important modification of a milk constituent that is common to all of these products is fermentation of the major milk sugar, lactose. Lactose is a disaccharide made up of D-galactose and D-glucose residues held together by a $1 \rightarrow 4$-β-linkage. The chemical designation for lactose is 4-O-β-D-galactopyranosyl-D-glucose pyranose. Approximately 4.75 to 4.9% of the weight of the lacteal secretion is contributed by lactose. There are other minor sugars and sugar phosphates in milk but these occur only in trace amounts. Hence, lactose is the major source of carbon and energy for micro-organisms growing in milk. This means that micro-organisms that possess the enzymic mechanisms for the use of lactose have a competitive growth advantage in milk (Vedamuthu, 1978a). In lactic-acid fermentations involved in cultured milk-product manufacture, a maximum of 1.5% acidity as lactic acid can be attained. For production of 1.5% lactic acid, only about 30% of the lactose content of milk is used. So, there is a great excess of unspent substrate available after fermentation is complete.

Utilization of any substrate by microbes involves its transport across the cell membrane and its breakdown through one of the major metabolic pathways in the cell system. Transportation of the substrate may require a permease or other energy-consuming reactions. Breakdown of the substrate within the cell generally consists of cleaving of specific linkages or other modifications required to route the substrate into specific metabolic pathways. The bond energy derived from degradation or oxidation of complex carbon substrates through these pathways is used for driving energy-requiring cellular synthetic mechanisms using short-chain carbon intermediates. Such energy- and short carbon-chain intermediate-yielding pathways are made up of a sequential series of enzymic reactions generally accommodating simple sugars or their derivatives, primarily the aldohexose glucose. Lactose must first be cleaved into its monosaccharide components, glucose and galactose or their derivatives. The galactose moiety is then converted into a glucose derivative and fed into the relevant energy-yielding pathway(s).

We have a good understanding of the mechanisms concerning lactose transportation across the cell membrane in bacteria used as starters in cultured dairy products. Lactic streptococci (McKay *et al.*, 1969), *Streptococcus thermophilus* (Reddy *et al.*, 1973), and the lactobacilli (Premi *et al.*, 1972) use the phosphoenolpyruvate–sugar phosphotransferase system (PTS) first described by Kundig *et al.* (1964) for translocation of lactose to the interior of the cell. In this system, lactose is phosphorylated at C-6 of the galactose moeity of the disaccharide and the phosphory-lated sugar is transferred through the various components in the system. The reactions involved are as follows (Anderson and McKay, 1977):

$$\text{PEP} + \text{HPr} \underset{\text{Mg}^{2+}}{\overset{\text{E}_{\text{I}}}{\rightleftharpoons}} \text{P-HPr} + \text{Pyruvate}$$

$$\text{F}_{\text{III}}\text{-Lac} + 2\text{P-HPr} \rightleftharpoons \text{F}_{\text{III}}\text{-Lac-2P} + 2\text{HPr}$$

$$\text{F}_{\text{III}}\text{-Lac-2P} + 2 \text{ Lactose} \underset{\text{Mg}^{2+}}{\overset{\text{E}_{\text{II}}\text{-Lac}}{\longrightarrow}} 2 \text{ Lactose-P} + \text{F}_{\text{III}}\text{-Lac}$$

In this scheme, HPr is a heat-stable soluble protein and E_I (enzyme 1) is also a soluble protein. These components are constitutively produced. Enzyme 2 (E_{II}-Lac) is a lactose-specific, membrane-bound component. Factor 3 (F_{III}-Lac) is lactose-specific and is soluble. Components E_{II}-Lac and F_{III}-Lac are inducible only in the presence of lactose or galactose. Recently, Thompson (1978) presented evidence to show that *in vivo* metabolic control of the lactose–PTS system is exercized by the activity of pyruvate kinase, an allosteric glycolytic enzyme. The feed-forward activation of pyruvate kinase in turn is controlled by the concentration of phosphorylated intermediates formed before the glyceraldehyde 3-phosphate dehydrogenase reaction in the glycolytic sequence. Using phenotypic variants (for the *lac* character) of *Strep. cremoris* B_1 and *Staphylococcus aureus* mutants deficient in specific lactose–PTS components in complementation experiments, Anderson and McKay (1977) showed that the genetic determinants for E_{II}-Lac and F_{III}-Lac in *Strep. cremoris* B_1 are located on a plasmid (an extrachromosomal genetic element) with a molecular weight of $36 \cdot 10^6$. The results presented by LeBlanc *et al.* (1979) also suggest a plasmid genetic determinant for the lactose–PTS system in *Strep. lactis* DR1251.

The phosphorylated lactose (phosphate group on C-6 of the galactose moeity) is cleaved by β-D-phosphogalactoside galactohydrolase into glucose and galactose 6-phosphate. Glucose is primarily metabolized by the hexose diphosphate pathway (HDP) although the presence of high concentrations of the key enzymes of the hexose monophosphate (HMP) pathway have been detected in lactic streptococci (Sandine and Elliker, 1970). Although there is sufficient evidence to show that lactose and glucose are translocated into lactic streptococcal cells via specific PTS systems (Thompson, 1978), the picture is not entirely clear with regard to galactose (LeBlanc et al., 1979; Thompson, 1978).

There is evidence to show that galactose 6-phosphate is metabolized through the D-tagatose 6-phosphate pathway. Free galactose in the presence of a lactose–PTS system, is transported through this system. In cells lacking a lactose–PTS system, free galactose may be transported by a specific permease and metabolized via the Leloir pathway (LeBlanc et al., 1979).

Knowledge of enzymic and genetic mechanisms for carbohydrate metabolism in lactic streptococci has exploded over the past decade. For detailed discussions of the enzymic mechanisms, there are excellent reviews by McKay et al. (1971), Lawrence et al. (1976) and Sandine and Elliker (1970). Genetic aspects are covered in papers by McKay (1978), McKay and Baldwin (1978) and Kempler and McKay (1979).

B. Citrate Metabolism

The second most important transformation of a milk component into an essential flavour compound in cultured dairy products involves citrate metabolism. Milk contains an average of 0.2% citrate. Among the ash components of milk, the citrate content exhibits the greatest seasonal fluctuation (a range of 0.07 to 0.4%). Citrate is converted into diacetyl by 'flavour bacteria' included in starter mixtures for cultured dairy products. The flavour bacteria are also called 'aroma bacteria' and citric acid fermenters. Diacetyl is the single most important and essential flavour compound that imparts the characteristic 'buttery', nut-meat like aroma and flavour of dairy products. In purified form, diacetyl is one of the most potent fragrances known to man. At high concentrations, the fragrance of diacetyl is very overpowering, coarse and unpleasant but at very low concentrations (about 3–5 μg g^{-1}), it has a fine, delicate odour. The important flavour bacteria are *Leuconostoc cremoris* and *Streptococcus lactis* subsp. *diacetylactis*. *Leuconostoc cremoris* growing in broth or milk consists of cocci arranged in very long chains. These bacteria grow poorly in milk and are relatively inert in this substrate. They produce very little, if any, acid from lactose in milk. To stimulate their growth in milk, addition of dextrose and yeast extract is necessary. Leuconostocs ferment milk citrate only when there is sufficient acid development. This requires associative growth with acid-producing lactic-acid bacteria. The need for acid development for active citrate fermentation is explained by the observation that the citrate permease of aroma bacteria is active only below pH 6.0 (Harvey and Collins, 1962). The normal pH value of milk is between 6.6 and 6.7. *Streptococcus lactis* subsp. *diacetylactis* on the other hand, grows well in milk, can ferment lactose and accumulate up to 0.4–0.65% lactic acid in 24 hours at 30°C. There is strain variation in the amount of acid produced in milk. Because of its lactose fermenting ability, in contrast to leuconostocs, *Strep. lactis* subsp. *diacetylactis* also can produce diacetyl from milk citrate in pure culture.

The literature contains conflicting reports on the biosynthetic pathway for diacetyl among aroma bacteria. The major area of disagreement among the various researchers relates to the role of α-acetolactate as an intermediate in formation of the diketone. Early work by DeMan (1956), Pette (1949) and Seitz *et al.* (1936a) showed that diacetyl is formed by oxidative decarboxylation of α-acetolactate. Later, Speckman and Collins (1968), using column chromatographic techniques capable of precisely separating diacetyl and its reduction product acetoin, showed that, while α-acetolactate is an intermediate in the formation of acetoin, it is not involved in synthesis of diacetyl. Speckman and Collins (1968) proposed that the dicarbonyl is formed as a result of the reaction between active acetaldehyde (acetaldehyde–thiamin pyrophosphate complex) and acetyl coenzyme-A. Later, they also reported that α-acetolactate only exists in a low enzyme-bound steady-state concentration within the cells of aroma bacteria (Collins and Speckman, 1974). Jonsson and Pettersson (1977) re-examined this question. They found that free extracellular α-acetolactate is found in cultures of *L. cremoris* and *Strep. lactis* subsp. *diacetylactis*, but accumulation of the compound occurs at a low oxidation–reduction potential that is not conducive to spontaneous oxidative decarboxylation of α-acetolactate to diacetyl. Therefore, these workers suggested that the biosynthetic mechanism for diacetyl proposed by Speckman and Collins (1968) is valid.

Harvey and Collins (1963) addressed the question of the physiological significance for aroma bacteria of the production of diacetyl and its reduction products from citrate. They suggested that diacetyl production is probably a detoxification mechanism to convert accumulated pyruvic acid into neutral C_4 compounds. Pyruvic acid is a key intermediate in carbon metabolism of lactic-acid bacteria. The central role of pyruvate in carbon metabolism of these bacteria is very well depicted in a recent paper by Coventry et al. (1978). In dissimilation of carbohydrates through the HMP and HDP pathways, pyruvic acid is the key intermediate in the final steps of partial oxidation of carbohydrates. Pyruvic acid formed through these pathways, however, does not accumulate, because regeneration of NAD^+ necessary for the oxidation of glyceraldehyde 3-phosphate earlier in the sequence of reactions, is coupled with reduction of pyruvate to lactic acid or other fermentation products. In lactic-acid bacteria that ferment both lactose and citrate, metabolism of both of these substrates contributes to the pyruvic acid pool. It is the pyruvate derived from citrate that would accumulate to toxic concentrations were it not converted into diacetyl and other neutral C_4 compounds. For a thorough survey of research done on the biosynthesis of diacetyl by starter bacteria, the papers by Seitz et al. (1963a), Collins (1972) and Jonsson and Pettersson (1977) are excellent.

Diacetyl synthesized by flavour bacteria in dairy products does not accumulate indefinitely. Once the concentration of diacetyl precursor, citrate, falls below a critical value, the diketone is rapidly converted into acetoin, a flavourless compound. Conversion of diacetyl into acetoin is catalysed by diacetyl reductase, an enzyme that is widely distributed among flavour bacteria and other contaminating psychrotrophic bacteria commonly found in dairy environs. Because the reduction of diacetyl to acetoin results in loss of desirable flavour in cultured dairy products, measures should be taken to obtain a high concentration of diacetyl and prevent or slow down the reductase-catalysed reaction. The simplest way to increase the amount of diacetyl synthesized is to boost the level of available substrate by fortification with citric acid or sodium citrate. As pointed out earlier, the citrate content of milk is limited and is subject to wide seasonal fluctuations. Fortification of milk therefore is justified to ensure biosynthesis of uniformly high concentrations of diacetyl. Fortification of milk with citrate also delays onset of the reductive phase of the diacetyl synthesis–destruction cycle. Loss of accumulated diacetyl can be controlled by the following methods. (a) Selecting for flavour bacteria with very low diacetyl reductase activity or for mutants lacking diacetyl reductase (Arora *et al.*, 1978). (b) Immediate chilling of the cultured product after the desirable degree of acidity and flavour is reached. Rapid cooling of the cultured product at this stage considerably slows down diacetyl destruction probably because enzyme activity is affected. Also, holding at refrigeration temperature results in a gradual increase in diacetyl concentration. The reason for this phenomenon is not understood (Pack *et al.*, 1968a). (c) Gentle agitation of the fermented product to incorporate air. (d) Hydrogen peroxide–catalase treatment of milk before culturing (Pack *et al.*, 1968b). Reduction of diacetyl to acetoin results in simultaneous regeneration of NAD^+. Aroma bacteria possess a NAD^+-oxidase system, whereby hydrogen from NADH can be directly transferred to oxygen (Bruhn and Collins, 1970). Oxygen introduced into the product during aeration caused by gentle agitation, or liberated as a result of cleavage of hydrogen peroxide by catalase, provides an alternative (competitive) mechanism for oxidation of NADH. This partially spares the diacetyl from acting as the hydrogen acceptor for regeneration of NAD^+. (e) Preventing contamination of the fermented product with psychrotrophic bacteria which possess very high diacetyl reductase activities (Seitz *et al.*, 1963b).

C. Miscellaneous Metabolic Products

1. Acetaldehyde

Among the miscellaneous metabolic products, acetaldehyde is an important flavour compound. In cultured creamery butter, cultured buttermilk and cultured sour cream, acetaldehyde is undesirable because it imparts a flavour defect referred to as 'green' or 'yogurt' flavour. Certain strains of *Strep. lactis* and a great majority of *Strep. lactis* subsp. *diacetylactis* strains produce relatively high concentrations of acetaldehyde (Harvey, 1960; Lindsay *et al.*, 1965). Such strains should be eliminated from starter mixtures for these products. On the other hand, *L. cremoris* which possesses relatively high concentrations of alcohol dehydrogenase scavenges acetaldehyde by converting it into ethanol (Keenan and Lindsay, 1966). In the manufacture of cultured dairy products in which acetaldehyde is undesirable, it is advisable to use *L. cremoris* rather than *Strep. lactis* subsp. *diacetylactis* for flavour production.

Acetaldehyde is, however, a very important flavour component in yogurt and related products. For excellent yogurt production selected strains of *Strep. thermophilus* and *L. bulgaricus* capable of good acetaldehyde production during symbiotic growth are necessary (Bottazzi *et al.*, 1973; Hamden *et al.*, 1971). Bottazzi and his coworkers (Bottazzi and Dellaglio, 1967; Bottazzi and Vescova, 1969) conducted an extensive survey of acetaldehyde production by several strains of *Strep. thermophilus* and *L. bulgaricus* in milk cultures.

Acetaldehyde is primarily derived from lactose, although other mechanisms for production of the carbonyl residue are found among lactic-acid bacteria. For example, metabolism of threonine and deoxyribonucleic acid by those bacteria also results in formation of acetaldehyde. For a comprehensive discussion of acetaldehyde metabolism and its role in cultured dairy products, the reader should refer to the reviews by Lees and Jago (1978a,b).

2. Other Carbonyl Compounds

Lactic streptococci produce a wide variety of neutral and acidic carbonyl compounds (Harper and Huber, 1961; Vedamuthu *et al.*, 1966) besides acetaldehyde. These compounds are primarily derived from carbohydrate although other mechanisms involving nitrogen-containing substrates are also known.

Among the carbonyl compounds, the acetone-producing potential of starter bacteria was evaluated by Keenan *et al.* (1967). One very important carbonyl compound that causes flavour problems is 3-methylbutanal. This compound imparts a 'malty' flavour to cultured milks and ripened cream butter. This aldehyde is produced by deamination followed by decarboxylation of leucine by certain strains of lactic-acid bacteria, especially *Strep. lactis* subsp. *maltaromicus* and *L. maltaromicus* (Sheldon *et al.*, 1971; Miller *et al.*, 1974). Differences in the ability of lactic-acid bacteria to form aldehydes from various amino acids were noted by McLeod and Morgan (1958). A summary of various reactions that can lead to formation of carbonyl compounds in cultured dairy products is given by Kilara and Shahani (1978).

3. Alcohol and Volatile Fatty Acids

Ethanol is produced in small amounts by aroma bacteria and in heterolactic fermentations. Bills and Day (1966) studied the alcohol dehydrogenase activities of lactic-acid bacteria and found that *Leuconostoc* spp. possess high concentrations of this enzyme.

In kefir and koumiss, alcohol is an important component characteristic of the product. The alcohol in kefir is produced by the yeasts *Saccharomyces kefir* and *Torula kefir*. The concentration of ethanol found in a good quality kefir is 1.0%. Torula yeasts are responsible for alcohol production in koumiss. The alcohol concentration in koumiss varies from 0.1 to 1.0% (Kosikowski, 1977).

Among the volatile fatty acids, formic, acetic and propionic acids are important in cultured dairy products. Sufficient amounts of these acids are produced by lactic starter bacteria (especially the flavour bacteria) to bring out the total flavour profile in cultured milks. Nakae and Elliot (1965) found that *Strep. lactis* subsp. *diacetylactis* strains produce more volatile fatty acids from casein hydrolysate than other lactic streptococci.

4. Carbon Dioxide

Carbon dioxide plays an essential role in the flavour impact of cultured buttermilk, kefir and koumiss. The gas is entrapped in the thickened milk and provides a lift, fizz, or effervescence to these cultured products. The flavour impact contributed by carbon dioxide is similar to the carbonation in soft drinks. A top-quality kefir or koumiss when agitated should foam and fizz and form a head of carbon dioxide gas as does beer.

Carbon dioxide is derived from lactose by heterolactic bacteria. Fermentation of citrate by aroma bacteria also yields considerable amounts of carbon dioxide in milk. Sandine *et al.* (1958) made an extensive study of high carbon dioxide-producing lactic streptococci, and devised a simple apparatus for testing the gas-producing potential of culture mixtures (Sandine *et al.*, 1957).

The minor metabolic products, although found only in small or even trace concentrations, may be important in maintaining a desirable flavour balance in cultured dairy products. Any shift in the flavour balance may result in the organoleptic perception of off-flavours. These aspects are covered in an excellent review by Sandine *et al.* (1972a). For a comprehensive survey of starter cultures and their role in dairy products, the reviews by Keogh (1976) and Sharpe (1979) should be consulted.

III. RIPENED CREAM BUTTER

Cultured creamery butter or ripened cream butter is the product obtained by churning cream that has been cultured with selected strains of lactic streptococci and leuconostocs. At present, very little ripened cream butter is produced in the U.S.A. The bulk of the butter manufactured in the U.S.A. is sweet cream butter, because ripened cream butter undergoes chemical deterioration faster than sweet cream butter. Hence, ripened cream butter has a shorter shelf-life. Chemical deterioration of the butter is brought about by the synergistic and catalytic effect of the combination of salt and increased acidity. In Europe and in parts of Canada, Australia and New Zealand, ripened cream butter is made in large quantities.

Ripened cream butter has a much better flavour than sweet cream butter because of the pleasant lactic-acid flavour and diacetyl derived from the ripened cream. In Europe, two types of starters are used for manufacturing ripened cream butter, namely B starters and B-D starters. The B starters consist of selected strains of *S. lactis* and/or *Strep. cremoris* mixed with high flavour-producing strains of *L. cremoris* (*Betacoccus* abbreviated as B). The B-D starters contain, in addition to acid-producing lactic streptococci, selected strains of *Leuconostoc cremoris* ('B' component) and *Strep. lactis* subsp. *diacetylactis* (*diacetylactis* abbreviated as D). In B-D starters, the mixture of leuconostoc and *Strep. lactis* subsp. *diacetylactis* probably ensures a good diacetyl content as well as a clean butter flavour without any harsh 'green-acetaldehyde notes' because the leuconostocs scavenge the acetaldehyde accumulated by the D component.

The major contribution of the culture to ripened cream butter is flavour. The mild pleasant lactic acid flavour is derived from the fermentation of lactose in the serum of the cream. Lowering of the pH value by lactic acid facilitates uptake of citrate by aroma bacteria. Relatively high concentrations of diacetyl are needed in the churned butter to obtain sufficiently high residual concentrations of flavour compound in the finished product. Diacetyl is water soluble and there is some loss of flavour during the drawing of buttermilk and washing of the butter granules.

Pasteurized cream usually cooled to 18.3°C is inoculated with starter and held for 5 to 6 hours until the pH value drops to 4.9. The cream is then rapidly chilled to 12.8°C and kept overnight. By the morning, the pH value has declined to 4.6. It is then cooled to 7.2°C and held until churning. Low-temperature ripening of the cream results in a slow uniform rate of acid development and helps to retain the fat globules in the cream in the proper physiochemical state for efficient churning (the ratio of solid fat to liquid fat approaching 1:1). Kosikowski (1977) mentioned a modified process where 1 to 3% by weight of an active butter culture in milk is added directly to sweet cream butter during working to obtain a flavour reminiscent of the ripened cream product. For further details on the manufacturing procedures and microbiology of ripened cream butter, the reader should consult Kosikowski's book on cheese and fermented milk foods (1977) and other standard texts (e.g. Foster *et al.*, 1957).

IV. CULTURED BUTTERMILK

Cultured buttermilk may be defined as "a skimmed or partially skimmed, pasteurized milk that has been fermented by a suitable mixture of lactic culture and aroma organisms". The product obtained is a drinkable viscous fluid containing characteristic pleasing aroma and flavour (Chandan *et al.*, 1969). The coagulum that is formed is an acid curd but the resulting product is not simply sour milk. It possesses the distinctive characteristics of a good butter starter.

Cultured buttermilk is very popular in the U.S.A. because it is nutritious and refreshing. It also is used in cooking and as an ingredient in pancakes, dinner rolls and biscuits, and other dishes as a substitute for sour cream because of its low contents of calories and cholesterol (milk fat) relative to sour cream. Consumption of cultured buttermilk is much higher in the south and south-eastern states than in other parts of the U.S.A., and there is regional variation in the viscosity, acidity and flavour contents to suit the tastes of the local populace (Richter, 1977). Chandan *et al.* (1969) reported that cultured buttermilk is also gaining market acceptance in Western Europe.

Cultured buttermilk is generally made from fresh skim-milk. The proposed standards for cultured buttermilk as specified by the United States Food and Drug Administration (United States Department of Health, Education and Welfare, Food and Drug Administration, 1977) states that cultured buttermilk should contain less than 0.5% milk fat and not less than 8.25% milk solids-not-fat and should have been cultured with lactic acid-producing bacteria. The dairy and non-dairy ingredients allowed are cream, milk, partially skimmed milk, skim-milk, concentrated skim-milk, non-fat dry milk, or other milk-derived ingredients to increase the non-fat solids content provided that the ratio of protein to total non-fat solids and the protein-efficiency ratio of the finished product remain unaltered. Other optional ingredients include aroma bacteria, salt, citric acid (or sodium citrate) up to a level of 0.15%, butterfat granules or flakes, stabilizers, nutritive carbohydrate sweetners, characterizing flavour ingredients, colouring that does not simulate the colour of milkfat or butterfat, and vitamins A and D (not less than 2000 International Units per 946.35 ml) dispersed in suitable 'carriers'.

The bacteria used in buttermilk fermentation are selected strains of *Strep. lactis* and/or *Strep. cremoris* in combination with *L. cremoris*. The acid-producing lactic streptococci are selected on the basis of the following criteria: (a) rapid acid production at 21–24°C, (b) production of clean, pleasant lactic-acid flavour with no trace of maltiness or 'green' acetaldehyde flavour, and, (c) compatibility with the aroma bacteria strain(s) included in the mixture.

Although the optimum temperature for growth of lactic streptococci is near 30°C, in the manufacture of cultured buttermilk the preferred incubation temperature is between 21°C and 24°C. This is because at 21–24°C a balanced growth of acid producers and aroma bacteria (*Leuconostoc* spp.) is obtained. Although the aroma bacteria do grow well at 30°C (Pack *et al.*, 1968a), at temperatures higher than 24°C, the acid producers are progressively favoured because they multiply at a much faster rate than aroma bacteria. Buttermilk made at temperatures higher than 24°C usually lacks the fine diacetyl flavour because of insufficient growth of aroma bacteria. Also, incubation between 21°C and 24°C permits overnight holding to fit convenient work schedules, and close monitoring at frequent intervals is not necessary. Also, at 21–24°C there is energy saving in maintenance of the desired incubation temperature and in rapid cooling to 4.4°C.

Because of high acetaldehyde production by *Strep. lactis* subsp.

diacetylactis, these bacteria are not generally included in buttermilk cultures. There are, however, certain strains of *Strep. lactis* subsp. *diacetylactis* that produce smaller amounts of acetaldehyde but greater amounts of diacetyl (desirable ratio of diacetyl:acetaldehyde is greater than 3:1 but less than 4.5:1) that can be used (Lindsay *et al.*, 1965).

In the manufacture of cultured buttermilk, skim-milk or low-fat milk is heat-treated at 82–87°C for 30 to 45 minutes, cooled to 21–24°C and inoculated with starter. The heat treatment of milk destroys undesirable pathogenic and spoilage microflora, eliminates any biological competition for the starter bacteria, destroys natural inhibitors in milk, and facilitates partial denaturation of whey proteins which makes them susceptible to precipitation and coagulation as the pH value drops during fermentation. Coprecipitation of whey proteins with the acid casein provides a good body and viscosity in the finished buttermilk. The length of incubation varies from 14 to 16 hours depending on the incubation temperature and the acidity desired in the final product. A top-quality buttermilk has a titratable acidity of 0.75 to 0.85% (expressed as lactic acid). The delicate flavour associated with top-quality cultured buttermilk is contributed primarily by lactic acid, traces of acetic and formic acids, alcohol, diacetyl and carbon dioxide. For a fine-flavoured buttermilk, the diacetyl content should preferably be between 2 and 4 mg per kg. To attain the proper acidity and flavour content, strict control over heat treatment of milk, preparation of starter, incubation conditions, and prompt cooling to 5°C should be exercized. Addition of 0.1% citric acid or sodium citrate helps to maintain a sufficient concentration of diacetyl. Preventing post-fermentation contamination with psychrotrophs (which have high diacetyl reductase activity) is also important.

Cultured buttermilk should have a thick, smooth and fairly viscous body. In certain markets, uniformly distributed lumps in the buttermilk are preferred. A slimy or ropy body indicates improper selection of starter strains or contamination with psychrotrophic bacteria. There should be very little separation of whey or 'wheying-off'. The consistency should be such that when poured, the product flows evenly, but it should not exhibit an uneven or runny quality. Good body characteristics can be obtained if tight control is exercised over the use of a proper milk-solids level (at least 9.0% solids-not-fat), heating sufficiently high over a definite length of time to obtain whey protein denaturation, incubation conditions, and rapid cooling to arrest excessive acid production which causes curd shrinkage and wheying-off, and gentle agitation to break the coagulum so that the fine strands of protein holding the moisture in the gel will not be disrupted. For a detailed discussion of cultured buttermilk manufacturing and merchandising, the reader should consult the publications of Emmons and Tuckey (1967), Kosikowski (1977), Richter (1977), Vedamuthu (1978b) and White (1978). A good discussion on cultured buttermilk also is found in the *Manual for Milk Plant Operators* (Milk Industry Foundation, 1957) and *Dairy Microbiology* by Foster *et al.* (1957).

V. SOUR CREAM

The Code of Federal Regulations of the Food and Drug Administration of the United States (United States Department of Health, Education and Welfare, Food and Drug Administration, 1978) defines sour cream thus: 'Sour cream results from the souring, by lactic acid-producing bacteria, of pasteurized cream. Sour cream contains not less than 18% milkfat'. When flavourings are used, the milkfat level should not be less than 14.4%. Sour cream is used as a topping on baked potatoes, on vegetables, salads, flavoured gelatin and fruits, as a base for making dips, and as an ingredient in bakery goods. Cultured sour cream is a companion product for cultured buttermilk because the starter bacteria and culturing conditions for the two products are similar.

Good quality sour cream has a viscous, smooth, silky body, with an absence of grainy texture and very little or no separation of whey, a pleasant tart lactic-acid taste and the delicate aroma of diacetyl. Over-stabilization of sour cream often affects uniform melting of the product when applied on warm baked potatoes, and should be avoided. The use of rennet to firm up body usually results in a grainy texture. To avoid this, other hydrocolloid stabilizers should be used.

Emmons and Tuckey (1967) listed the following factors as important in obtaining desirable body, texture, and flavour characteristics in cultured sour cream: amount of milkfat, proportions of solids-not-fat, heat treatment, homogenization temperature, pressure and number of stages used, incubation time and temperature, pH value developed at packaging, temperature of sour cream at filling, and method of packaging. To facilitate breaking up of clumps, an in-line screen is used between the incubation tank and the filler. After filling, the product should be held in the cooler (4.5°C) for at least 72 hours to obtain complete solidification of the milkfat. For further details on the production of sour cream and defects commonly encountered in this product, the publications of Kosikowski (1977) and Emmons and Tuckey (1967) are recommended.

VI. YOGURT

Among the various cultured dairy products, yogurt is unique. It is the only product in which acetaldehyde in relatively high concentrations is desired as an essential flavour component. Yogurt has gained tremendous popularity over the last 15 years in Europe, Canada and the United States of America. In the U.S.A., yogurt sales in 1955 were 17 million pounds. By 1976, the sales had boomed to 500 million pounds. The phenomenal increase in the consumption of yogurt is related to the addition of fruits, flavours, and sweeteners to plain yogurt. Although consumers in Western Europe, Canada and the U.S.A. were familiar with yogurt as a cultured product, the high acidity, tartness, and the 'green' acetaldehyde flavour of plain yogurt were not appealing to them. The introduction of sweetened flavoured yogurt transformed the image of the product, and its appeal as a nutritious quick snack gathered momentum with the use of attractive packaging.

According to a survey made in 1977, over 90% of the fresh yogurt marketed in the U.S.A. was flavoured. At present, yogurt is sold in different forms and is used for different applications in the U.S.A. Apart from plain yogurt, one can purchase flavoured yogurt without fruit (e.g. lemon or vanilla) yogurt with fruit purées, whole or sliced fruits mixed uniformly throughout (Swiss style), or filled in the bottom of the cup or on top of the yogurt to be mixed in by the consumer (Sundae style), soft or hard-frozen yogurt in various flavours, frozen yogurt sticks, fluid yogurt drinks, and fruit-topped pies filled with yogurt. Spray-dried yogurt products also are sold for making up baking mixes, soups, salad dressing and confectionary (Vedamuthu, 1979b).

Microbiologically, yogurt may be defined as the end product of a controlled fermentation of high-solids whole milk with a symbiotic mixture of *Strep. thermophilus* and *L. bulgaricus*. For the manufacture of high-grade yogurt, the ratio of *Strep. thermophilus* to *L. bulgaricus* should be close to 1:1. In the dairy industry, the *Strep. thermophilus* component is commonly referred to as coccus and *L. bulgaricus* as rods. The fermentation is carried out within a temperature range of 35°C to 45°C, and a median temperature of 40°C is generally preferred. During fermentation, the coccus and the rod work together and in succession to produce the characteristic tartness and the 'green' flavour of yogurt. Early in the incubation, *Strep. thermophilus* grows rapidly and dominates the fermentation. During this period the oxidation–reduction potential of the system is depressed. The rods grow rather slowly during the early fermentation but their weak proteolytic activity liberates sufficient amounts of peptides and amino acids to stimulate the coccus. Vigorous growth of *Strep. thermophilus* results in accumulation of moderate amounts of lactic acid, acetic acid, acetaldehyde, diacetyl and formic acid. When the pH value of the yogurt mix (initially between 6.3 to 6.5) drops below 5.5, the rapid growth of *Strep. thermophilus* is arrested and that of *L. bulgaricus* is favoured. Depletion of oxygen from the system and the availability of formate stimulate the rods. Until the pH value reaches 4.2, *Strep. thermophilus* grows very slowly. Below pH 4.2, the entire fermentation is dominated by the lactobacilli. The major portion of the lactic acid and the acetaldehyde necessary for the characteristic flavour of yogurt is contributed by the lactobacilli, but the second part of the fermentation is greatly aided by the initial metabolic activity of the coccus component. Yogurt should be cooled at a pH value of 4.2 to 4.3 to obtain a high-grade product.

The body of yogurt should possess a relatively high viscosity and should be firm and cohesive enough to be removed and eaten with a spoon. There should be very little wheying off during normal handling for mixing, cooling, pumping and packaging. Factors that affect the body characteristics of yogurt are the fat and solids-not-fat concentrations in the mix, stabilizers used, control over weighing and blending of ingredients in the mix, heat treatment of the mix, concentration of protein by new processes such as ultrafiltration, concentration of Ca^{2+} and Mg^{2+} if retentate from ultrafiltration is used, starter cultures used (especially strains that produce capsular slime in milk). Also important are standardization of the incubation conditions (temperature, time and length), accuracy of pH value and/or titratable acidity measurements, care exercized to ensure gentle handling during post-incubation operations, and the sugar concentration in fruit preserves used in flavoured yogurts.

Top-quality yogurt should have a smooth, rich texture free from lumps, granules or graininess. There should be no gas pockets, fissures or gassy effervescence. Several explanations for grainy texture are found in the literature. If the coagulum becomes excessively firm before stirring, the finished yogurt tends to get grainy. Also, any disturbance of the yogurt mix just before the gelling stage gives rise to a coarse texture. The use of rennet to obtain a good body invariably leads to graininess. Sometimes, inadequate mixing of the powdered ingredients causes a granular feeling in the mouth. The butterfat content in the mix and its homogenization contribute to the smoothness and richness of yogurt.

Gas pockets and fissures in yogurt are caused by trapped carbon dioxide and (or) hydrogen produced by contaminant flora. Coliform bacteria, spore-forming *Bacillus* spp. and yeasts are usually involved. Contamination of the starter with carbon dioxide-producing aroma bacteria also may cause defects. Slime-producing strains of yogurt bacteria are useful in obtaining a good body, but excessive sliminess would result in a mouth-feel similar to the white of raw eggs. Such strains should be avoided.

The typical yogurt flavour can only be detected in plain yogurt. Because yogurt is a product of bacterial symbiosis, the flavour depends on the proportion of rods and cocci and the combined metabolic activity. The distinctive flavour of yogurt is contributed by lactic acid, which is odourless, and by trace amounts of acetaladehyde, diacetyl and acetic acid, which are volatile and have strong odours. Milk components and other ingredients in the mix also play a role in providing a background flavour. Off-tastes and odours are usually by-products of faulty fermentation, provided the ingredients used in the mix were of good quality. To obtain good flavoured yogurt, the starter strains should be selected with a view to obtaining good symbiosis (Moon and Reinbold, 1976), the conditions of culturing should be adjusted to obtain near equal proportions of cocci and rods, and the fermentation mixture should be promptly and rapidly cooled when the pH value reaches 4.2–4.3. If the pH value falls below 4.2, in large tanks that require a long time to cool down to $10°C$, the desired coccus to rod ratio and the flavour balance are lost. Then, the product obtained will be too sour and coarse. A high concentration of sucrose in the mix also affects fermentation because *Strep. thermophilus* is generally inhibited by sucrose concentrations above 4%.

The flavour impact of fruit or flavoured yogurt is mainly dependent on the quality of the fruit mixture or flavouring that has been added.

Proper selection of these materials with regard to flavour, colour, and microbial contamination is necessary. Colour stability of natural pigments in fruits and berries should be experimentally established under ranges of pH and Eh values encountered in yogurt systems before selection of fruit and fruit juices are made.

Yogurt is a high-acid food. Hence, spoilage of yogurt is caused by aciduric microflora, generally yeasts and moulds. Fungal contamination of yogurt could come from filling equipment, air in the packaging room, fruits and syrups, and contaminated lids and cups. Fungal problems can be controlled by proper sterilization of filling equipment, careful storage of cups and lids, installation of filtered, laminar air-flow facilities in the filling room, periodic fumigation of the filling room with formaldehyde, use of sorbate in the product, and by careful selection of fruits, fruit preserves and syrups used for making flavoured yogurt. For detailed discussion of the manufacture and microbiology of yogurt, the publications of Rasic and Kurmann (1978), Tamime and Greig (1979), Robinson and Tamime (1975, 1976), Kroger (1973) and Humphreys and Plunkett (1966) should be consulted.

VII. KEFIR

Traditional kefir is an historic and old product from the Caucasus Mountain region in Russia. In the Soviet Union, *per capita* annual consumption of this product is over 2 kg. The unique feature of kefir is the use of kefir grains as a starter to make a batch of the product. The grains can be re-used several times if proper sanitation is observed in recovering, drying, and storing the grains from batch to batch.

Kefir grains are gelatinous white or cream-coloured irregular granules varying in size from that of a wheat grain to a walnut. The granules are largely made up of a polysaccharide called kefiran, although some denatured milk protein may be associated with the polysaccharide matrix. The granules are insoluble in water and swell up when soaked in water to form a slimy, jelly-like product. Within the folds or involutions of the granules, bacteria and yeasts that form the characteristic flora of kefir are found. There appears to be a symbiotic relationship between the yeasts and the bacteria in this specific ecological niche.

The flora of traditional kefir consists of two species of yeasts, *Sacch. kefir* and *Torula kefir*, and certain lactobacilli named *Lactobacillus caucasicus*.

Lactic streptococci and *Leuconostoc* spp. are also associated with kefir granules. Other micro-organisms that have been isolated from kefir granules have been identified as coliforms, micrococci and spore-forming bacilli, but these bacteria are undesirable and may be contaminants. Kefir is made by adding kefir grains to cow's milk that has been cooled to 23–25°C after heat treatment at 85°C for 30 minutes. The inoculated milk is incubated at 23°C overnight. By morning, a smooth soft curd forms which, when agitated, foams and fizzes like beer. The major end-products of the fermentation are lactic acid (about 0.8%), ethanol (about 1.0%) and carbon dioxide. Other minor components are traces of acetaldehyde, diacetyl and acetone. The flavour of kefir may be described as mildly alcoholic, yeasty-sour with a tangy effervescence.

When kefir curd is agitated, the kefir grains are carried upwards to the surface of the curd by escaping carbon dioxide bubbles. The grains are strained out, washed in clean cold water, and stored in cold water at 4°C. The grains also may be dried in a warm oven and stored in foil pouches. Grains stored in water are active for 8–10 days. Properly dried grains are active for 12–18 months. Dried grains should be transferred at least three times in milk before they are fully activated. Currently, freeze-dried preparations also are available. For further information on kefir, the reader should refer to Keogh (1976), Kosikowski (1977) and Vaissier (1977, 1978a,b).

VIII. KOUMISS

Koumiss is similar to kefir, but is made from mare's milk. The product originated in the Asiatic Steppes of Soviet Russia and derives its name from a tribe called Kumanes, who lived along the river Kumane. Mare's milk does not coagulate at the isoelectric point of casein and hence koumiss, which contains about 0.7–1.8% lactic acid and 1–2.5% ethanol, is not a curdled product. Finished koumiss is a greyish-white drink. The micro-organisms involved in koumiss fermentation are *L. bulgaricus* and lactose-fermenting yeasts. The major end-products are lactic acid, ethanol and carbon dioxide. Koumiss also fizzes and effervesces like kefir. In earlier days, if the fermentation slowed down, pieces of horse flesh were added to accelerate the fermentation. Koumiss is considered to be a therapeutic drink.

In the modern manufacture of koumiss in Russia, cow's milk is used.

The starter used consists of *L. bulgaricus*, *L. acidophilus* and *Saccharomyces lactis*. For more details, the reader should refer to the publications of Kosikowski (1977), Chandan *et al.* (1969) and Keogh (1976).

IX. BULGARIAN MILK

Bulgarian milk, as the name suggests, originated in Bulgaria. It is an extremely high acid and acrid product. It is made from whole milk. The milk is heated at 85°C for 30 minutes, cooled to 37–38°C and inoculated with 2% milk starter made with *L. bulgaricus*. The milk is incubated until the acidity reaches at least 1.4% and is then cooled to 7°C. In some cases, the acidity of the product may be as high as 4%.

X. ACIDOPHILUS MILK

Acidophilus milk or 'reform yogurt' is the product obtained by fermenting milk with an authentic culture of *L. acidophilus*. The 8th edition of the *Bergey's Manual of Determinative Bacteriology* (Buchanan and Gibbons, 1974) details the entire biochemical and genetic character-istics of this micro-organism. One of the unique features of this species is its ability to survive the severe conditions found in the intestinal tract of man, animals and birds. Some research workers have proposed that the ability to initiate growth in, or form colonies on, media containing bile salts should be used as a distinguishing characteristic for *L. acidophilus*. This bacterium forms part of the normal intestinal flora, but there appears to be some variation between strains isolated from different sources. In a survey designed to establish if there are differences among *L. acidophilus* strains isolated from faeces of humans, pigs and chickens, Gilliland *et al.* (1975) found that the deoxyribonucleic acid of the human isolates generally had a lower guanine plus cytosine ratio than did the pig and chicken biotypes.

The therapeutic value of milk containing viable *L. acidophilus* in controlling intestinal disorders has been known for a long time. The precise mode of action(s) of these bacteria in maintaining health in the gut, however, has yet to be elucidated. For extensive discussion of these aspects, the publications of Speck (1976, 1978) and Sandine *et al.* (1972b) are recommended. The major prerequisites for a viable bacterial preparation to be useful in intestinal therapy may be summarized as follows. (a) The micro-organism should be a part of the normal flora of the gut. *Lactobacillus acidophilus* can be consistently isolated from faeces of normal, healthy individuals. (b) The micro-organism should be capable of withstanding the ecological and environmental conditions in the gut. When larger numbers of *L. acidophilus* are ingested, these bacteria being aciduric, survive in sufficient numbers when exposed to the low pH value of the stomach. The bacteria also thrive under the conditions of the low redox potential in the gut. Additionally, these bacteria can survive and grow in the presence of concentrations of bile salts found in the enteric system. (c) The micro-organism should be capable of implanting in the intestines. Among the various biotypes of *L. acidophilus*, there appears to be host specificity in implantation. Although experimental evidence in the human system is currently lacking, it is suggested that, for effective intestinal therapy, it is necessary to use authentic *L. acidophilus* strains isolated from normal healthy humans.

Traditional acidophilus milk is an extremely sour product. The finished product contains very little, if any, metabolic by-products other than lactic acid. Acidophilus milk is made from partially skimmed milk. To stimulate growth of *L. acidophilus*, the milk needs to be heated at 120°C for 15 minutes which denatures and releases some peptides from proteins in the milk. After heat treatment, the milk is cooled to 37–38°C, inoculated with 5% milk starter, and incubated at the same temperature for 18 to 24 hours. It is cooled to 7°C after the acidity reaches 1.0% and bottled.

Because of the extreme sourness and the lack of other balancing flavours in acidophilus milk, it has not been a popular drink. Because there are sufficient data to show that ingestion of viable *L. acidophilus* may be beneficial in maintaining the health of the enteric system, attempts have been made to devise ways of making milk containing large numbers of viable *L. acidophilus* palatable (Myers, 1931; Duggan *et al.*, 1959). These studies recommended addition of a concentrated cell suspension of *L. acidophilus* to pasteurized milk in the cold (5°C), which is then mixed thoroughly, bottled and distributed under refrigerated storage. The *L. acidophilus* cells added to the milk retained their viability for 10–15 days, but did not multiply at the refrigeration temperatures. Also, their metabolic activity was arrested at 5°C. So, the milk retained the fresh 'sweet' taste (Foster *et al.*, 1957). This concept was recently revived and commercially exploited in the U.S.A. under the label

'Sweet Acidophilus Milk'. Frozen concentrated preparations of *L. acidophilus* are used for preparing this product (Speck, 1978).

A product closely related to acidophilus milk is a product called 'Biogarde' made and marketed in Germany. In the production of this product, *L. acidophilus*, *Strep. thermophilus* and selected strains of bifidobacteria (another bacterium usually isolated from faeces of normal, healthy individuals) are cultured separately and added to milk and propagated. In the finished product, only 1% of the flora consists of *L. acidophilus* and bifidobacteria. The inability to obtain a greater proportion of *L. acidophilus* and bifidobacteria, which are the essential components in products intended for therapeutic purposes, is attributed to the incompatibility of culture mixtures used in 'Biogarde' (Speck, 1978). For a detailed discussion of acidophilus milk and cultured foods containing bifidobacteria, the reader should refer to Rasic and Kurmann (1978).

Another cultured product prepared with an intestinal isolate of *Lactobacillus casei* is marketed in Japan under the label 'Yakult'. In the preparation of 'Yakult', skim-milk, glucose, and Chorella extract are dissolved in hot water, filtered, sterilized and cooled to 37°C. A bulk starter of *L. casei* (strain Shirota) is added and the fermentation tank held at 37°C for 4 days or until the acidity reaches 2.7%. At this point, the product is cooled and sweetners and flavours are added. It is then bottled and distributed door-to-door by housewives. Studies in Japan have revealed that 'Yakult' is beneficial in maintaining the health of the enteric system (Speck, 1978).

XI LEBEN AND DAHI

Leben is one of the oldest fermented milks that originated in the Tigris–Euphrates Valley. It is now consumed all over the Middle East. Most of the leben in these countries is made in individual households or on a cottage-industry scale. There are, however, modern dairy plants making this product. There are minor variations in production of this product, and the name also varies from country to country.

Leben is a concentrated yogurt product and the condensing may be accomplished by hanging the fermented curd in a cloth bag which allows the whey to drain out as the curd shrinks and is squeezed out by its own weight. In Turkey, a goat or sheep's skin bag is used. In Egypt, earthenware is used and moisture is removed by surface evaporation from the porous container. Removal of moisture, concentration of solids, and the high acidity of the product provide the limited shelf stability of leben until it is consumed. In certain areas the concentrated curd mass is rolled into balls, sun-dried and sold.

The predominant flora of leben consists of *Strep. lactis*, *Strep. thermophilus*, *L. bulgaricus* and lactose-fermenting yeasts. The product is made from sheep's, goat's, cow's or buffalo's milk, or a mixture of two or more of these milks. A portion saved from the previous days, production is used as a starter. For detailed discussion of the production, and chemical, microbiological and marketing aspects of leben, the publications of Kosikowski (1977), Abo-Elnaga et al. (1977) and Tamime and Robinson (1978) should be consulted.

Dahi is the Indian equivalent of yogurt. About 7.07% of the total milk production in India is converted into dahi. It may be produced from cow's milk or buffalo milk or from a mixture of both. In the manufacture of dahi, milk is brought to the boil to destroy pathogenic and spoilage micro-organisms, cooled to about 37°C, and inoculated with a small portion of the product saved from the previous day. The inoculated milk is kept near a warm place (close to a cooking oven) overnight. By morning, a coagulum is formed. The acidity of the product varies from region to region in the Indian subcontinent. Also, the curd tension varies depending upon the type of milk used. Dahi made from buffalo milk has a higher curd tension than the product made from cow's milk because of the higher solids content in buffalo milk.

The micro-organisms involved in dahi fermentation include *Strep. lactis*, *Strep. thermophilus*, *L. bulgaricus*, *L. helveticus*, *L. plantarum* and lactose-fermenting yeasts. Dahi is eaten with rice in the South and with flat wheat bread (chapathi) in the North. In some places it is sweetened with sugar, flavoured with nutmeg and consumed as a dessert. Dahi is also mixed with ice-water, sweetened and mixed with aromatic spices, and made into a refreshing drink called lassi. A good discussion on dahi can be found in the book by Rangappa and Achaya (1975).

XII. VILIA

Vilia or filia is a Finish fermented milk. In Finland it is known as viili or filli. On the farms, this product is made from fresh unheated cow's milk.

In the dairies, pasteurized milk standardized to 2.5% milkfat is used. The starters used consist of a mixture of capsule-producing strains of *Strep. lactis* and *Strep. cremoris*. The milk is tempered to 17–18°C, inoculated with starter, and held at this temperature until a smooth uniform coagulum is formed. The product has a stringy texture, a pleasant acid taste and good diacetyl flavour. In Eastern Finland, the ropy texture is not popular. The ropy characteristic is retained if the incubation temperature is held between 17°C and 22°C. Above 22°C, the ropy characteristic is not consistently produced and may be lost. Top-quality villi also has a surface growth of *Geotrichium candidum*, which lowers the acidity by utilizing some of the lactic acid. Viili may be eaten plain or with sugar or cinnamon.

REFERENCES

Abo-Elnaga, I.G., El-Aswad, M. and Moqi, M. (1977). *Milchwissenschaft* **32**, 521.
Anderson, D.G. and McKay, L.L. (1977). *Journal of Bacteriology* **136**, 465.
Arora, B.C., Dutta, S.M., Sabharwal, V.B. and Ranganathan, B. (1978). *Acta Microbiologica Polanica* **27**, 353.
Bills, D.D. and Day, E.A. (1966). *Journal of Dairy Science* **49**, 1473.
Bottazzi, V. and Dellaglio, F. (1967). *Journal of Dairy Research* **34**, 109.
Bottazzi, V. and Vescovo, M. (1969). *Netherlands Milk and Dairy Journal* **23**, 71.
Bottazzi, V., Bottistatti, B. and Montescani, G. (1973). *Le Lait* **53**, 295.
Bruhn, J.C. and Collins, E.B. (1970). *Journal of Dairy Science* **53**, 857.
Buchanan, R.E. and Gibbons, N.E. (1974). 'Bergey's Manual of Determinative Bacteriology', 8th edition Williams and Wilkins Co., Baltimore.
Chandan, R.C., Gordon, J.F. and Walker, D.A. (1969). *Process Biochemistry* **4** (2), 13.
Collins, E. B. (1972). *Journal of Dairy Science* **55**, 1022.
Collins, E.B. and Speckman, R.A. (1974). *Canadian Journal of Microbiology* **20**, 805.
Coventry, M.J., Hillier, A.J. and Jago, G.R. (1978). *Australian Journal of Dairy Technology* **33**, 148.
DeMan, J.C. (1956). *Netherlands Dairy and Milk Journal* **10**, 38.
Duggan, D.E., Anderson, A.W. and Elliker, P.R. (1959). *Food Technology* **13**, 465.
Emmons, D.B. and Tuckey, S.L. (1967). 'Cottage Cheese and other Cultured Milk Products'. Pfizer Cheese Monographs No. 3. Chas. Pfizer and Co., Inc., New York.
Foster, E.M., Nelson, F.F., Speck, M.L., Doetsch, R.N. and Olson, J.C., Jr. (1957. 'Diary Microbiology'. Prentice-Hall Inc., Englewood Cliffs, N. J.
Gilliland, S.E., Speck, M.L. and Morgan, C.G. (1975). *Applied Microbiology* **30**, 541.
Hamden, I.Y., Kansman, J.E. and Deane, D.D. (1971). *Journal of Dairy Science* **54**, 1080.
Harper, W.J. and Huber, R.M. (1961). *Applied Microbiology* **9**, 184.
Harvey, R.J. (1960). *Journal of Dairy Research* **24**, 41.
Harvey, R.J. and Collins, E.B. (1962. *Journal of Bacteriology* **83**, 1005.
Harvey, R.J. and Collins, E.B. (1963). *Journal of Bacteriology* **86**, 1301.
Humphreys, C.L. and Plunkett, M. (1966). *Dairy Science Abstracts* **31**, 607.
Jonsson, H. and Pettersson, H.E. (1977). *Milchwissenschaft* **32**, 587.

Keenan, T.W. and Lindsay, R.C. (1966). *Journal of Dairy Science* **49,** 1563.
Keenan, T.W., Bills, D.D. and Lindsay, R.C. (1967). *Canadian Journal of Microbiology* **13,** 1118.
Kempler, G.M. and McKay, L.L. (1979). *Applied and Environmental Microbiology* **37,** 1041.
Keogh, B.P. (1976). *Commonwealth Scientific and Industrial Research Organization (Australia) Food Research Quarterly* **36,** (2), 35.
Kilara, A. and Shahani, K.M. (1978). *Journal of Dairy Science* **61,** 1793.
Kosikowski, F.V. (1977). Cheese and Fermented Milk Foods. 2nd edition. Edwards Bros. Inc., Ann Arbor, Michigan.
Kroger, M. (1973). *Dairy and Ice Cream Field* **156**(1), 38.
Kundig, W., Ghosh, S. and Roseman, S. (1964). *Proceedings of the National Academy of Sciences of the United States of America* **52,** 1067.
Lawrence, R.C., Thomas, T.D. and Terazaghi, B.E. (1976). *Journal of Dairy Research* **43,** 141.
LeBlanc, D.J., Crow, V.L., Lee, L.N. and Garon, C.F. (1979). *Journal of Bacteriology* **137,** 878.
Lees, G.J. and Jago, G.R. (1978a). *Journal of Dairy Science* **61,** 1205.
Lees, G.J. and Jago, G.R. (1978b). *Journal of Dairy Science* **61,** 1216.
Lindsay, R.C., Day, E.A. and Sandine, W.E. (1965). *Journal of Dairy Science* **48,** 863.
McKay, L.L. (1978). *Food Technology* **32,** 181.
McKay, L.L. and Baldwin, K.A. (1978). *Applied and Environmental Microbiology* **36,** 360.
McKay, L.L., Walter, L.A., Sandine, W.E. and Elliker, P.R. (1969). *Journal of Bacteriology* **99,** 603.
McKay, L.L., Sandine, W.E. and Elliker, P.R. (1971). *Dairy Science Abstracts* **33,** 493.
McLeod, P. and Morgan, M.E. (1958). *Journal of Dairy Science* **41,** 908.
Milk Industry Foundation. (1957). Manual for Milk Plant Operators, 2nd edition. MIF, Washington.
Miller, A., Morgan, M.E. and Libbey, L.M. (1974). *International Journal of Systematic Bacteriology* **24,** 346.
Moon, N.J. and Reinbold, G.W. (1976). *Cultured Dairy Products Journal* **9,** 10.
Myers, R.P. (1931). *American Journal of Public Health* **21,** 861.
Nakae, T. and Elliot, J.A. (1965). *Journal of Dairy Science* **48,** 287.
Pack, M.Y., Vedamuthu, E.R., Sandine, W.E. and Elliker, P.R. (1968a). *Journal of Dairy Science* **51,** 339.
Pack, M.Y., Vedamuthu, E.R., Sandine, W.E. and Elliker, P.R. (1968b). *Journal of Dairy Science* **51,** 345.
Pette, J.W. (1949). *Proceedings of the 12th International Dairy Congress* **2,** 572.
Premi, L., Sandine, W.E. and Elliker, P.R. (1972). *Applied Microbiology* **24,** 51.
Rangappa, K.S. and Achaya, K.T. (1975). 'Indian Dairy Products', 2nd edition. Asia Publishing House, Bombay, India.
Rasic, J.L. and Kurmann, J.A. (1978). 'Yoghurt. Scientific Grounds, Technology, Manufacture and Preparations'. Technical Dairy Publishing House, Copenhagen, Denmark.
Reddy, M.S., Williams, F.D. and Reinbold, G.W. (1973). *Journal of Dairy Science* **56,** 634.
Richter, R.L. (1977). *Cultured Dairy Products Journal* **12,** 22.
Robinson, R.K. and Tamime, A.Y. (1975). *Journal of the Society of Dairy Technology (England)* **28,** 149.
Robinson, R.K. and Tamime, A.Y. (1976). *Journal of the Society of Dairy Technology (England)* **29,** 149.

Sandine, W.E. and Elliker, P.R. (1970). *Journal of Agricultural and Food Chemistry* **24**, 1106.

Sandine, W.E., Elliker, P.R. and Anderson, A.W. (1957). *Milk Products Journal* **48**(1), 12.

Sandine, W.E., Elliker, P.R. and Anderson, A.W. (1958). *Journal of Dairy Science* **41**, 706.

Sandine, W.E., Daly, C., Elliker, P.R. and Vedamuthu, E.R. (1972a). *Journal of Dairy Science* **55**, 1031.

Sandine, W.E., Muralidhara, K.S., Elliker, P.R. and England, D.C. (1972b). *Journal of Milk and Food Technology* **35**, 691.

Seitz, E.W., Sandine, W.E., Elliker, P.R. and Day, E.A. (1963a) *Canadian Journal of Microbiology* **9**, 431.

Seitz, E.W., Sandine, W.E., Elliker, P.R. and Day, E.A. (1963b) *Journal of Dairy Science* **46**, 186.

Sharpe, E.M. (1979). *Journal of the Society of Dairy Technology (England)* **32**(1), 9.

Sheldon, R.M., Lindsay, R.C., Libbey, L.M. and Morgan, M.E. (1971). *Applied Microbiology* **22**, 263.

Speck, M.L. (1976). *Journal of Dairy Science* **59**, 338.

Speck, M.L. (1978). *Developments in Industrial Microbiology* **19**, 95.

Speckman, R.A. and Collins, E.B. (1968). *Journal of Bacteriology* **95**, 174.

Tamime, A.Y. and Robinson, R.K. (1978). *Milchwissenschaft* **33**, 209.

Tamime, A.Y. and Greig, R.I.W. (1979). *Dairy Industries International* **44**(9), 3.

Thompson, J. (1978). *Journal of Bacteriology* **136**, 465.

United States Department of Health, Education and Welfare, Food and Drug Administration (1977). *Federal Register* **42**(12), 29919.

United States Department of Health, Education and Welfare, Food and Drug Administration. (1978). *Code of Federal Regulations* **21**, Parts 100 to 199, pp. 135.

Vaissier, Y. (1977). *Revue Laitière Francaise* **915**, 1.

Vaissier, Y. (1978a) *Revue Laitière Francaise* **361**, 1.

Vaissier, Y. (1978b). *Revue Laitière Francaise* **362**, 1.

Vedamuthu, E.R. (1978a). *Journal of Food Protection* **41**, 654.

Vedamuthu, E.R. (1978b). *Dairy and Ice Cream Field* **161**(1), 66.

Vedamuthu, E.R. (1979a). *Developments in Industrial Microbiology* **20**, 187.

Vedamuthu, E.R. (1979b). *Indian Dairyman* **31**, 289.

Vedamuthu, E.R., Sandine, W.E. and Elliker, P.R. (1966). *Journal of Dairy Science* **49**, 151.

White, C.H. (1978). *Cultured Dairy Products Journal* **13**(1), 16.

Note Added in Proof

Since the completion of this chapter, several promising developments have occurred that have a bearing on genetic engineering (recombinant DNA) of starter bacteria used in fermented milks. Currently, we have experimental evidence for transduction, conjugation and transformation among lactic streptococci. Genetic recombination through protoplast fusion of lactic streptococci has also been reported (Gasson, 1980). For recent developments on genetics governing the metabolic activities of lactic streptococci, the reader should refer to Davies and Gasson (1981) and Kempler and McKay (1981). Genetic determinants for host–phage restriction modification systems in lactic streptococci have also been described (Sanders and Klaenhammer, 1981). At present, active research on the genetics of yogurt starter bacteria is being carried out in the U.S.A. (Somkuti *et al.*, 1981) and Italy (Vescovo *et al.*, 1982). These developments undoubtedly will have far-reaching effects on scientific control of dairy fermentations and in new-product development.

In this chapter, manufacturing aspects of fermented milks were discussed only briefly. For more detailed treatments of these aspects, the recent review by Chandan (1982) is recommended.

References

Chandan, R.C. (1982). *In* 'Industrial Microbiology', 4th edn. Avi Publishing Co., Westport, Connecticut.
Davies, L.F. and Gasson, M.J. (1981). *Journal of Dairy Research* **64**, 1527.
Gasson, M.J. (1980). *Federation of European Microbiological Societies Microbiology Letters* **9**, 99.
Kempler, G.M. and McKay, L.L. (1981). *Journal of Dairy Science* **64**, 1527.
Sanders, M.E. and Klaenhammer, T.R. (1981). *Applied and Environmental Microbiology* **42**, 944.
Somkuti, G.A., Steinberg, D.H. and Bencivengo, M.M. (1981). Abstracts of American Society for Microbiology International Conference on Streptococcal Genetics P-4. American Society for Microbiology, Sarasota, Florida.
Vescovo, M., Morelli, L. and Bottazzi, V. (1982). *Applied and Environmental Microbiology* **43**, 50

7. Fermented Vegetables

H. P. FLEMING

Food Fermentation Laboratory, U.S. Department of Agriculture, Agricultural Research Service, Southern Region, and North Carolina Agricultural Research Service, Department of Food Science, North Carolina State University, Raleigh, North Carolina 27650, U.S.A

I. INTRODUCTION

Preservation of foods by fermentation is thought to have originated in the Orient before recorded history (Pederson, 1960). Salting (brining) is a requisite for preserving vegetables and certain fruits, such as olives, by fermentation because it helps to direct the course of the fermentation and prevent softening and other degradative changes in plant tissues. Brining probably preceded or occurred simultaneously with fermentation as a food preservation method. The type and extent of microbial action in salted vegetables is highly dependent on the concentration of salt. Microbial fermentation may be rapid at low concentrations of salt, but extremely slow or non-existent at high concentrations. Other chemical and physical factors are also important in regulating the rate and extent of fermentation.

Brining and fermentation were primary methods for preserving vegetables throughout the World prior to the advent of canning and freezing. Although secondary to modern preservation methods in Western civilization, brining and fermentation remain important methods for preserving certain vegetables in highly developed countries because they: (1) impart certain desired organoleptic qualities in the products; (2) provide a means for extending the processing season for fruits and vegetables; and (3) require comparatively little mechanical energy input, a fact that enhances the potential for these age-old methods of preservation in our modern energy-sensitive World.

Probably most vegetables have been preserved by brining and/or fermentation. Comprehensive reviews are available on the fermentation of sauerkraut (Pederson, 1960, 1979; Stamer, 1975), olives (Vaughn *et al.*, 1943; Vaughn, 1954, 1975) and cucumbers (Etchells *et al.*, 1951, 1975). Examples of other vegetables and fruits that have been brined for home and commercial purposes include carrots (Niketic-Aleksic *et al.*, 1973), celery (Bates, 1970), various vegetable blends (Orillo *et al.*, 1969), green beans, lima beans, green peas, corn, okra and green tomatoes (Etchells *et al.*, 1947), cauliflower (Dakin and Milton, 1964), and whole peppers, pepper mash, onions and citron. In addition, fermented beets, turnips, radishes, chard, Brussels sprouts, mustard leaves, lettuce, fresh peas and vegetable blends have been produced in considerable quantity in the Orient (Pederson, 1979). The purpose of this review is to summarize general principles governing brine fermentation of vegetables and certain fruits. Fermentations of cabbage, olives and cucumbers are emphasized to illustrate those principles.

II. BRINING OPERATIONS FOR NATURAL FERMENTATIONS

The general procedure for preserving vegetables by natural (i.e. by naturally occurring micro-organisms) fermentation is outlined in Figure 1. Specific treatments vary among vegetables, some of which will be indicated.

A. Prebrining Treatments

Vegetables may be treated in various ways before brining, depending on the nature of the fresh vegetable and the product desired. Examples of such prebrining treatments include grading, washing, cutting, piercing and exposing to alkali.

Cabbage for sauerkraut is wilted, cored, trimmed and shredded prior to placement in a fermentation tank. Olives destined for table use receive various prebrining treatments, depending on desired qualities of the end product (Vaughn, 1954; Fernandez-Diez, 1971). Olives contain oleuropein, an extremely bitter phenolic glucoside, and varieties rich in this compound must be debittered. In the preparation of Spanish-style green olives, the olives undergo a prebrining treatment with 1 to 2% sodium hydroxide to destroy the bitter principle. The alkali is allowed to penetrate about three-quarters of the distance to the pit, and air is

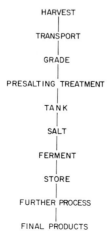

Fig. 1. Flow chart for preservation of vegetables by brining.

excluded during this operation to prevent darkening. The olives are then washed several times to remove alkali. During washing, some of the fermentable sugars and other nutrients are also removed from the olives. Sicilian-style green olives are not debittered before brining, since the varieties of olive used contain relatively low concentrations of the bitter principle. Greek-style ripe olives are prepared from naturally tree-ripened olives and are not debittered. The final product retains a level of bitterness desirable to some people.

Cucumbers are usually graded to size before brining because size influences the brining procedure. Cucumbers mostly are brined whole, although residual cut pieces from the manufacture of fresh-pack cucumbers (prepared by direct acidification and pasteurization of fresh cucumbers) are brined for use in relishes. Cucumbers may be pierced to prevent internal accumulation of gas (bloater formation) during fermentation, but this practice is limited primarily to specialty products such as overnight and genuine dills by a few briners. Cucumbers destined for natural fermentation are not normally washed before brining. It has been suggested, however, that the initial brine be drained from small cucumbers, which usually retain fungal-laden flowers, in order to lower the concentration of softening enzymes (Etchells *et al.*, 1955, 1958).

B. Brining Procedures

Vegetables may be brined at various salt concentrations by either the dry-salt method or the brine-solution method (Table 1). By the dry-salt method, salt is added with the vegetables, such as cabbage, as they are placed in the tanks; the tanks are then headed with timber. By the brine-solution method, as used with cucumbers and olives, a small amount of brine is added to the tanks to cushion the fall of the vegetables as they are loaded into the tanks. Then, the tanks are headed with timber and additional brine is added to cover the vegetables and heading timber. The added salt and water-soluble constituents of the vegetables interdiffuse until equilibrium of all constituents is attained. The equilibration process continues until some time after the fermentation is complete. With proper salting and under correct holding conditions, the fermented vegetables may be stored for several months or years without serious loss in texture or other quality factors.

Table 1

Brining procedures for vegetables

	Concentration (%, w/v) of salt during		
Method of salting	Fermentation	Storage	Vegetable
Dry salting	2–3	2–3	Cabbage
Brine solution	5–8	8–16	Cucumbers
	4–7	4–7	Green olives

The concentrations of salt indicated generally are used for commercial brining of cabbage, cucumbers and olives in the United States. Wide variations exist in the salt concentration used for peppers, onions and cauliflower.

Salt for brining purposes may include solar, rock and granulated types. Borg *et al.* (1972) observed no significant differences in microbial populations of these three types of salt and found that lactic-acid bacteria grew equally well in the presence of all types.

For sauerkraut fermentation, dry salt is distributed uniformly into the shredded cabbage as it is added to the tank. The rate of salt addition is such that the final concentration will be 2 to 3% (preferably 2.25%) sodium chloride. That concentration is maintained during fermentation and storage. The preferred temperature during fermentation is around 18°C (Pederson, 1960).

Alkali-treated and washed green olives to be prepared into Spanish-style olives are brined and kept at 4 to 7% sodium chloride during fermentation and storage. Sicilian-style green olives are brined by the solution method, and dry salt added as needed to maintain the brine at 6 to 7% sodium chloride (Vaughn, 1954). Greek-style, naturally ripe olives may be dry-salted or placed in a brine solution, with salt concentrations ranging from 7 to 19% (Balatsouras, 1966). Olives for eventual processing as California-style, black-ripe olives are temporarily stored in brines of 5 to 7% sodium chloride (Vaughn, 1954). These olives undergo fermentation. Alternatively, olives for black-ripe processing may be held in a salt-free solution containing 0.67% lactic acid, 1% acetic acid, 0.3% sodium benzoate and 0.3% potassium sorbate (Vaughn *et al.*, 1969b) or in similar proprietary solutions. These olives do not ferment. Salt-free storage is now used primarily for Sevillano and Ascolano varieties which are more likely to shrivel in salt solutions;

brining is still an important method of storing smaller varieties (Manzanillo and Mission) of olives (J.R. Webster, personal communication).

Cucumbers are brined in a solution of 5 to 8% sodium chloride, and dry salt is added to maintain the brine at this concentration during fermentation. Etchells and Hontz (1972) suggested that the concentration of sodium chloride at equilibrium be adjusted according to brine temperature, i.e. lower salt concentrations at lower temperatures (e.g. about 5% sodium chloride at 20°C and below) and higher salt concentrations at higher temperatures (e.g. about 8% sodium chloride at 30°C). Brines containing large-size cucumbers are purged with nitrogen gas during the fermentation period to remove carbon dioxide from solution and thereby prevent bloater damage (Etchells et al., 1973) as will be discussed later.

C. Brining Vessels

Vegetables are brined in vessels varying in size from a few litres for home and small-scale commercial use to over 75,000 litres for large-scale commercial operations. The size is dictated by the volume of fresh intake, physical properties of the vegetable and economic factors. Materials used for brining vessels include earthenware (as in stone crocks for home use), wood (e.g. cypress, fir, redwood), food-grade plastics, glazed tile, fibre-glass, and concrete. When fibre-glass or concrete is used, the interior of the vessel is coated with a food-grade material. Metal deteriorates rapidly in the saline and acidic environment and is not used within the brining vessels, but steel straps or rods may be placed externally for tank reinforcement.

Brine depth is the primary dimension that governs tank size for vegetable fermentations. Fresh vegetables contain air, many having a specific gravity of less than one, and are therefore buoyant when placed in brine solutions that have specific gravities above one. Headboards are mounted in the tops of brine tanks to keep freshly tanked vegetables submerged in the brine. The buoyancy of freshly brined cucumbers can cause physical damage to those cucumbers in the top section (Fleming et al., 1977). Tank depths of about 2.5 metres are typical for brined vegetables, but may reach 4 metres for some cucumber tanks. The buoyancy problem is especially serious with peppers because they are hollow, necessitating placement of baffles (false heads) at about 1.25-metre intervals in deep tanks to prevent the peppers at the top of the tank from being crushed.

The manner in which the vessel is headed and the environment in which it is held are extremely important relative to the attention that the brined vegetables must receive during fermentation and storage. Cucumbers, for example, are brined in cylindrical tanks with the brine surface exposed to the atmosphere (Fig. 2). Growth of film yeasts, moulds, and spoilage bacteria on the surface of the brine must be restricted. This growth is restricted if the brine surface is exposed to sunlight; but evaporation, rainwater and extraneous contamination from the atmosphere create problems. The tanks may be sheltered, but then the surface must be periodically skimmed to remove surface growth.

Sauerkraut tanks or vats are covered with plastic that seals the kraut surface from the atmosphere. The salted cut cabbage creates its own brine. When the plastic cover that is draped over the tank sides is properly weighted with water, the brine is kept level with the surface of the sauerkraut. Thus, the brine surface is anaerobic, and is not susceptible to growth by surface yeasts, moulds, and spoilage bacteria (Pederson and Albury, 1969).

Surface growth on olive brines exposed to air is similar to that on cucumber brines. No problem with surface growth occurs when the olives are fermented and stored in air-tight wooden or plastic drums. Bulk storage in large tanks has been of recent interest in Spain (Borbolla y Alcala et al., 1969) and the United States (Vaughn, 1975). The use of 'anaerobic' bulk tanks in the olive brining industry is an important development which offers a means of eliminating the surface-growth problem (Vaughn, 1975). Some olive companies have designed their own special tanks to maintain anaerobiosis (Figs. 3 and 4). One company uses 45,000-litre capacity, fibre-glass tanks (Fig. 3) normally used for petroleum storage. The tanks are coated internally for food use and fitted with two manholes, each 0.91 m in diameter, for filling and emptying. These tanks are then buried with only the manholes protruding above ground. The tanks are filled with olives, brine is added until the manhole is partially filled, and a plastic disc is floated on the brine surface to further restrict exposure of the brine surface to the atmosphere. Underground storage of vegetables offers the advantage of uniform and moderate temperatures. This advantage has been put to

Fig. 2. Photograph of the surface of cucumber brining tanks during an active stage of fermentation. (a) Surface of a 23,000-litre, wooden tank that is being purged of carbon dioxide. Note that the headboards are not visible, giving no evidence of excessive buoyancy pressure. Dissolved carbon dioxide being removed from the brine by purging with nitrogen gas with the side-arm apparatus shown causes the frothy surface. The tank in the background is being flume-filled. (b) Surface of a 23,000-litre tank that is not being purged of carbon dioxide. Note the heavy timbers that are broken due to excessive buoyancy pressure created by bloated cucumbers. The tank must be reheaded to prevent air exposure and eventual spoilage of the cucumbers. The photographs are reproduced with compliments of Mount Olive Pickle Company, Inc., Mount Olive, N.C.

Fig. 3. Photographs of 45,000-litre, fibre-glass tanks for underground storage of Spanish-style green olives. The tanks will be placed underground, leaving only the two manholes protruding above ground. The manholes will be used for filling and unloading, and are fitted with plastic floats to further restrict air exposure of the brine surface. (a) Shows tanks aligned in their burial site. (b) Shows site of the buried tanks. The photographs are reproduced with the compliments of J.R. Webster, Lindsay Olive Growers, Lindsay, California, U.S.A.

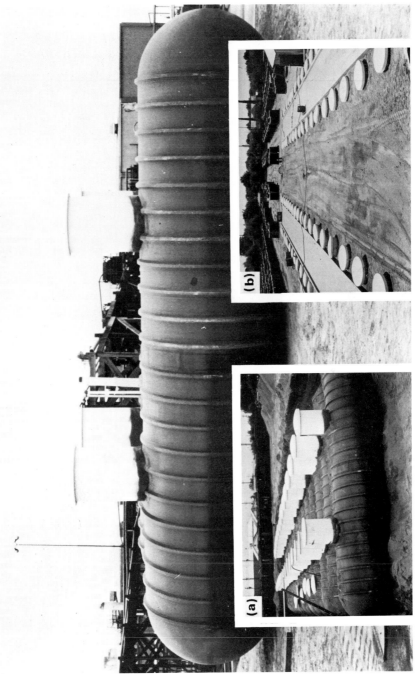

Fig. 3. (For legend see facing page.)

Fig. 4. (For legend see facing page.)

use by storage of olives underground in southern California to avoid high brine temperatures that occur during summer months and which result in softening of the olives.

Important changes in design, size and materials of brining vessels have occurred in recent years. Further changes should be expected because of the trend toward larger tanks, the quest for higher quality and more economical products, and greater regulatory restrictions on disposal of salt and other wastes.

III. MICRO-ORGANISMS DURING NATURAL FERMENTATION

A. Growth Sequence of Microbial Groups

Fresh vegetables contain a numerous and varied epiphytic microflora, including many potential spoilage micro-organisms, and an extremely small population of lactic-acid bacteria. Fresh pickling cucumbers, for example, may contain over $1 \cdot 10^7$ total aerobic micro-organisms per g but only $5 \cdot 10^3$ acid-forming bacteria per g of cucumber (Table 2; Etchells *et al.*, 1961, 1975). Less than 10 lactobacilli per g of plant material were enumerated in the majority of various plant materials analysed by Mundt and Hammer (1968). Streptococci and *Leuconostoc mesenteroides* occurred about 14 times as frequently as lactobacilli in a survey of 109 fresh plant samples, which included vegetables, grasses, fodders, legumes and cereals (Mundt and Hammer, 1968). Stirling and Whittenbury (1963) found that fresh grasses and legumes contained extremely low numbers of lactic-acid bacteria, but that the numbers were much higher on the harvesting equipment. They suggested that nutrients made available through cutting of the plant material provided a medium for growth of the few lactic-acid bacteria, which apparently were latent on the intact plant.

When vegetables are properly brined at salt concentrations of up to about 8% and allowed to undergo natural fermentation, the brine

Fig. 4. Photograph of a brine yard with 1500-litre plastic storage containers for Spanish-style green olives. (a) Shows the cap and plastic float assembly used to restrict the surface of the brine to air. (b) Shows the top of a storage container with a plastic float assembly in place. After olives are placed in the container, brine is added until it reaches a specified level in the reservoir of the cap. The plastic float is then installed and the brine level is monitored by the indicator stick extending from the float through the cover plate. The photographs are reproduced with the compliments of J.R. Webster, Lindsay Olive Growers, Lindsay, California, U.S.A.

Table 2

Microbial populations on cucumber fruit and blossoms

| Microbial group | Colony counts (thousands) | | | |
| | Cucumber fruit | | Cucumber blossoms | |
	Per g	Per unit	Per g	Per unit
Bacteria				
Aerobes				
Total	16,000.0	182,320.0	18,200,000	476,000
Spores	17.0	218.0	67,800	1,940
Anaerobes				
Total	1,830.0	19,800.0	3,092,000	78,760
Spores	0.8	9.8	2,100	191
Coliforms	3,940.0	49,125.0	6,400,000	167,530
Acid-formers	5.0	60.0	26,000	765
Yeasts	1.6	18.0	3,030	82
Moulds	3.4	44.0	11,300	295

Data shown are based on results of Etchells *et al.* (1961) as recapitulated by Etchells *et al.* (1975).

solution supports growth or fermentation by a sequence of various types of micro-organisms. This sequence may be categorized into four stages, namely *initiation, primary fermentation, secondary fermentation,* and *post-fermentation* (Table 3). These four stages are based on changes in the chemical and physical environments during fermentation and storage of cabbage, olives, cucumbers and other vegetables. These environments

Table 3

Sequence of microbial types during natural fermentation of brined vegetables

Stage	Prevalent micro-organisms
Initiation	Various Gram-positive and -negative bacteria
Primary fermentation	Lactic-acid bacteria, yeasts
Secondary fermentation	Yeasts
Post-fermentation	Open tanks[a], surface growth of oxidative yeasts, moulds and bacteria
	Anaerobic tanks, none

[a] This refers to tanks with the brine surface exposed to the atmosphere. Exposure of the brine surface to sufficient sunlight will restrict surface growth, but surface growth may be great if the brine surface is shaded.

dictate the type and extent of microbial growth. The rapidity of transition between stages varies among vegetables due to properties of the vegetables as well as the chemical and physical conditions under which they are held.

The *initiation* stage may include growth by many of the faculative and strictly anaerobic micro-organisms originally present on the fresh material, but as lactic-acid bacteria become established, the pH value is lowered and growth of undesirable micro-organisms such as Gram-negative and spore-forming bacteria is inhibited. The quality of the final product depends largely on the rapidity with which the lactic-acid bacteria are established and the undesirable bacteria are excluded.

During the *primary fermentation* stage, lactic-acid bacteria and fermentative yeasts are the predominant active microflora. They grow in the brine until the fermentable carbohydrates are exhausted or until the lactic-acid bacteria are inhibited by low pH values, resulting from production of lactic and acetic acids. Buffering capacity and the fermentable carbohydrate content of the plant material are important factors which govern the extent of fermentation by lactic-acid bacteria and the extent of subsequent fermentation by yeasts.

Secondary fermentation is essentially due to fermentative yeasts. These yeasts may become established during the primary fermentation, are acid tolerant and, if fermentable sugars remain after the lactic-acid bacteria are inhibited by low pH values, continue to grow until the fermentable carbohydrates are exhausted.

During the *post-fermentation* stage, when fermentable carbohydrates are exhausted, microbial growth is restricted to the surface of brines exposed to air. When the surface of brines is so exposed, oxidative yeasts, moulds, and ultimately spoilage bacteria may become established on the surface of improperly managed tanks. No surface growth occurs in anaerobic tanks.

Numerous chemical and physical factors influence the rate and extent of growth of various micro-organisms, as well as their sequence of appearance during fermentation. Acidity and pH value greatly influence establishment and extent of growth of lactic-acid bacteria. Salt concentration, temperature, natural inhibitory compounds of plant origin, chemical additives, exposure of the brine surface to air and sunlight, fermentable carbohydrate content of the vegetable and availability of nutrients in the brine are other important factors that affect fermentation. Most of these factors are dealt with to some extent in this review.

B. Lactic-Acid Bacteria

The lactic acid bacteria that predominate initiation and primary fermentation of brined vegetables include *Streptococcus faecalis*, *Leuconostoc mesenteroides*, *Lactobacillus brevis*, *Pediococcus cerevisiae* (probably *Pediococcus pentosaeceus*, according to recent classification; Buchanan and Gibbons, 1974), and *Lactobacillus plantarum* (Pederson, 1960; Vaughn, 1954, 1975; Etchells *et al.*, 1975). These species are listed in order of increasing total acid production and in the order of their appearance during the sauerkraut fermentation (Pederson, 1960). Only the last three species appear to grow to any appreciable extent in olives or cucumbers, however, apparently due to the higher brine strengths used for these products than for sauerkraut. All factors that govern sequential growth of lactic-acid bacteria are not known, but the size of the initial populations, rapidity of growth in the brine solution, and limiting pH values are highly important. Limiting pH value, or the lowest pH value at which the organism will grow, and corresponding acidities for these bacteria in pure culture fermentations are given in Table 4. *Streptococcus faecalis* and *L. mesenteroides*, not being as salt and acid tolerant as the other species, are probably not of major significance in vegetables brined above about 5% sodium chloride (Pederson, 1960; Etchells *et al.*, 1964).

Table 4

Terminal pH values and acidities tolerated by lactic-acid bacteria involved in pure-culture vegetable fermentations

| | Final pH values and acidities[a] | | | |
| | Cabbage juice[b] | | Cucumbers[c] | |
Organism	pH Value	Acid (%)	pH Value	Acid (%)
Leuconostoc mesenteroides	3.9	1.04	—	0.23
Lactobacillus brevis	3.9	1.06	3.7	0.54
Pediococcus cerevisiae	3.5	0.90	3.4	0.63
Lactobacillus plantarum	3.5	1.40	3.2	0.91

[a] Acidity expressed as lactic acid.
[b] From Stamer *et al.* (1971) for fermentation of filter-sterilized cabbage juice.
[c] From Etchells *et al.* (1964). The salt concentration was 5.5% for all fermentations, except *L. mesenteroides* (4.2% salt; these data are from Etchells *et al.*, 1975).

Lactobacillus plantarum is the most acid-tolerant species, and terminates the lactic fermentation of vegetables. Even when other species were introduced initially into otherwise natural fermentations of cucumbers, *L. plantarum* was the final organism to grow (Pederson and Albury, 1961).

Heterofermentative lactic-acid bacteria produce sizeable quantities of acetic acid, ethanol, carbon dioxide, mannitol, dextrans, and traces of other compounds in addition to lactic acid; homofermentative lactic-acid bacteria produce mainly lactic acid (Table 5). Heterofermentative lactic-acid bacteria are desirable in sauerkraut fermentation because the volatile acids and other compounds they produce impart desired flavours to the final product (Pederson, 1960; Stamer, 1975). In cucumbers, heterofermentative lactic-acid bacteria have been implicated in bloater damage due to their large production of carbon dioxide (Etchells *et al.*, 1968b). However, even the relatively small amount of carbon dioxide produced by the homofermentative *L. plantarum*, when combined with that produced by the cucumber tissue, was found to induce bloater damage (Fleming *et al.*, 1973a).

Although the contributions of lactic-acid bacteria and other organisms to the flavour of many fermented foods are well known, comparatively little is known about their importance to the flavour of fermented vegetables. The concentration of volatile acid is known to be

Table 5

Products of lactic-acid bacteria during vegetable fermentations[a]

Fermentation type or species	Major products from fermentation of vegetable sugars[b]	Lactic acid configuration
Heterofermentative	Lactic acid, acetic acid, ethanol, carbon dioxide, mannitol, dextran	
Leuconostoc mesenteroides		D(−)
Lactobacillus brevis		DL
Homofermentative	Lactic acid	
Pediococcus cerevisiae		DL
Lactobacillus plantarum		DL

[a] Primary catabolism of hexoses is by the phosphoketolase pathway for heterofermentative and the Embden-Meyerhof-Parnas pathway for homofermentative lactic-acid bacteria (Doelle, 1975).

[b] The major fermentable sugars in vegetables are glucose, fructose and sucrose.

important to the flavour of sauerkraut (Pederson, 1960). Acetaldehyde and diacetyl may be present early in the fermentation (Hrdlicka *et al.*, 1967). During the pure-culture fermentations of cucumbers, changes that occur in the concentrations of ethanol and various aldehydes may influence flavour (Aurand *et al.*, 1965). Fleming *et al.* (1969a) concluded that the most important contribution of *L. plantarum* to the flavour of olives in pure-culture fermentation studies was to produce preservative concentrations of acid and exclude undesirable micro-organisms which produce off-flavours.

C. Yeasts

Yeasts are probably active in most vegetable fermentations. Surface or film (oxidative) and subsurface (fermentative) yeasts may populate vegetable brines, depending on various chemical and physical conditions of the brines during fermentation and storage. Yeasts may be present in relatively low numbers on fresh vegetables such as cucumbers (Table 2, p. 238). Usually, they are present in very low numbers in brines of cucumbers (Etchells and Jones, 1943) and olives (Mrak *et al.*, 1956) during the first few days after brining. Fermentative yeasts grow throughout the fermentation of cucumbers and olives, including the primary and secondary stages, until fermentable carbohydrates are exhausted. Fermentative yeasts apparently are not found in high numbers during sauerkraut fermentations. A partial explanation is that the low concentrations of salt used favour a primary fermentation predominated by lactic-acid bacteria.

J. L. Etchells and his coworkers have made extensive studies on the yeasts of cucumber brines, including their identification and factors affecting their growth (Etchells and Bell, 1950a,b; Etchells *et al.*, 1952, 1953, 1961). The principal species of fermentative yeasts (the recent classification of yeasts by Lodder, 1970, is given in parentheses), listed in the approximate order of their occurrence in commercial cucumber brines, were: *Brettanomyces versatilis* (now *Torulopsis versatilis*); *Hansenula subpelliculosa*; *Torulopsis caroliniana* (now *T. lactis-condensi*); *Torulopsis holmii*; *Saccharomyces rosei*; *Saccharomyces elegans* (now *S. baillii*); *Saccharomyces delbrueckii*; *Brettanomyces sphaericus* (now *Torulopsis etchellsii*); and *Hansenula anomala* (Etchells *et al.*, 1961). The principal oxidative yeasts were: *Debaryomyces membranaefaciens* var. *Holl.*; *Endo-*

mycopsis ohmeri (now *Pichia ohmeri*); *Zygosaccharomyces halomembranis* (now *Saccharomyces rouxii*); and *Candida krusei* (Etchells and Bell, 1950b). Several other fermentative and oxidative yeasts were present in smaller numbers. *Rhodotorula* species, for example, occurred infrequently, but were recognizable by the pink film they formed on brine surfaces.

Mrak *et al.* (1956) found predominantly fermentative yeasts in commercial Spanish-style green olives during the first seven weeks after brining, which approximated the period of lactic acid fermentation. The principal species were: *Candida krusei*; *C. tenuis*; *C. solani*; *Torulopsis sphaerica*; *T. holmii*; and *Hansenula subpelliculosa*. During the next several weeks they found, in addition to the fermentative yeasts, the following principal species of oxidative yeasts: *Pichia membranaefaciens*; *C. mycoderma* (now *C. valida*); and *C. rugosa*.

In the fermentation of cucumbers, high concentrations (e.g. 10 to 16%) of sodium chloride favour growth of yeasts, whereas low concentrations (e.g. 5 to 8%) favour growth of lactic-acid bacteria (Etchells and Jones, 1943). The concentration of sodium chloride also influences species of yeasts in cucumber (Etchells and Bell, 1950a; Etchells *et al.*, 1952) and olive brines (Mrak *et al.*, 1956). Of the oxidative yeasts, species of *Debaryomyces* and *Zygosaccharomyces* (now *Saccharomyces*) grew in brines at sodium chloride concentrations of up to 20%, *Pichia* species up to 15%, and *Candida* species up to 10%, as shown by luxuriant film formation (Etchells and Bell, 1950a). Brine acidity also influences the species of yeasts. For example, *T. lactis-condensii* occurs early in cucumber fermentation when the acidity is low, but apparently is eliminated as the acidity increases (Etchells *et al.*, 1952). On the other hand, *Torulopsis* species may be active throughout brine storage (Etchells *et al.*, 1952). Yeasts generally have been considered as spoilage organisms, or at best have been viewed with indifference, in regard to cabbage, cucumber, and olive fermentations. Fermentative yeasts produce large amounts of carbon dioxide which can cause gaseous deterioration in brined cucumbers (Jones and Etchells, 1943; Etchells *et al.*, 1952). Pink discolouration in sauerkraut is caused by *Rhodotorula* species which may grow if the cabbage contains over 3% salt (Pederson, 1960).

Oxidative yeasts can utilize lactic acid and lower the brine acidity to allow other spoilage micro-organisms to grow (Etchells *et al.*, 1975; Mrak *et al.*, 1956). Utilization of brine acidity by these yeasts is limited to aerobic conditions, however, as they are unable to utilize organic acids under anaerobic conditions (Mrak *et al.*, 1956).

Yeasts have not been implicated in the softening of brined cucumbers. Although *Saccharomyces fragilis* and four strains of *Sacch. cerevisiae* have been found to hydrolyse pectin, none of these yeasts has been isolated from cucumber brines (Bell and Etchells, 1956). However, species of *Debaryomyces*, *Pichia* and *Candida* isolated from films on cucumber brines were able to de-esterify pectin (Bell and Etchells, 1956). *Rhodotorula glutinis* var. *glutinis*, *R. minuta* var. *minuta* and *R. rubra* have been found to cause softening in brined olives (Vaughn *et al.*, 1969a).

Sorbic acid has been used to discourage growth of fermentative and oxidative yeasts (Phillips and Mundt, 1950; Costilow *et al.*, 1957; Bell *et al.*, 1959). Although sorbic acid lowers the size of the yeast population, some concern exists that surviving yeasts may develop sufficient tolerance to the compound to yield a subsequent problem in the brine yard (Etchells *et al.*, 1961). Furthermore, fermentable sugars must be removed sooner or later if full advantage of fermentation as a method of preservation is to be obtained.

Yeasts also have a beneficial role in brined vegetables during bulk fermentation by utilizing fermentable sugars, since lactic-acid bacteria may be inhibited by low pH before the fermentation is complete. This is particularly true of fermentations involving vegetables containing a high sugar content such as carrots. Vaughn *et al.* (1976) reported that certain yeasts may be responsible for acid production in brined olives. Also, certain yeasts may impart a desirable flavour, although this possibility has not been fully explored in regard to vegetable fermentations.

D. Moulds

Moulds, especially highly pectinolytic species, can cause serious spoilage problems due to enzymic softening of vegetable tissue and should be avoided on the fresh vegetable and during brine storage and further processing. Pectinolytic enzymes may be present on the vegetable at the time of brining or may be produced by moulds that are permitted to grow on the brine surface. Raymond *et al.* (1960) found 73 species in 34 genera of moulds to be present on the blossoms, ovaries and fruit of cucumbers. Seven species in five of the genera accounted for most of the isolates and included, in decreasing order of occurrence, *Penicillium oxalicum*, *Ascochyta cucumis*, *Fusarium roseum*, *Cladosporium cladosporioides*, *Alternaria tenuis*, *Fusarium oxysporum* and *Fusarium solani*.

IV. SPOILAGE PROBLEMS

Preservation of vegetables by brine fermentation is highly dependent on the brining treatment and the resulting type and extent of microbial action during and after fermentation in bulk containers. Complete conversion of fermentable carbohydrates into acids and other end products renders the vegetables stable to subsequent fermentation. Fully fermented cucumbers (Etchells *et al.*, 1951) and olives (Fernandez-Diez, 1971), for example, may be made into finished products without the need for heat processing. The presence of residual fermentable sugar in such products can lead to gas pressure and unsightly brine turbidity in the final package from growth of yeasts and lactic-acid bacteria. Fermentation, if properly done, offers advantages as a method for preserving vegetables, as indicated in the Introduction to this chapter, but improperly brined vegetables can result in serious spoilage.

A. Softening

Softening can be a very serious problem with brined vegetables and may be caused by enzymes of plant or microbial origin. The concentration of salt needed to prevent softening varies widely among vegetables and is a major reason for differences in salt concentrations used for the vegetables listed in Table 1 (p. 231). For example, 2% sodium chloride is sufficient to prevent softening in properly attended sauerkraut, but bell peppers must be saturated with salt (about 26%) to maintain firm texture. A low concentration of salt favours a rapid fermentation in sauerkraut, but a high concentration of salt precludes microbial fermentation in bell peppers. This difference in tendency to soften among vegetables apparently is related to the activity of natural softening enzymes in the plant tissue, or possibly to differences in resistance of the tissue to attack by microbial softening enzymes.

Soft kraut is caused by insufficient salt and by yeast and mould growth at the surface, where air has been allowed to contact the kraut. These problems can be largely eliminated by insuring that the proper concentration of salt is distributed uniformly throughout the kraut at the time of tanking, and by preventing air from contacting the kraut surface by proper heading such as with the use of an air-impermeable plastic cover over the kraut surface (Pederson and Albury, 1969).

Softening of olives also may result from pectinolytic enzymes produced by moulds (Vaughn, 1975) and yeasts (Vaughn *et al.*, 1969a) which are allowed to grow on brine surfaces (Vaughn, 1954, 1975). Methods for controlling surface growth of moulds and yeasts have already been discussed. Softening can also result from growth of undesirable bacteria in the early stages of brining, especially in Spanish-style olives, before the pH value is lowered by lactic-acid bacteria. Softening can be especially serious if the salt concentration is too low. Fortifying brines with additional salt and with acid (to lower the pH value below that at which spoilage bacteria can grow) will prevent bacterial softening.

Softening of cucumbers may occur due to the action of pectinolytic enzymes which originate in the cucumber fruit itself and from microbial sources. The pectin methylesterase content of cucumbers is approximately constant regardless of fruit size (Bell *et al.*, 1951), but polygalacturonase activity increases dramatically in the larger-size fruit as they ripen. The increase in polygalacturonase activity is thought to give rise to the problem of tissue softening in the seed area, a problem commonly termed 'soft centres' (McFeeters *et al.*, 1980). Softening may occur due to bacterial growth on the fruit prior to brining, but the major microbial source of polygalacturonase softening in brined cucumbers is fungi that populate the fruit and especially the fungal-laden flowers, which are frequently retained on small fruit (Etchells *et al.*, 1958). Bacterial softening is not a serious problem if the freshly harvested fruit is brined promptly after harvest because most bacterial polygalacturonases are not active in the pH range encountered in properly fermenting cucumbers (Etchells *et al.*, 1952). Mould enzymes present more of a problem, however, due to their activity at low pH values, and they should be removed. Draining the brine from tanks of small cucumbers about 36 hours after brining or washing the fruit prior to brining as in the controlled fermentation process (Etchells *et al.*, 1973) will effectively lower the level of these enzymes. Softening enzymes can be inhibited by tannin-like substances extracted from grape leaves (Bell and Etchells, 1958) and other plants (Bell *et al.*, 1962, 1965a,b), but such compounds have not been approved for use in brined vegetables.

Although polygalacturonase activity is suppressed by high concentrations of sodium chloride in the brine (Bell and Etchells, 1961), it is preferable to exclude polygalacturonase enzymes from the tank as much as possible. Current and projected restrictions on salt disposal demand concerted efforts to minimize salt usage in brining. Buescher *et al.* (1979) found that addition of calcium chloride to cucumber brines resulted in firm brine-stock pickles, even in the presence of added polygalacturonase enzymes from a mould. They suggested that the use of calcium may offer a practical means of lowering the sodium chloride concentration required for cucumber brining. Salt-free storage of olives (Vaughn *et al.*, 1969b) and cucumbers (Shoup *et al.*, 1975) has been proposed as a means of storing these products. The storage solution is acidified but does not contain salt; the product does not undergo fermentation.

B. Gaseous Deterioration

Fisheye spoilage in Spanish-style green olives is characterized by formation of gas blisters under the skin. The disorder is due to growth of gas-forming bacteria, especially coliforms, which grow when the brine contains less than 5% salt or has a high pH value (4.8–8.5). Brines may be fortified with the proper concentration of salt and acid to prevent this type of spoilage (Vaughn, 1954).

Bloater damage in brined cucumbers results from an increase in gas pressure inside the cucumbers during fermentation. This gas pressure is due to the combined effects of nitrogen, which is trapped inside the cucumbers when they are brined, and carbon dioxide (Fleming and Pharr, 1980). The carbon dioxide originates from the cucumber tissue (Fleming *et al.*, 1973b) and from gas-forming micro-organisms active in the brine such as yeasts (Etchells *et al.*, 1952), heterofermentative lactic-acid bacteria (Etchells *et al.*, 1968b), and even the homofermentative *L. plantarum* (Fleming *et al.*, 1973a). Bloater damage in brined cucumbers can be greatly decreased if carbon dioxide is purged from the brine, as is later discussed (p. 252).

C. Other Spoilage Problems

Discolouration is an important concern with some brined vegetables. Pink kraut may be caused by pink yeasts, which can grow when excess salt is used (Pederson and Albury, 1969). This problem can be avoided by proper concentrations and distribution of the salt. *Lactobacillus brevis* has been implicated in causing pink to brown off-colours (Stamer, 1975). Laboratory tests showed that increased quantities of an unidentified pink compound were formed when *L. brevis*-inoculated cabbage juice was buffered with calcium carbonate (Stamer, 1975). Bleaching may occur when brined vegetables are exposed to sunlight. Thus, it is important that the product not be excessively exposed to sunlight during brine storage or further processing. Spanish-style green olives may darken during the alkali prebrining treatment if air is allowed to contact the olives.

Undesirable flavours and odours may result from growth of undesirable micro-organisms during fermentation and storage of brined vegetables. Malodorous fermentations of olives occur when undesirable bacteria grow before the lactic-acid bacteria become established. Certain rancid odours are caused by butyric-acid bacteria, and 'zapatera' (characterized by a sagey offensive odour) is caused by unidentified bacteria (Vaughn, 1954). These problems can be prevented by proper lye and washing treatments and by use of brine concentrations that favour early growth of lactic-acid bacteria.

Off-flavours and odours in brined cucumbers originate from growth of coliform and other undesirable bacteria before lactic-acid bacteria become established, from oxidative yeasts, fungi and bacteria which grow on the surface of improperly maintained brines, and from oxidative changes brought on by exposure of the cucumbers to sunlight. Most of these problems occur because of negligence and can be avoided by proper brining, avoidance of surface growth, and by not allowing cucumbers to be exposed to sunlight for extended periods after headboards have been removed from the tank.

V. USE OF PURE CULTURES OF LACTIC-ACID BACTERIA

A. Pure-Culture Inoculation

Numerous studies have been made on pure-culture inoculation of sauerkraut, including those of Gruber (1909), LeFevre (1919, 1920, 1928), Pederson (1930) and Engelland (1962). Pederson (1960) concluded that inoculation was impractical and unnecessary, since the organisms responsible for the fermentation occur naturally in adequate numbers, and proper fermentation will occur if temperature and salt concentrations are suitable. Stamer (1968) indicated that a proper ratio of heterofermentative to homofermentative lactic-acid bacteria is probably necessary to ensure superior kraut, and that this ratio is unknown.

Use of lactic starter cultures for Spanish-style green olives was tested by Cruess (1937). Subsequently, starters of *L. plantarum* were used commercially in the California olive industry from 1937 to 1955; they have not been used extensively since (Vaughn, 1975). Addition of cultures increased the rate of acid production during the first two months after brining and resulted in less spoilage (Vaughn *et al.*, 1943). More recently, lactic acid has been added to the brines of lye-treated and washed olives, decreasing the pH value to 7.0 or below. When the need is indicated, normal brine is used to inoculate abnormal fermentations (Vaughn, 1975).

Etchells *et al.* (1966) found that heat shocking of lye-treated, green Manzanillo olives greatly increased their brine fermentation by pure cultures of lactic-acid bacteria. When the olives were neither lye-treated nor heated, added lactic cultures caused essentially no fermentation; only yeasts grew. They suggested that the heat treatment destroyed a naturally occurring inhibitor of lactic-acid bacteria, and that the inhibitor may account for the occurrence of stuck fermentations, which are characterized by the absence of lactic-acid bacteria and the presence of yeasts in the brine. Subsequent studies revealed that hydrolysis products of oleuropein, including its aglycone and elenolic acid (Fig. 5), are inhibitory to lactic-acid bacteria but not to yeasts (Fleming and Etchells, 1967; Fleming *et al.*, 1969b, 1973c). Fleming *et al.* (1973c) found that oleuropein was not greatly inhibitory to lactic-acid bacteria, contrary to earlier suggestions by Vaughn (1954) and Juven *et al.* (1968). Indeed, lactic-acid bacteria were found to utilize oleuropein (Garrido-Fernandez and Vaughn, 1978). Fleming *et al.* (1973c) suggested that oleuropein is degraded to its aglycone when unheated olives are brined, perhaps by the natural β-glucosidase which Cruess and Alsberg (1934) reported, and by further degradation to elenolic acid. Juven and Henis (1970) found that oleuropein became more inhibitory when it was treated with β-glucosidase. Perhaps heat inactivates β-glucosidase in the olives, preventing breakdown of oleuropein to yield the inhibitory aglycone when the olives are brined.

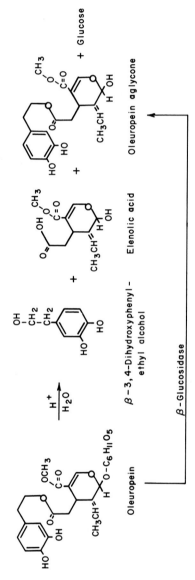

Fig. 5. Structures of oleuropein and its hydrolysis products. From Walter *et al.* (1973) and based on the work of Panizzi *et al.* (1960).

Pederson and Albury (1956) altered the course of natural cucumber fermentations by the addition of pure cultures of heterofermentative and homofermentative lactic-acid bacteria. No attempt was made to remove or inactivate the natural microflora. They found that *L. plantarum* completed all fermentations, regardless of the species of bacteria used for the inoculum, apparently because of its greater acid tolerance and presence among the natural microflora of cucumber (Pederson and Albury, 1961). Etchells *et al.* (1964, 1968a) obtained pure-culture fermentation by hot-water blanching or by gamma radiation of cucumbers prior to inoculation with lactic-acid bacteria. *Lactobacillus plantarum* produced the highest concentration of acid and grew at the lowest pH value at 8% sodium chloride. *Pediococcus cerevisiae* and *L. brevis* also grew well at 8% salt, but several thermophilic lactic-acid bacteria were limited in salt tolerance to about 2.5 to 4% sodium chloride.

B. Controlled Fermentation of Cucumbers

A controlled fermentation procedure for brined whole cucumbers was

Fig. 6. Flow chart for brine fermentation of cucumbers. Steps that have been added to the overall conventional procedure (natural fermentation) to render 'controlled fermentation' are indicated in bold face. See Etchells and Hontz (1972) and Etchells *et al.* (1973) for details of procedures for natural and controlled fermentation of cucumbers.

Fig. 7. Photograph of a modern cucumber brine yard with fibre-glass brining tanks and a 19,000-litre, liquid nitrogen storage tank for purging of carbon dioxide from fermenting brines. Nitrogen gas is piped to each of the brining tanks. The photograph is reproduced with the compliments of Straub Food Products, Stockton, California, U.S.A.

developed by Etchells *et al.* (1973, 1976), and has been in commercial use for several years (Fig. 6). In addition to certain steps taken previously with natural fermentations of brined cucumbers, the steps as outlined in Figure 6 foster a 'controlled fermentation'. The cucumbers are *washed* to lower the concentrations of softening enzymes, dirt and undesirable micro-organisms. After heading and addition of the cover brine (about 6.6% sodium chloride, w/w), the brine is immediately *acidified* with glacial acetic acid or vinegar (about 0.16% acetic acid, concentration at equilibration) to an initial pH value of around 2.8, which prevents growth of the natural microflora prior to addition of the culture. During the next 18 to 24 hours, nutrients diffuse from the cucumbers into the brine, and salt and acid diffuse into the cucumbers. The brine is *purged* continuously with nitrogen gas (Figs 2 and 7), which sweeps carbon dioxide from the brine until the fermentation is completed (Fleming *et al.*, 1973a, 1975; Etchells *et al.*, 1973). Salt is added in the dry form onto the top of the tank in two to three increments over the next few days to compensate for water in the cucumber (about 95%). This additional salt maintains the equilibrated brine strength at

5 to 8% sodium chloride. About 18–24 hours after brining, the brine pH value is altered from below 4.0 to around 4.6 by addition of sodium acetate, a *buffer* (Etchells *et al.*, 1973). Alternatively, commercial firms more commonly raise the pH value of the acidified brine to around 4.6 by carefully adding a predetermined amount of sodium hydroxide, hence, forming a buffer (Lingle, 1975). The sodium hydroxide is added, preferably, as pellets contained in 2.3 kg plastic bags. The bags are punctured for gradual dissolution and release of the alkali into the stream of acidified brine, which continuously exits from the side-arm purger. The lactic *culture* (*L. plantarum* alone or with *P. cerevisiae*) is then added to an initial population of around one billion cells per 5 litres of brined cucumbers. Fermentation by the added culture at 25°C to 30°C is completed within 7–12 days (Etchells *et al.*, 1973). The brine-stock cucumbers are then stored under ambient conditions until needed for further processing into dills, sweets, sours and relishes.

The foregoing procedure does not result in a pure culture fermentation. Rather, the key features of the procedure, as emphasized in Figure 6, serve to establish the conditions that favour growth of the added culture rather than the naturally occurring micro-organisms. Although the entire controlled fermentation procedure is in limited use, purging brines of carbon dioxide to prevent bloater formation has been widely adopted and has had a significant impact on the pickling cucumber industry through increased yields and improved quality of brine-stock pickles (Fleming *et al.*, 1975; Fleming, 1979; Costilow *et al.*, 1977). The effects of purging are further obvious in less buoyancy pressures on headboards of tanks, since the cucumbers do not become bloated by buildup of internal gas pockets (Fig. 2, p. 234).

A controlled fermentation procedure also has been described for sliced cucumbers (Fleming *et al.*, 1978). The cucumbers are sliced, heated, cooled and brined in a solution of sodium chloride and calcium acetate and inoculated with *L. plantarum*. Calcium acetate serves as a buffer and to firm the tissue.

Concentrated cultures specifically prepared for brine fermentations of cucumbers and other vegetables are available commercially. The cultures are shipped to the briner in frozen form with dry ice and may be kept for several weeks on dry ice or in a −40°C freezer (Porubcan and Sellars, 1979). Success has been claimed in the use of improved lyophilized cultures, which may be shipped at ambient temperature and then held under refrigeration (Porubcan and Sellars, 1975).

Although nitrogen was originally recommended as the purging gas, inert gas (obtained by combustion of the oxygen in air) and air have been used (Fleming, 1979). Air will sweep carbon dioxide from brines and thereby effectively prevent bloater formation, but may encourage growth of undesirable aerobic micro-organisms such as moulds, which may induce enzymic softening, or oxidative yeasts, which may utilize excessive amounts of lactic acid (Fleming *et al.*, 1975; Potts and Fleming, 1979 and unpublished work; Fleming, 1979).

VI. SUMMARY

Most vegetables can be preserved by brine fermentation. Major principles involved in successful vegetable fermentations include the presence of the proper amount of salt, conversion of fermentable carbohydrates into acids and other end products by certain lactic-acid bacteria and yeasts, and the preclusion of spoilage micro-organisms. Properties of the individual vegetable, particularly its susceptibility to softening, dictate the concentration of salt needed for preservation. In turn, the salt concentration greatly influences the type and extent of microbial action. Other factors that affect microbial growth include temperature, degree of exposure to air, and properties of the vegetable and the brine in which it is held (e.g. fermentable carbohydrate level, buffer capacity, pH value, acidity, natural inhibitory compounds, and availability of nutrients in the brine).

Considerable knowledge on vegetable fermentations has been accumulated during the past several decades and is available for further technological application. Anaerobic post-fermentation storage, controlled fermentation methods, purging of carbon dioxide from brines, and use of calcium chloride to lower salt requirements are recent developments which may be only the beginning of a new era in vegetable fermentations. Further research will undoubtedly yield new insights which may advance vegetable fermentations into the modern technological age that has been attained by many other fermentation industries.

VII. ACKNOWLEDGEMENTS

I am indebted to J. L. Etchells for his many helpful suggestions over the past 15 years and for his continuing concern for the scientific and technological advancement of the vegetable fermentation industry. I also thank M. A. Daeschel for assistance in updating yeast nomenclature. This publication is paper no. *6471* of the journal series of the North Carolina Agricultural Research Service, Raleigh, U.S.A.

Mention of a trademark or proprietary product does not constitute a guarantee or warranty of the product by the U.S. Department of Agriculture or North Carolina Agricultural Research Service, nor does it imply approval to the exclusion of other products that may be suitable.

REFERENCES

Aurand, L.W., Singleton, J.A., Bell, T.A. and Etchells, J.L. (1965). *Journal of Food Science* **30**, 288.

Balatsouras, G.D. (1966). *Grasas y Aceites* **17**, 83.

Bates, R.P. (1970). *Journal of Food Science* **35**, 476.

Bell, T.A. and Etchells, J.L. (1956). *Applied Microbiology* **4**, 196.

Bell, T.A. and Etchells, J.L. (1958). *Botanical Gazette* **119**, 192.

Bell, T.A. and Etchells, J.L. (1961). *Journal of Food Science* **26**, 84.

Bell, T.A., Etchells, J.L. and Jones, I.D. (1951.) *Archives of Biochemistry and Biophysics* **31**, 431.

Bell, T.A., Etchells, J.L. and Borg, A.F. (1959). *Journal of Bacteriology* **77**, 573.

Bell, T.A., Etchells, J.L., Williams, C.F. and Porter, W.L. (1962). *Botanical Gazette* **123**, 220.

Bell, T.A., Etchells, J.L., Singleton, J.A. and Smart, W.W.G., Jr. (1965a). *Journal of Food Science* **30**, 233.

Bell, T.A., Etchells, J.L. and Smart, W.W.G., Jr. (1965b). *Botanical Gazette* **126**, 40.

Borbolla Y Alcala, J.M.R., Fernandez-Diez, M.J., Gonzalez Cancho, F. and Cordon Casanueva, J.L. (1969). *Grasas y Aceites* **20**, 55.

Borg, A.F., Etchells, J.L. and Bell, T.A. (1972). *Pickle Pak Science* **2**, 11.

Buchanan, R.E. and Gibbons, N.E., eds. (1974). 'Bergey's Manual of Determinative Bacteriology,' 8th edition. Williams and Wilkins Company, Baltimore, Maryland.

Buescher, R.W., Hudson, J.M. and Adams, J.R. (1979). *Journal of Food Science* **44**, 1786.

Costilow, R.N., Coughlin, F.M., Robbins, E.K. and Hsu, W.T. (1957). *Applied Microbiology* **5**, 373.

Costilow, R.N., Bedford, C.L., Mingus, D. and Black, D. (1977). *Journal of Food Science* **42**, 234.

Cruess, W.V. (1937). *Fruit Products Journal* **17**, 12.

Cruess, W.V. and Alsberg, C.L. (1934). *Journal of the American Chemical Society* **56**, 2115.

Dakin, J.C. and Milton, J.M. (1964). *Food Processing and Marketing* **33**, 432.

Doelle, H.W. (1975). 'Bacterial Metabolism,' 2nd edition. Academic Press, New York.

Engelland, G.C. (1962). United States Patent 3,024,116.

Etchells, J.L. and Bell, T.A. (1950a). *Food Technology* **4**, 77.

Etchells, J.L. and Bell, T.A. (1950b). *Farlowia* **4**, 87.

Etchells, J.L. and Hontz, L.H. (1972). *Pickle Pak Science* **2**, 1.

Etchells, J.L. and Jones, I.D. (1943). *Food Industries* **15**, 54.

Etchells, J.L., Jones, I.D. and Lewis, W.M. (1947). United States Department of Agriculture Technical Bulletin 947.

Etchells, J.L., Jones, I.D. and Bell, T.A. (1951). *In* 'Crops in Peace and War. Yearbook of Agriculture, 1950–1951.' pp. 229–236. U.S. Department of Agriculture, U.S. Government Printing Office, Washington, DC.

Etchells, J.L., Costilow, R.N. and Bell, T.A. (1952). *Farlowia* **4**, 249.

Etchells, J.L., Bell, T.A. and Jones, I.D. (1953). *Farlowia* **4**, 265.

Etchells, J.L., Bell, T.A. and Jones, I.D. (1955). *Research and Farming* **13**, 14.

Etchells, J.L., Bell, T.A., Monroe, R.J., Masley, P.M. and Demain, A.L. (1958). *Applied Microbiology* **6**, 427.

Etchells, J.L., Borg, A.F. and Bell, T.A. (1961). *Applied Microbiology* **9**, 139.

Etchells, J.L., Costilow, R.N., Anderson, T.E. and Bell, T.A. (1964). *Applied Microbiology* **12**, 523.

Etchells, J.L., Borg, A.F., Kittel, I.D., Bell, T.A. and Fleming, H.P. (1966). *Applied Microbiology* **14**, 1027.

Etchells, J.L., Bell, T.A. and Costilow, R.N. (1968a). United States Patent 3,403,032.

Etchells, J.L., Borg, A.F. and Bell, T.A. (1968b). *Applied Microbiology* **16**, 1029.

Etchells, J.L., Bell, T.A., Fleming, H.P., Kelling, R.E. and Thompson, R.L. (1973). *Pickle Pak Science* **3**, 4.

Etchells, J.L., Fleming, H.P. and Bell, T.A. (1975). *In* 'Lactic Acid Bacteria in Beverages and Food.' (J.G. Carr, C.V. Cutting and G.C. Whiting, eds.), pp. 281–305. Academic Press, New York.

Etchells, J.L., Bell, T.A., Fleming, H.P. and Thompson, R.L. (1976). United States Patent 3,932,674.

Fernandez-Diez, M.J. (1971). *In* 'The Biochemistry of Fruits and Their Products' (A.C. Hulme, ed.), vol. 2, pp. 255–279. Academic Press, New York.

Fleming, H.P. (1979). *Pickle Pak Science* **6**, 8.

Fleming, H.P. and Etchells, J.L. (1967). *Applied Microbiology* **15**, 1178.

Fleming, H.P. and Pharr, D.M. (1980). *Journal of Food Science* **45**, 1595.

Fleming, H.P., Etchells, J.L. and Bell, T.A. (1969a). *Journal of Food Science* **34**, 419.

Fleming, H.P., Walter, W.M., Jr. and Etchells, J.L. (1969b). *Applied Microbiology* **18**, 856.

Fleming, H.P., Thompson, R.L., Etchells, J.L., Kelling, R.E. and Bell, T.A. (1973a). *Journal of Food Science* **38**, 499.

Fleming, H.P., Thompson, R.L., Etchells, J.L., Kelling, R.E. and Bell, T.A. (1973b). *Journal of Food Science* **38**, 504.

Fleming, H.P., Walter, W.M., Jr. and Etchells, J.L. (1973c). *Applied Microbiology* **26**, 777.

Fleming, H.P., Etchells, J.L., Thompson, R.L. and Bell, T.A. (1975). *Journal of Food Science* **40**, 1304.

Fleming, H.P., Thompson, R.L., Bell, T.A. and Monroe, R.J. (1977). *Journal of Food Science* **42**, 1464.

Fleming, H.P., Thompson, R.L., Bell, T.A. and Hontz, L.H. (1978). *Journal of Food Science* **43**, 888.

Garrido-Fernandez, A. and Vaughn, R.H. (1978). *Canadian Journal of Microbiology* **24**, 680.

Gruber, T. (1909). *Centralblat Bakteriologie ParasitKde* **2**, 555.

Hrdlicka, H., Curda, D. and Pavelka, J. (1967). *Sbornik Vysoke Skoly Chemićko-Technologicke v. Praze Potraviny* **15**, 51.

Jones, I.D. and Etchells, J.L. (1943). *Food Industries* **15**, 62.

Juven, B. and Henis, Y. (1970). *Journal of Applied Bacteriology* **33**, 721.

Juven, B., Samish, Z. and Henis, Y. (1968). *Israel Journal of Agricultural Research* **18**, 137.

LeFevre, E. (1919). *Canner* **48**, 176.

LeFevre, E. (1920). *Canner* **50**, 161.

LeFevre, E. (1928). United States Department of Agriculture Circular 35.

Lingle, M. (1975). *Food Production Management, July*, 10.

Lodder, J. (1970). 'The Yeasts. A Taxonomic Study.' North-Holland Publishing Company, London.

McFeeters, R.F., Bell, T.A. and Fleming, H.P. (1980). *Journal of Food Biochemistry*, **4**, 1.

Mrak, E.M., Vaughn, R.H., Miller, M.W. and Phaff, H.J. (1956). *Food Technology* **10**, 416.

Mundt, J.O. and Hammer, J.L. (1968). *Applied Microbiology* **16**, 1326.

Niketic-Aleksic, G.K., Bourne, M.C. and Stamer, J.R. (1973). *Journal of Food Science* **38**, 84.

Orillo, C.A., Sison, E.C., Luis, M. and Pederson, C.S. (1969). *Applied Microbiology* **17**, 10.

Panizzi, L.M., Scarpati, J.M. and Oriente, E.G. (1960). *Nota II. Gazzeta Chimica Italiana* **90**, 1449.

Pederson, C.S. (1930). New York Agricultural Experiment Station Bulletin 169.

Pederson, C.S. (1960). *In* 'Advances in Food Research' (C.O. Chichester, E.M. Mrak and G.F. Stewart, eds.), vol. 10, pp. 233–291, Academic Press, New York.

Pederson, C.S. (1979). 'Microbiology of Food Fermentations.' Avi Publishing Company, Westport, Connecticut.

Pederson, C.S. and Albury, M.N. (1956). *Applied Microbiology* **4**, 259.

Pederson, C.S. and Albury, M.N. (1961). *Food Technology* **15**, 351.

Pederson, C.S. and Albury, M.N. (1969). New York State Agricultural Experiment Station Bulletin 824.

Phillips, G.F. and Mundt, J.O. (1950). *Food Technology* **4**, 291.

Porubcan, R.S. and Sellars, R.L. (1975). United States Patent 3,897,307.

Porubcan, R.S. and Sellars, R.L. (1979). *In* 'Microbial Technology' (H.J. Peppler and D. Perlman, eds.), vol. 1, pp. 59–92. Academic Press, New York.

Potts, E.A. and Fleming, H.P. (1979). *Journal of Food Science* **44**, 429.

Raymond, F.L., Etchells, J.L., Bell, T.A. and Masley, P.M. (1960). *Mycologia* **51**, 492.

Shoup, J.L., Gould, W.A., Giesman, J.R. and Crean, D.E. (1975). *Journal of Food Science* **40**, 689.

Stamer, J.R. (1968). *Proceedings of Frontiers in Food Research, June*, 46.

Stamer, J.R. (1975). *In* 'Lactic Acid Bacteria in Beverages and Food' (J.G. Carr, C.V. Cutting and G.C. Whiting, eds.), pp. 267–280. Academic Press, New York.

Stamer, J.R., Stoyla, B.O. and Dunckel, B.A. (1971). *Journal of Milk and Food Technology* **34**, 521.

Stirling, A.C. and Whittenbury, R.J. (1963). *Journal of Applied Bacteriology* **26**, 86.

Vaughn, R.H. (1954). *In* 'Industrial Fermentations' (L.A. Underkofler and R.J. Hickey, eds.), vol. 2, pp. 417–478. Chemical Publishing Company, Inc., New York.

Vaughn, R.H. (1975). *In* 'Lactic Acid Bacteria in Beverages and Food' (J.G. Carr, C.V. Cutting and G.C. Whiting, eds.). pp. 307–323. Academic Press, New York.

Vaughn, R.H., Douglas, H.C. and Gilliland, J.R. (1943). University of California Agricultural Experiment Station Bulletin 678.

Vaughn, R.H., Jakubczyk, T., MacMillan, J.D., Higgins, T.E., Dave, B.A. and Crampton, V.M. (1969a). *Applied Microbiology* **18**, 771.

Vaughn, R.H., Martin, M.H., Stevenson, K.E., Johnson, M.G. and Crampton, V.M. (1969b). *Food Technology* **23,** 124.

Vaughn, R.H., Joe, T., Crampton, V.M., Lieb, B. and Patel, I.B. (1976). *Abstracts of the Annual Meeting of the American Society for Microbiology* 188.

Walter, W.M., Jr., Fleming, H.P. and Etchells, J.L. (1973). *Applied Microbiology* **26,** 773.

8. Coffee

ROBERT O. ARUNGA

Kenya Industrial Research and Development Institute, Nairobi, Kenya

I. INTRODUCTION

Coffee is a general term which refers to the fruits, seeds and products of plants of the genus *Coffea*. Although there are over 40 species of this genus, those cultivated commercially are *Coffea arabica*, *Coffea canephora* (robusta), *Coffea arabusta*, *Coffea liberica* and *Coffea excelsa*. The coffee plant belongs to the family Rubiaceae. A mature coffee fruit is a fleshy spheroidal berry about 15–20 mm in diameter. The term berry is not botanically accurate because the coffee fruit is a drupe. It changes from a green to a cherry-red colour while ripening. The fruit normally contains two beans surrounded by a thin membrane known as the silver skin. The beans and the silver skin are protected by a hard, horny endocarp generally referred to as the parchment. Adhering firmly to the outside of the parchment is a pulpy, mucilaginous mesocarp which is covered by the fruit skin or pulp (exocarp) shown schematically in Figure 1.

Coffee is generally marketed as green beans. Green beans are subsequently blended, roasted and ground to make the well-known coffee grounds which are marketed for making the coffee beverage. Preparation of instant coffee involves further processing which incorporates dissolution of ground coffee in water followed by dehydration of the soluble component. Coffee processing to the green bean stage has as its primary objective the dehydration of the bean to a point where biological activity is minimized. Coffee beans may be dehydrated either in the cherry state or after removal of the pulp and mucilage (parchment state). The ripe coffee cherry can therefore be processed by two distinct methods to obtain the green coffee of commerce. It may be dried directly and milled or it may be pulped, fermented, dried and milled. The latter method, incorporating pulping and fermentation and generally referred to as 'wet processing,' accounts for about one-third of the World green coffee production (Menchu and Rolz, 1973) but is normally confined to the *C. arabica* species.

Pulping involves removal of the coffee skin and is accomplished by suitable mechanical methods described by Sivetz and Foote (1963). After pulping, coffee is fermented. Fermentation of coffee is the process by which the mucilaginous mesocarp adhering to the coffee parchment is degraded by enzymes. The mucilage is subsequently washed off to

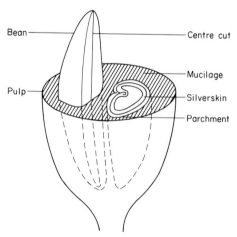

Fig. 1. Partial cross-section of coffee cherry.

leave parchment coffee which is subjected to a drying regime to obtain a moisture content of 10–11%. Coffee fermentation accomplishes two important objectives. It removes the sticky mucilage layer allowing for quick drying of the parchment coffee, and improves the raw appearance of the beans. Both of these have a direct bearing on the final coffee-cup quality (Wootton, 1963).

II. METHODS OF REMOVAL OF COFFEE MUCILAGE

A. Natural Fermentation

Coffee fermentation has been described as the removal of mucilage from parchment coffee effected by enzymes. Natural fermentation refers to the process of mucilage removal by enzymes naturally occurring in the coffee cherry and/or elaborated by the natural microflora picked by the coffee from the environment (Frank et al., 1965; Hall, 1967). Pulped coffee is placed in concrete or wooden tanks and left to ferment under water or with constant drainage of water and mucilage liquors. The latter process known as 'dry fermentation' is preferred. Under-water fermentation is slower and results in a higher production of volatile acids which may result in tainting of the final coffee beverage (Wootton, 1963). The natural fermentation takes anything from 20 to 100 hours. The duration of fermentation varies with the stage of ripeness, temperature, pH value, concentration of ions, coffee variety, microflora population and aeration. It has been demonstrated that lowering the temperature and pH value retards the rate of fermentation (Wootton, 1963) and that aerobic fermentations are faster than anaerobic fermentations (Menchu and Rolz, 1973). It would be expected that the availability of oxygen under water is restricted by the amount that can dissolve in the water at any given time.

There are various factory practices which are calculated to increase the rate of fermentation. These include dry feeding of pulped coffee into fermentation tanks, and using recirculated water which is rich in enzymes. Sophisticated factories aerate or use other additives which enhance enzyme activity. The addition of lime to provide calcium ions which activate specific enzymes is a good example (Haarer, 1962). After the mucilage has been degraded, parchment coffee is washed and graded by water in concrete canals (Sivetz and Foote, 1963).

In East Africa, Wootton (1963) recommended a two-stage fermentation procedure which includes a quick-dry fermentation stage, washing off the mucilage, followed by a 24-hour under-water soak. The advantages of this procedure were shown to be improvement of raw-bean appearance through outward diffusion of undesirable browning compounds from the beans, specifically from the centre cut and the silver skin. Other workers have asserted that the whole fermentation procedure does not improve quality but rather leads to unnecessary and expensive weight losses of up to 3% (Bannell, 1954). Coffee fermented under water or processed by the two-stage fermentation procedure tends to deteriorate in quality during drying because of the preponderance of cracked parchment. This may be avoided by subsequent carefully controlled drying (Kulaba, 1978/1979).

Natural fermentation of coffee has to be carefully controlled, otherwise off-flavours can develop and be reflected in the final liquor quality. Wootton (1960) has reported that onion flavour develops in coffee as a result of the production of propionic acid. He concluded that production of propionic and butyric acids during the final stages of fermentation is greater during an under-water fermentation and is also dependent on a heavy initial washing before fermentation. Northmore (1968) has indicated that the incidence of an off-flavour referred to as 'stinkers' may be associated with high temperatures obtainable during fermentation. Gibson et al. (1971/1972) have shown that 'sourness' and 'stinkers' in coffee are caused during fermentation under anaerobic conditions created by high proportions of reducing agents in the fermentation waters. Off-flavours in coffee are caused by various factors which need proper investigation based on a correct understanding of the biochemistry involved in the fermentation process. This has led to the introduction of various methods of coffee processing which are not dependent on natural fermentation and are therefore easier to control. However, the delicate nature of the coffee-bean tissue defies any attempts to rid it completely of off-flavours as detected by a subjective human palate.

B. Commercial Enzymes

Several commercial enzymes are now available for coffee fermentation. The earliest one was marketed under the trade name Benefax (Sivetz and Foote, 1963). Later brands have included Pectozyme, Cofepec

(Brownbridge and Wootton, 1966), and Ultrazym (Arunga, 1973). These are mould-enzyme preparations with appropriate inert fillers. The commercial enzymes are generally mixtures of pectic enzymes but may contain hemicellulases and cellulases. Because of financial constraints, these enzymes have not been widely used. Most factories have kept the commercial enzymes for use during peak periods or when natural fermentations are slow. Conditions created by overproduction and slow fermentations usually upset the smooth running of a factory. Congestion can occur either in fermentation and soaking tanks or on drying tables. These conditions affect coffee quality adversely because of the concomitant physiological activities by wild micro-organisms and those in the bean. Since commercial enzymes are applied by mixing with coffee in fermentation tanks, only a little saving of space is afforded by their use in the normal factory routine.

C. Mechanical Demucilaging

Several mechanical machines have been developed for demucilaging coffee beans. The Raoeng or Aquapulper (Sivetz and Foote, 1963) is a combined pulper and demucilager. It has been reported that it consumes a lot of power and water (Wootton, 1970). Sivetz and Foote (1963) have, however, indicated that if it is used only as a demucilager the power requirements are decreased. Recent trials with Aquapulper (Kamau, 1978) have confirmed these observations and recommended further investigations. The Hess Coffee Washer was patented by Hess and Hato (1955) as a low-pressure low-power machine for mucilage removal from coffee beans. A simple machine designed in Hawaii known as the HAES Demucilaging Machine (Fukunaga, 1957) can be operated easily with a two horse power motor and is claimed to eliminate most of the shortcomings of other mechanical devices. Wootton (1964) has tested this Hawaian machine and reports that it is difficult to adjust.

The drawbacks of the mechanical demucilaging machines are the high water and power consumptions, bean damage and possible adverse quality factors (Wootton, 1970). If mechanical demucilaging is followed by an under-water soak, then the quality is not adversely affected. This has to be balanced against weight losses, and space and time requirements of the natural fermentation procedure.

D. Chemical Demucilaging

Coffee mucilage can be removed chemically using alkalis, acids or warm water. These treatments essentially involve hydrolysis and depolymerization. Carbonell and Vilanova (1952) proposed the use of sodium hydroxide for mucilage removal from coffee beans. Sivetz and Foote (1963) have described a method proposed by P.B. Tester for demucilaging coffee using hot water which has the disadvantage of high fuel costs. Alkalis and acids require careful handling and appropriate disposal methods. These chemicals may therefore not find wide applicability in factories at the farm level.

III. BIOCHEMISTRY OF COFFEE FERMENTATION

A. Composition of Mucilage

The chemical and physical characteristics of coffee mucilage are basic to an understanding of coffee fermentation. Mucilage forms 20–25% (wet basis) of coffee cherry. Freshly pulped, ripe coffee cherry has a mucilage layer of 0.5–2.0 mm thickness. Coffee-mucilage thickness varies with fruit variety, stage of ripeness and environmental cultivation conditions. The mucilage is colourless but, on exposure to air, turns brownish as a result of oxidative enzymic reactions. It has an open cellular structure which on examination appears to consist of long cells with occasional broken cell walls.

Chemically, coffee mucilage consists of all of the higher plant cell materials including water, sugars, pectic substances, holocellulose, lipids and proteins. Some of the available data on coffee mucilage composition are given in Table 1. Several other workers have reported conflicting data (Coleman et al., 1955; Nadal, 1959; Wilbaux, 1963). These differences reflect problems associated with isolation procedures, different coffee varieties and stages of maturity. It is clear from these analyses that there are various components still to be accounted for. More attention has been directed towards components implicated in the fermentation process. The most important chemical components of mucilage are pectic substances together with carbohydrates and their breakdown products. These substances are expected to be found in

Table 1

Chemical composition (%), on a wet and dry basis, of coffee mucilage

Mucilage components	Chemical composition (%)
(a) Wet basis. From Menchu and Rolz (1973)	
Moisture	85.0
Total carbohydrates	7.0
Nitrogen	0.15
Acidity (as citric acid)	0.08
Alcohol-insoluble compounds	5.0
Pectin (as galacturonic acid)	2.6
(b) Dry basis. From Picado (1934)	
Pectic substances	33
Reducing sugars	30
Non-reducing sugars	20
Cellulose and ash	17

mucilage either in the soluble form or as insoluble complexes forming various cell structures.

The constituents of the cell wall, especially the middle lamella and the primary wall, must be considered of greatest importance in the breakdown of mucilage. Pectic substances and hemicellulose are the constituents of particular interest. Pectic substances are made up of chains of $1 \rightarrow 4$-linked α-galacturonic acid units with partially methylated carboxyl groups and acetylated hydroxy groups. Units of α-D-galacturonic acid (Pilnik and Voragen, 1970) are linked by glycosidic bonds (Fig. 2). Pectic substances are classified as protopectin, pectinic acids, pectin and pectic acids. The chemistry of protopectin has been comprehensively reviewed by Joslyn (1962). Protopectin is the insoluble pectic substance consisting of completely esterified and cross-linked galacturonic acid chains bonded to other plant materials. Other pectic substances are classified according to the degree of

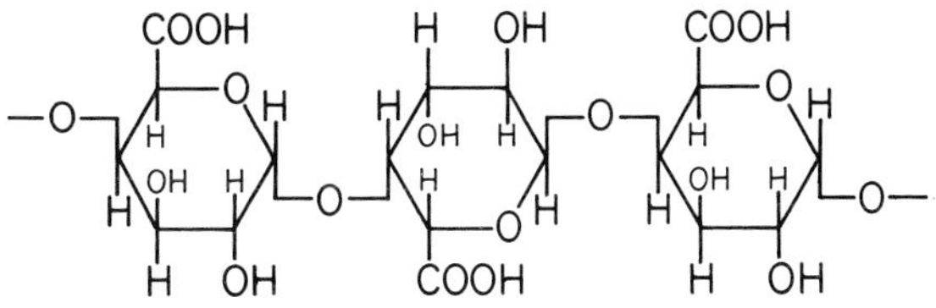

Fig. 2. Structure of polygalacturonic acid.

esterification and solubility, with pectic acids being essentially a soluble polygalacturonic acid free from methyl ester groups.

Wootton (1970) has postulated that the important component in the coffee fermentation is the insoluble fraction of mucilage, mainly the cell-wall and intercellular material characteristic of the parenchymatous cells of fruits. He confirmed that the middle lamella of coffee mucilage cells is primarily pectinic, and that the primary cell contains pectin and cellulose materials. The insoluble fraction of coffee mucilage is expected to consist mainly of pectic substances in close association with other cell-wall and intercellular materials including hemicelluloses and phospho- and galactolipids. Breakdown of this cellular material and its detachment from coffee parchment are the important biochemical processes in coffee fermentation. Although breakdown components of coffee mucilage are well known, the way in which they are linked to form the complex structure known as mucilage is not properly understood. Separation of mucilage from parchment and its subsequent fractionation invariably result in artifacts (Anderson et al., 1961).

B. Changes During Coffee Fermentation

When coffee is pulped and left in a dry heap or under water, fermentation occurs. After a period which takes anything from 20 to 100 hours, depending mainly on the environmental temperature, the mucilage detaches from the parchment and can be readily washed with water. On completion of fermentation, a few beans when rubbed in the hand feel gritty. Various chemical changes have been reported to occur during the process of fermentation. Arunga (1970/1971) studied, using conventional fractionation methods, the gross compositional changes that occur during fermentation (Table 2). Production of carboxylic acids which change the pH value of fermentation liquor from 5.9 to 4.0 have been confirmed (Wootton, 1970; Menchu and Rolz, 1973). Wootton (1960) demonstrated that acetic and lactic acids are produced early in coffee fermentation and that propionic and butyric acids are elaborated later. Menchu and Rolz (1973), however, observed that acetic and propionic acids appear early in the process and confirmed the appearance of butyric acid at the end of fermentation. Wootton (1960) has reported that there is a close positive correlation between the appearance of propionic acid in the fermentation stage and the

Table 2

The composition of coffee mucilage before and after complete fermentation. From Arunga (1970/1971).

Component	Percentage and dry basis	
	Before fermentation	After fermentation
Water-soluble	35.3	50.7
Lipid	6.0	4.0
Pectin	47.0	36.2
Holocellulose	9.4	8.0
Unaccounted	2.3	1.1

incidence of 'onion' flavour in coffee beverage. These carboxylic acids are produced through degradation of sugars by micro-organisms. Loew (1907) explained the fermentation as the result of yeast attack on mucilage leading to production of alcohol and acetic acid.

Menchu and Rolz (1973) have given quantitative data to show that ethanol is one of the products of coffee fermentation. The evolution of hydrogen and carbon dioxide has been demonstrated during both dry and under-water fermentations by Hall (1967/1968) who further postulated that hydrogen is produced through breakdown of sugars by bacteria of the coliform group. *Escherichia coli* metabolizes glucose by a mixed acid fermentation at pH 7.8:

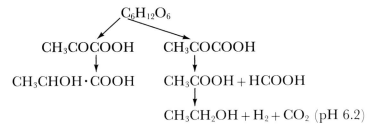

Aerobacter aerogenes gives a lower yield of mixed acids, particularly of lactic acid, because some pyruvic acid is converted into acetylmethyl-carbinol and butanediol (Hall, 1967/1968).

Several investigators have confirmed the presence of reducing and non-reducing sugars in soluble mucilage fractions after complete fermentation (Wootton, 1970; Wilbaux, 1963). Some of the sugars forming part of the structure of mucilage are arabinose, xylose, galactose, fructose and glucose. Wootton (1970) suggested that arabinose, xylose and galactose are part of the insoluble structure of mucilage. The soluble sugars form an excellent medium for growth of micro-organisms. A lipid fraction isolated by Arunga (1970/1971) from fermented mucilage when analysed by thin-layer chromatographic methods indicated the presence of an esterified sterol glycoside. It was suggested that, since pectic acids with four or fewer galacturonic acid units are not found in natural fermentation liquors (Wootton, 1970, Arunga, 1970/1971), mucilage degradation may involve breakages in cross-linkages which may implicate lipids and hemicellulose materials.

Although Case (1935) reported that there is practically no temperature rise during the course of the fermentation, it is most likely that this process is exothermic. Temperature rises of 2–8°C under certain experimental conditions have been observed by Northmore (1968). In a factory where heat loss is at a maximum, this phenomenon is not easily detectable.

Changes in the quality of the coffee bean are fundamental to the continued practice of naturally fermenting coffee. In the two-stage fermentation developed by Wootton (1963) in East Africa, it is reported that the raw-bean quality improves and that this improvement is reflected in the roast and final beverage quality. The improvement in raw appearance is dependent on diffusion of various compounds from the bean which also result in weight losses. While Wootton (1970) claims a weight loss of the order of 1.0–1.5%, other workers (Carbonell and Vilanova, 1952; Bannell, 1954; Sivetz and Foote, 1963) have reported losses ranging from 3 to 12%. The higher weight losses were observed in under-water fermentations. This magnitude of loss would make fermentation a very expensive exercise thus nullifying gains in raw-bean quality. Despite these observations, natural fermentation of *Coffea arabica* is the preferred dimucilaging method.

IV. MICROBIOLOGY OF COFFEE FERMENTATION

It has been repeatedly shown by various workers that breakdown of mucilage during fermentation initially occurs as a result of natural enzymes in the coffee cherry but that the major factors in natural fermentations are the extracellular enzymes elaborated by micro-organisms. Since mucilage contains simple sugars, polysaccharides, minerals, protein and lipids, it forms a good medium for microbial growth. Micro-organisms to be expected in the pulped coffee include the natural flora consisting of plant pathogens, and microbial populations from the surface of fruit, the soil and pulping water. It is therefore not surprising that many investigators have reported the occurrence of a number of bacteria, moulds and yeasts normally associated with the appropriate environments.

The presence of bacteria in fermenting coffee was observed by several early workers including Beckley (1930), and Pederson and Breed (1946). They mainly implicated lactic-acid producing bacteria of the genera *Leuconostoc* and *Lactobacillus*. Working with Brazilian coffee cherries, Vaughn *et al.* (1958) isolated coliform bacteria resembling species of the genera *Aerobacter* and *Escherichia*. Pectinolytic species in the genus *Bacillus* were also isolated from fermenting coffee. Although bacteria of the genus *Leuconostoc* and *Streptococcus* have been isolated from fermenting coffee, they have not shown the ability to degrade pectic enzymes (Van Pee and Castelein, 1972). Frank and Dela Cruz (1964), Frank *et al.* (1965) and Hall (1967) studied bacterial isolates from fermenting coffee in some detail using highly sophisticated microbiological methods. These workers identified bacterial cultures whose principal characteristics were outlined as pectolytic, Gram-negative, non-spore forming and lactose-fermenting. More recently, Van Pee and Castelein (1972), working with Congo coffee, isolated bacteria belonging to the family Entrobacteriaceae very similar to those isolated from fermenting coffee in Brazil (Vaughn *et al.*, 1958), Hawaii (Frank *et al.*, 1965) and Kenya (Hall, 1967). These bacteria were reported to resemble closely *Erwinia dissolvens* and may also correspond to *Erwinia atroseptica*.

Beginning with the early work of Loew (1907), yeasts were reported in fermenting coffee but have not demonstrated the ability to degrade pectin (Hall, 1967; Van Pee and Castelein, 1970). Agate and Bhat (1966) on the other hand isolated mucilage-degrading yeasts from the surface of *Coffea canephora*. Mould enzymes are known to speed up mucilage breakdown. Fungi of the genera *Aspergillus*, *Fusarium* and *Penicillium* were isolated by Vaughn *et al.* (1958) from depulped coffee.

Coffee naturally fermenting will pick up a number of micro-organisms from the environment which might be of no particular significance in the fermentation process. The importance of pectic substances in cell-wall structure leads to the hypothesis that pectinolytic micro-organisms may be of particular significance in the fermentation of coffee. These will include micro-organisms that have been isolated from coffee fermentations and shown to be pectinolytic belonging to the genera *Bacillus, Erwinia, Aspergillus, Penicillium* and *Fusarium*. Members of the fungal genera are used commercially for production of pectic enzymes but have not been detected in many coffee fermentations. Members of the genus *Erwinia* are suspected to be the important micro-organisms in production of extracellular pectic enzymes which break down coffee mucilage.

The genus *Erwinia*, belonging to the family Enterobacteriaceae, consists of 15 species and are characterized as pathogens which invade the tissues of living plants and produce dry necroses, galls, wilts and soft rots. Soft rot is caused by pectic enzyme attack of the middle lamella (Breed *et al.*, 1957). The genus is heterogeneous and is composed of *Erwinia sensu stricto* which does not produce visible gas from sugars and another group, for which the generic name *Pectobacterium* has been suggested, and which produces visible gas from sugars and usually causes soft rot. The type species of the genus *Erwinia*, recognized in Bergey's Manual of Determinative Bacteriology (Breed *et al.*, 1957) as *Erwinia amylovora*, liquefies gelatin but does not cause rot. Of particular significance to coffee fermentations are species of the second group, especially *Erwinia dissolvens* and *Erwinia atroseptica*. Van Pee and Castelein (1972) concluded that the bacterial isolates important in coffee fermentation belong to the species *E. dissolvens*.

Erwinia dissolvens is considered atypical in the genus *Erwinia*. The species belongs to the group of pathogens that cause soft rot in plants, produce gas from sugars, do not liquefy gelatin, and hydrolyse starch. It has been suggested that they be placed in the genus *Aerobacter*. *Erwinia dissolvens* is a Gram-negative rod, occurring in pairs, non-motile, with an optimum growth temperature of 30°C. Bacterial isolates from coffee closely correspond to *E. dissolvens* although Van Pee and Castelein (1972) indicated that their isolates did not produce indole; moreover, all of the reported coffee bacterial isolates are non-spore formers.

The complexity of the fermentation process defies any firm conclusions as to the specific micro-organism responsible for the whole phenomenon. It is more likely that a number of micro-organisms contribute enzymes which degrade specific components of mucilage.

V. ENZYMES IN COFFEE FERMENTATION

It is now widely accepted that enzymes from the coffee cherry play a minor role in natural coffee fermentation contrary to the views of

Beckley (1930) and Case (1935). The hypothesis that coffee fermentation is effected by enzymes elaborated mainly by bacteria is supported by several studies (Carbonell and Vilanova, 1952; Vaughn *et al.*, 1958; Hall, 1967; Arunga, 1970/1971; Van Pee and Castelein, 1972). Hall (1967) in particular has demonstrated that there is a significant increase in pectic-enzyme concentration during the course of the fermentation which can be attributed to micro-organisms, whose numbers were also observed to increase.

On account of the importance of pectic substances in the structure of the middle lamella and the primary wall, pectic enzymes elaborated by bacteria would appear to be of particular significance in the coffee fermentation process. Some of these enzymes have been detected in coffee fermentations. One of these is pectinesterase, a saponifying enzyme that splits the methyl ester group of pectin and pectinic acids:

$$\text{Pectin} + H_2O \rightarrow \text{pectic acid} + CH_3OH$$

This enzyme was detected in fermenting coffee but appears to be a coffee-cherry enzyme (Hall, 1967). Pectinesterase occurs in several fruits including currants, pears, apples and tomatoes (Pilnik and Voragen, 1970). Hall (1967) extracted pectinesterase from mucilage using the methods of Hobson (1963) and assayed it by titrating with sodium hydroxide, thus measuring the increase in free carboxyl groups. During these studies, it was observed that the activity of the enzyme increased during the course of fermentation suggesting that some of the enzyme may be elaborated by micro-organisms. Increase of pectinesterase activity during maturation and ripening of tomatoes has been reported by Hobson (1963, 1964). It is, however, important to note that it has been observed that plant pectinesterases have pH optima between 7 and 8 (Hultin and Levine, 1963), while pectinesterases from moulds and fungi have pH optima from 4 to 5. The latter pH range may be more relevant to the coffee fermentation. Pectinesterase is inactivated by heating coffee beans at 95°C for 20 minutes (Hall, 1967).

Polymethylgalacturonase and polygalacturonase are depolymerizing enzymes which attack pectin and pectic acid respectively by splitting the α-(1→4) glycosidic bonds. They may be further classified as endo- or exo- to designate random or terminal cleavaging enzymes respectively. The depolymerizing enzymes are mainly elaborated by yeasts, moulds and bacteria, but are also found in plants such as tomatoes, peaches and pears (McCready and McComb, 1954; Hobson, 1962). Bacterial isolates from coffee fermentations have been shown to elaborate depolymerizing enzymes (Hall, 1967; Van Pee and Castelein, 1972). The enzymes were similarly detected in fermentation liquors (Arunga, 1970/1971). Bacterial and fruit depolymerizing enzymes have pH optima between 3.0 and 5.5 which is within the coffee-fermentation liquor range. Depolymerase activity can be detected in coffee-fermentation liquors and bacterial cultures by measuring viscosity and estimating the increase in reducing groups. These assay methods are also applicable to the transeliminases.

Pectin transeliminase and pectic acid transeliminase which depolymarize pectin and pectic acid, respectively, with formation of a double bond can also attack randomly or terminally (Fig. 3). Hall (1967) has demonstrated transeliminative activity in some pure bacterial cultures isolated from coffee fermentations. This activity has not been detected in coffee-fermentation liquors (Arunga, 1970/1971). Transeliminases are not expected to play an important role in coffee fermentation because their pH optimum is much higher than that obtainable during fermentation.

During the coffee fermentation, arabinose is constantly produced (Wootton, 1970). Hall (1967) demonstrated the presence of the enzyme α-L-arabinofuranosidase in extracts of fermenting coffee, and concluded that this is a coffee-cherry enzyme which may play some role in fermentation.

A careful study of the structure of plant tissues, especially of the primary cell wall and middle lamella, will indicate that several enzymes probably play a role in coffee fermentation by degrading various mucilage components including pectic substances and hemicelluloses. Arunga (1970/1971) observed that commercial pectic enzymes and hemicellulases rapidly degraded coffee mucilage. Any other enzyme

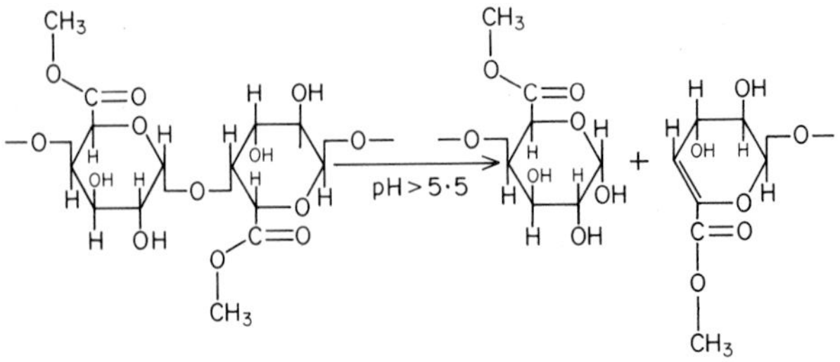

Fig. 3. Transeliminative cleavage of glycosidic linkages by pectin transeliminase.

attacking other components may be implicated in this complex process. The important enzymes from this analysis appear to be pectinesterase, endo- and exopolymethylgalacturonase, together with endo- and exopolygalacturonase. The role of these pectic enzymes in the structural weakening of cell wall has been conclusively investigated by Hobson (1964).

The elaboration of a protopectinase or a macerase which solubilizes protopectin has been proposed, but it is now clear that this activity is performed by a mixture of enzymes including pectinesterase and the depolymerizing enzymes (Pilnik and Voragen, 1970). The supposition that cell-wall material is composed of homopolysaccharides is not supported by present knowledge (Isherwood, 1970). The cell wall and middle lamella, it appears, are composed of complex heteropolymers which on hydrolysis will give different monosaccharides. In coffee mucilage, this hypothesis is supported by the varied breakdown products of the insoluble portion which usually include galacturonic acid, arabinose and, occasionally, galactose.

REFERENCES

Agate, A.D. and Bhat, J.V. (1966). *Applied Microbiology* **14,** 256.
Anderson, D.M.W., Bews, A.M., Garbutt, S. and King, J. (1961). *Journal of the Chemical Society*, 5230.
Arunga, R.O. (1970/1971). East African Industrial Research Organization, Annual Report, 69.
Arunga, R.O. (1973). *Kenya Coffee* **38,** 354.
Bannell, M. (1954). *Tea and Coffee Trade Journal* **106,** 34.
Beckley, V.A. (1930). Department of Agriculture, Kenya Bulletin No 8.
Breed, R.S., Murray, E.G.D. and Smith, N.R. (1957). Bergey's Manual of Determinative Bacteriology', 7th edition. The Williams and Wilkins Company, Baltimore.
Brownbridge, J.M.C. and Wootton, A.E. (1966). *Kenya Coffee* **31,** 253.
Carbonell, R.J. and Vilanova, M.T. (1952). Technical Bulletin No 13. Ministerio de Agricultura y Ganaderia Centro Nacional de Agronomia, Santa Tecla, El Salvador.
Case, E.M. (1935). Coffee Board of Kenya Monthly Bulletin No. 7.
Coleman, R.J., Lenney, J.F., Coscia, A.T. and Dicarlo, F.J. (1955). *Archives of Biochemistry and Biophysics* **59,** 127.
Frank, H.A. and Dela Cruz, A.S. (1964). *Food Science* **29,** 850.
Frank, H.A., Lum, N.A. and Dela Cruz, A.S. (1965). *Applied Microbiology* **13,** 201.
Fukunaga, E.T. (1957). Hawaii Agricultural Experiment Station, Bulletin 115.
Gibson, A., Arunga, R.O. and Butty, M. (1971/1972). East African Industrial Research Organization, Annual Report, p. 55.
Haarer, A.E. (1962). 'Modern Coffee Production', 2nd edition, pp. 492. Leonard Hill (Books) Ltd., London.

Hall, A.N. (1967). East African Coffee Processing Research Committee Report, 29.
Hall, A.N. (1967/1968). East African Industrial Research Organization, Annual Report, 27.
Hess, H.L. and Hato, R. (1955). United States Patent 2,722,226.
Hobson G.E. (1962). *Nature, London* **195**, 195.
Hobson, G.E. (1963). *Biochemical Journal* **88**, 358.
Hobson, G.E. (1964). *Biochemical Journal* **92**, 324.
Hultin, H.O. and Levine, A.S. (1963). *Archives of Biochemistry and Biophysics* **101**, 396.
Isherwood, F.A. (1970). In 'The Biochemistry of Fruits and their Products' (A.C. Hulme, ed.), vol. 1, pp. 33–52. Academic Press, London.
Joslyn, M.A. (1962). In 'Advances in Food Research' (C.O. Chichester and E.M. Mark, eds.), vol. II, pp. 1–107. Academic Press, New York.
Kamau, I.N. (1978). *Kenya Coffee* **43**, 383.
Kulaba, G.W. (1978/1979). Kenya Industrial Research and Development Institute, Annual Report, 39.
Loew, O. (1907). The Fermentation of Cacao, Annual Report, Puerto Rico Agricultural Experimental Station, 41–52.
McCready, R.M. and McComb, E.A. (1954). *Journal of Food Science* **19**, 530.
Menchu, J.F. and Rolz, C. (1973). *The Cafe Cacao* **17**, 53.
Nadal, N.G. (1959). *Coffee and Tea Industries* **82**, 17.
Northmore, J.M. (1968). East African Coffee Processing Research Committee Report, October.
Pederson, C.S. and Breed, R.S. (1946). *Food Research* **11**, 99.
Picardo, C. (1934). *Anquivos Instituto Biologica Vegetables* **2**, 67.
Pilnik, W. and Voragen, A.G.J. (1970). In 'The Biochemistry of Fruits and their Products' (A.C. Hulme, ed.), vol. 1, pp. 53–87. Academic Press, London.
Sivetz, M. and Foote, H.E. (1963). 'Coffee Processing Technology,' vol. 1, 71 pp. The Avi Publishing Company, Inc. Westport, Connecticut.
Van Pee, W. and Castelein, J.M. (1970). *East African Agricultural Journal* **36**, 308.
Van Pee, W. and Castelein, J.M. (1972). *Journal of Food Science* **37**, 171.
Vaughn, R.H., De Camango, R., Fallanghe, H., Mello-Ayres, G. and Sarzedello, A. (1958). *Food Technology, Supplement 4*, **12**, 57.
Wilbaux, R. (1963). UN/FAO Informal Working Bulletin, No. 20.
Wootton, A.E. (1960). East African Industrial Research Organization, Report, C.R. 11.
Wootton, A.E. (1963). East African Industrial Research Organization Report C.R. 12.
Wootton, A.E. (1964). Coffee Processing Sub-Committee, Coffee Board of Kenya.
Wootton, A.E. (1970). East African Industrial Research Organization, Report C.R. 14.

9. Cocoa

J. G. CARR

Long Ashton Research Station, University of Bristol, Long Ashton, Bristol, England

I. INTRODUCTION

Cocoa and chocolate are made from the seeds of *Theobroma cacao*, and this plant was cultivated by the Maya Indians of central America long before the Spanish invasion of 1519 (Urquhart, 1955). It was used by the Mayas and Aztecs as an item of food. It is difficult to see how these seeds became an article of diet since their flavour in the raw untreated state is not altogether pleasant, being rather bitter and astringent. This persists right through fermenting and drying. The Mayas made a drink from the beans pounded with maize and with added capsicum or other spices, which was not much liked by the Spaniards. This rather thick beverage was called *chocolatl* (Wood, 1975). However, it was they who first added sugar and used the mixture to prepare a drink. It is interesting to note that, in the majority of cocoa products available today, the addition of sugar, started by the Spaniards several centuries ago, persists.

Cocoa did not become established in Europe until the end of the Sixteenth Century, by which time the Spaniards were exporting it from Central America. It was not until later that cocoa was exported from Venezuela by the Dutch. Originally, cocoa drinking was restricted to Spain but, due to the activities of the Dutch, this habit spread amongst the wealthy of Europe and it was consumed in many royal courts, including that of Charles II.

Cocoa was introduced into Trinidad fairly soon after its discovery by the Spaniards and from there it spread to other parts of the Caribbean where the climate is suitable for its cultivation. In response to growing demands, the plants were cultivated in the Philippines, the East Indies and Sri Lanka. Later, cocoa spread through Brazil and, in the early part of the Nineteenth Century, to the islands of São Tomé and Fernando Póo. Towards the end of the Nineteenth Century, these islands were the source of material for the spread of cocoa throughout West Africa. This area of the world, which includes Ghana, Nigeria, Ivory Coast, Cameroons and São Tomé, still includes the World's largest producers of cocoa, followed by Brazil and other South American countries. Smaller amounts of cocoa come from other places including Asia. There is, for example, an expanding cocoa-growing industry in Malaysia which has its own special problems which will be mentioned later.

The nomenclature applied to cocoa is somewhat confusing, but there are three main names which occur in the literature. These are Criollo, meaning native, Forasteros, meaning foreign and Trinitario, appertaining to Trinidad (Wood, 1975). The first of these is taken from the characteristics of the old Venezuelan cocoa population. Their pods are either red or yellow, deeply furrowed, usually with ten grooves, very warty and pointed. The seeds are almost round in section and contain within them white or pale-violet cotyledons. Forasteros, or Amazonian Forasteros as they are usually called, are distributed naturally through the basin of this river and its tributaries, but may now be found in other parts of Brazil, Ecuador and West Africa. Their pods are yellow when ripe with a fairly smooth surface and round ends. The seeds are flattened and the cotyledons are dark purple or nearly black. One offshoot of these Forasteros is the fairly homogeneous population of West Africa which is termed Amelonado. Trinitario not only refers to Trinidad cocoa but to populations where Trinidad cocoa or Forasteros have been introduced into a substantially Criollo population and where the resulting hybrids have supplanted the original populace. There is a further distinction made in cocoa, namely the difference between fine and ordinary. Since these are grading terms of commerce, the reader is referred to Urquhart (1955) for an explanation.

The origins of eating chocolate date back to the middle of the last century and arise from the invention of the cocoa press by Van Houten in 1828. This enabled the removal of cocoa butter, thus producing a less fatty cocoa powder for use as drinking chocolate. At the same time the residual cocoa butter was available to be added back to a mixture of roasted, powdered cocoa beans together with sugar to make solid eating chocolate as we know it today. The manufacture of chocolate and its subsequent handling form a separate story, and it is only intended to describe in this chapter the fate of the cocoa beans from the stage where the pods are hanging on the trees to the end of drying, i.e. the various stages carried out in the producer country.

II. BOTANY OF COCOA

Theobroma cacao L. (the generic name means food of the gods; Chittenden, 1951) is, when mature, a woody tree some 8–10 metres tall belonging to the family Sterculiaceae. It is the only species of some 22 that bears seeds suitable for manufacture of chocolate. The seeds, of which there may be some 20–50 within the pod, germinate fairly readily and are epigeal when they grow. Eventually they produce a strong tap root which may penetrate the soil to a considerable depth if it is not waterlogged. If wet conditions obtain, the tap root may not elongate more than 45 cm. In addition to the tap root, there are laterals which usually develop some 15–20 cm below the surface. Cuttings do not develop a tap root; instead, several leaders grow down and this anchors the plant quite adequately.

The cocoa tree is a shade plant, growing in its natural habitat below larger forest trees. When cultivated, it is normally grown beneath taller trees and, on some plantations, coconut palms are used for this purpose. The plant's mode of growth is somewhat unusual, since it will grow on a single stem to a height of 1–2 metres and then fan into several lateral branches, usually three to five in number, called a 'jorquette' (jorqueta– South American Spanish for a fork). It increases its height by development and growth of a bud just below the jorquette, which produces a vertical shoot called a 'chupon'. The process of fanning out and then growing upwards may be repeated three or four times. Development of the uppermost branches usually leads to the disappearance of those lower down the tree. The leaves on the two types of branches differ; the early ones on the chupon have long petioles and are symmetrical, whereas those on the laterals have shorter petioles and tend to be asymmetrical.

The flowers of cocoa are quite small but are often present in large numbers. They are termed 'cauliflorous' because they arise from the old wood of the trunk and main branches. The inflorescence arises in a leaf axil, but this is not very obvious. After some time, secondary thickening occurs forming a bulge on the trunk or branch called the cushion. It is from this that several flowers may emerge simultaneously. Like other tropical plants, the cocoa tree bears flowers all year round and crops twice a year with an interval of about six months between each fruiting. Since the flowers are borne throughout the year, some fruit are also to be found, but the majority are borne at two well defined periods. The flowers are hermaphrodite and, if not pollinated, abciss within 24 hours. As an average tree is estimated to bear some 10,000 flowers in a year, the wastage rate is very high since only 10–50 ever develop into mature fruit. The flower is small and regular, having the formula $K_{(5)}C_5A_{(5+5)}G_{(5)}$ (Fig. 1). Pollination is by insects although Amelonado, as grown in West Africa, is self-compatible, whereas most others are self-incompatible.

According to Urquhart (1955) the fruit is a berry, but Wood (1975) describes it as an indehiscent drupe. Whatever the botanical niceties of the situation may be, the fruit is a pod containing up to 50 seeds which are somewhat smaller than broad beans. They are covered by a white mucilage which is said to contain an antigermination factor. In addition, it contains acids and sugar and, if tasted, has quite a pleasant fruity flavour. The function of this substance is not immediately obvious but it could be to insulate the beans in the event of the pod dropping

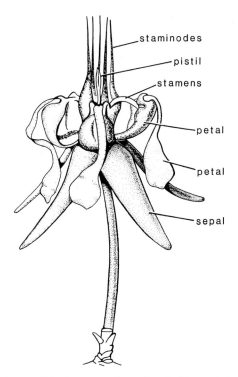

Fig. 1. Diagram of a flower of *Theobroma cacao*. Reproduced with permission from Wood (1975).

from the tree or it might simply be to hold each bean within its proper place inside the pod. Methods of harvesting vary in different parts of the world. For example, in West Africa it is customary to take pods off the tree and to transport them to where they are going to be fermented and then to extract the seeds. In contrast, bean gathering in Malaysia is done within the plantation. Pods are removed from the trees and the beans extracted and accumulated, placed in a box and transported back to the fermentary. The method of opening the pods varies from place to place. Often the pod is cut with a knife, care being taken not to damage any of the beans. The usual method, however, is not to cut the pod but to split it by banging it on a stone or a stake with a sharp point. By this method, the beans remain undamaged. The beans are extracted by hand, and it is probably this activity that inoculates them and causes the various microbial activities to occur. It is true to say that Malaysian beans extracted in the plantation ferment in the box in which they are transported to the fermentary. The initial inoculation with microbes may not occur from the hands of the harvesters. There are other agencies and these are the ever present fruit flies. In Malaysia, during the early days of cocoa bean fermentation, considerable difficulty was experienced in initiating fermentation and this was was eventually achieved by hanging bananas and other items attractive to fruit flies over the top of the fermentation boxes. No difficulty is encountered now, which presumably means that the boxes are covered with those microorganisms able to ferment cocoa beans.

III. METHODS OF FERMENTATION

There are three main methods of fermentation used in various parts of the World. The simplest is the one used in West Africa, which has the virtue of requiring no special apparatus and is therefore cheap. In this method, beans are piled up underneath plantain leaves which not only cover the surface but also the bottom of the pile (Fig. 2). To assist the sweatings to run away, the pile is made up over a series of radially arranged pieces of wood. The pile is kept together for six days during which time it is turned on the second and fourth days. This has the effect of making the aerobic parts anaerobic and *vice versa*. Piles vary in size and can be between two and four feet in diameter.

Another method that finds favour with many growers, and which is used extensively in South America, involves fermentation of beans in large hardwood boxes holding up to 1.5 tonnes. These either have slatted bottoms or holes in the sides and bottoms which have a two-fold function. They allow the sweatings to drain away and access of air. Often these boxes are stacked stepwise and have removable sides allowing easy transfer of beans to the box below. In this system, the first box often has twice the surface area of the other boxes and is half the depth. It is customary to place a covering over the surface of the beans, sometimes sacking and sometimes plantain leaves. The number of changes in this system is usually six and these take place once in 24 hours. Other box systems which are smaller are, perhaps, changed every 48 hours. A variation of the box system is to use stacked trays which give a series of thin layers of bean with plenty of air circulating between each layer.

Two other methods of bean fermentation, which are not used so

Fig. 2. Sampling a heap of cocoa beans. Note the plantain leaves ready to cover the surface.

frequently as those already described, are as follows. Beans may be placed in a plantain leaf-lined basket and left to ferment, or they may be placed in a hole in the ground. Both of these methods suffer from the disadvantage of low initial aeration and lack of drainage for the sweatings.

There are, in addition to the main methods of fermentation, many local variations. One of the variables is the length of time fermentation is allowed to proceed and, although six days has been stated as the usual duration, Roelofsen (1958) mentions two to twelve days. During the course of fermentation, the external appearance of the beans changes. Initially they are pinkish with a covering of white mucilage, but gradually they darken and the mucilage disappears. This colour change is oxidative for, when a heap is disarranged, the beans on the outside are darker than those on the inside. As the beans become mixed, their colour becomes a more uniform orange-brown and, towards the end of the fermentation, nearly all the mucilage has disappeared leaving the beans slightly sticky. At this stage they are ready for drying.

IV. DRYING

The aim of drying cocoa beans is to lower the moisture content from something over 50% to about 7–8%. This has the effect of preserving beans in storage from mould contamination. After drying, it may be necessary to store the beans for some time before shipping and, with only 7% moisture present, long periods of storage can be achieved. Although there are many ways to dry beans, these methods are either slow or fast. Beans dried slowly continue to undergo chemical changes initiated in fermentation until the moisture content is sufficiently low to halt them. In the quick-dried beans, these activities are rapidly stopped.

A. Sun Drying

Sun drying is carried out in those countries where the main crop is harvested during the dry season. Sun drying is widely employed in West Africa and the West Indies. All that is required is a large flat area on which the beans can be spread to a depth of about 5 cm, several operatives to rake them with wooden palettes, and a means of covering the drying beans rapidly when a sudden tropical storm occurs. The necessity for rapid action in the event of rain is met either by having a moveable roof or by spreading the beans on moveable trays that can be wheeled easily under a fixed roof. Even in fine weather, sun drying takes over a week and, in dull or wet weather, it may take a fortnight to reach the required moisture content. This method of drying has the one advantage that extraneous pieces of plant material and misshapen beans can be removed during the process.

B. Artificial Drying

The objective of all forms of artificial drying is to remove excess moisture from the beans in as short time as possible. This is done by subjecting the beans to a stream of hot air. Wood (1975) lists some seven different types of dryers namely, simple, platform, bin, rotary, lister extractor, tunnel and buthner. Simple dryers are usually fuelled by wood and, if the smoke leaks into the beans, it can give rise to a taint known as smoky cocoa. The rest of the dryers are mechanical, producing hot air, usually by burning oil, and often having devices for turning the beans to ensure even drying. For details of various dryers, the reader is referred to Wood (1975). Generally, the total drying time is about 48 hours. Mechanical dryers have a number of advantages. They have a large capacity, the control can be quite precise and they are less labour intensive than sun drying. Unwanted material has to be removed by hand. When this is done, the beans are bagged in jute bags and are then ready for storage prior to despatch to the chocolate manufacturers.

V. JUDGEMENT OF QUALITY

The final test of quality for any food product is whether it is pleasant in flavour and acceptable to the consumer. However, those responsible for the purchase of cocoa must have some scale of quality, and one of the tests applied is visual and called the 'cut test'. In this, 100 beans are removed at random from a sample and cut down the dividing line between the two cotyledons (at this stage referred to as the nib by manufacturers). There are small frames containing 100 compartments to make visual appraisal easier. The appraiser is attempting to estimate the percentage of beans that are fully fermented, partially fermented or not fermented at all. These are judged by the colour of the cotyledon. At the same time, the size of the beans and the thickness of the shell (dried testa) are also judged. Fully fermented Forasteros, the beans that have purple cotyledons, usually end up very dark brown, whereas the Criollo beans with the white cotyledons have a paler brown nib when fully fermented. At this stage, the nib is very hard and friable; it has a faint chocolate aroma and flavour, but the strong bitterness and astringency still dominate the taste.

VI. MICROBIOLOGICAL ACTIVITIES IN FERMENTING COCOA

So far, this chapter has been devoted to what happens as the beans develop, are harvested, fermented and dried. Nothing has yet been said about the microbiological and chemical changes that take place during fermentation and drying. Much of this information is based on research carried out in Ghana and Malaysia by the author and his colleagues on beans classified as Upper Amazon Hybrids which, having purple cotyledons, would be regarded as a form of Forastero (Carr *et al.*, 1979).

A. Yeasts

The first major microbial activity in cocoa fermentation, whether in a heap or box, is an alcoholic fermentation which can easily be detected by the aroma of ethanol. In heaps examined in Ghana, yeasts were present in fairly high numbers at the first examination and four hours later. After this time, they initially disappeared and were only detected sporadically until the experiments were ended after some 144 hours. In contrast, lactic-acid bacteria persisted right until the end of the experiment. The position from which samples were taken indicated that, for the brief time the yeasts were present, they were fairly evenly distributed from top to bottom. Lactic-acid bacteria, however, tended to occur more frequently in the middle and lower parts of the heap than in the upper part. In the Malaysian boxes, the yeasts tended to be less numerous in the early stages compared with other organisms present and sometimes they persisted rather longer than in the African heaps. Numbers indicated that fermenting yeasts were fairly evenly distributed throughout the boxes in the early stages. Non-fermenting yeasts could be isolated in very small numbers at the beginning and towards the end of the fermentation. The behaviour of the lactic-acid bacteria was similar in the Malaysian boxes and the African heaps.

The primary activity of the fermenting yeasts is dissimilation of sucrose, glucose and fructose to ethanol. Other activities are somewhat more speculative, e.g. breakdown of pectin in the mucilage to produce the watery sweatings (Dechau and Emeis, 1975). It is now known that some fermenting yeasts, such as *Kluyveromyces fragilis* (formerly *Saccharomyces*), can break down pectin. However, there are other activities, such as metabolism of organic acids, that may also occur when the yeasts are fermenting, but again, these have never been investigated fully, although Roelofsen (1958) describes some experiments in which it was shown that yeasts could dissimilate citric acid and cause maceration of the bean mucilage. Knapp (1937) makes the point that processes associated with fermentation produce energy as follows:

$$C_{12}H_{22}O_{11} + H_2O \rightarrow 2C_6H_{12}O_6 + 18.8 \text{ kJ}$$

Sucrose   Glucose
                     or
                     fructose

$$C_6H_{12}O_6 \rightarrow 2C_2H_5OH + 2CO_2 + 93.3 \text{ kJ}$$

Glucose or fructose   Ethanol

These activities raise the temperature but are not the main cause, as will be seen later. Table 1 shows the range of yeasts found in the heaps of Africa and boxes of Malaysia. According to Ciferri (1931) the following species were identified, although not necessarily given these names at that time: *Schizosaccharomyces*, *Saccharomyces*, *Hansenula*, *Debaryomyces* or *Candida* and *Kloeckera* spp. Similarly, Roelofsen (1958) lists *Candida*, *Torulopsis*, *Pichia*, *Saccharomyces*, *Kloeckera* and *Hansenula* spp. From these comparisons of others' results with our own, it would seem that the same type of yeasts can be found in fermenting cocoa beans wherever they are processed.

Table 1

Yeasts isolated from various cocoas

Name	Africa	Malaysia	Ability to ferment*
Hansenula spp.	+	+	+
Kloeckera spp.	+	+	+
Torulopsis spp.	+	+	±
Saccharomyces spp.	+	+	+
Candida spp.	+	+	±
Pichia spp.	+	−	W
Schizosaccharomyces spp.	+	−	+
Saccharomycopsis spp.	+	−	±
Rhodotorula spp.	−	+	−
Debaryomyces spp.	−	+	W
Hanseniospora spp.	−	+	+

* + indicates a positive reaction, ± some species positive, − negative, and W weak.

B. Lactic-Acid Bacteria

It is interesting to note that, according to Roelofsen (1958), the presence of lactic-acid bacteria was not proven until 1935 and then not confirmed until the work of Rombouts (1952). Indeed it is strange that this should be so since lactic-acid bacteria have been known since the pioneering work of Louis Pasteur. It might, perhaps, have been the type of medium used since Roelofsen (1958) refers to their cultivation aerobically on yeast-agar plates with 2% glucose and 2% calcium carbonate added. Such incubation conditions will certainly not lead to growth of lactic-acid bacteria very well, nor do many of them from acidic situations grow well, or even at all, at the pH value that 2% calcium carbonate would create. It is interesting to note that neither Knapp (1937) nor Hoynak *et al.* (1941) actually mentions lactic-acid bacteria, and even Roelofsen (1958) only mentions that he found in the fermenting cocoa of Java species of *Betabacterium* (Orla Jensen) but no *Betacoccus* spp. This, in part, confirms our findings since we did not find any betacocci, or, as they are now called, members of the genus *Leuconostoc*. This may be explained by the fact that these organisms have a low optimum growth temperature and would not be favoured by the elevated temperatures of fermenting cocoa beans. We not only found members of the genus *Betabacterium* (Orla Jensen, 1919) but also members of the genus *Streptobacterium* (Orla Jensen, 1919) in the fermenting beans of Africa and Malaysia. These, in modern terminology, correspond, respectively, to the hetero- and homofermentative lactobacilli, thus called because the former produce not only lactic acid but also acetic acid, carbon dioxide and ethanol from metabolism of glucose, whereas the latter produce almost exclusively lactic acid from the same substrate. The organisms found in African cocoa belonged to the ubiquitous species *Lactobacillus plantarum* and *L. fermentum* together with the two species described by Carr and Davies (1970, 1972, 1974) namely, *L. mali* and *L. collinoides*. Of these, *L. plantarum* and *L. mali* are homofermentative whereas *L. fermentum* and *L. collinoides* are heterofermentative. In addition to the four named species there were a few unidentified heterofermentative lactobacilli. In the Malaysian beans there were *Lactobacillus plantarum* and *L. collinoides* and, in addition, several unidentified homofermentative organisms. During an extensive investigation into the micro-organisms in Trinidad cocoa beans, Ostovar and Keeney (1973) found the following species in fermenting beans: *Lactobacillus acidophilus*, *L. bulgaricus*, *L. casei*, *L. fermentum*, *L. lactis*, *L. plantarum*. In addition, they found *Leuconostoc*, *Pediococcus* and *Streptococcus* spp., species that appeared not to occur in fermenting beans of Africa and Malaysia.

Production of lactic acid from hexose sugars by lactic-acid bacteria also adds to the energy which is lost as heat and helps to raise the temperature of the beans. This is not the only substrate that these organisms can metabolize. For example, most can ferment a wide range of sugars including pentoses. Most are able to attack malic and citric acids with production of end products such as lactic acid, acetic acid and carbon dioxide, as described by Whiting and Coggins (1964). All lactic-acid bacteria from Africa and Malaysia were tested for their ability to metabolize malic and citric acids and nearly all the strains could attack both acids. Dissimilation of these acids leads to an overall drop in acidity and rise in pH value. Thus it seems likely that this group of bacteria, which has been little studied in relation to cocoa, may contribute to the general drop in pulp acidity observed throughout the period of fermentation.

One other acid against which these were tested was quinic acid which of course is associated with so-called tannins in apples. Of the cocoa organisms tested, nearly all strains of *L. plantarum*, irrespective of country of origin, could attack this acid and bring about the changes described by Whiting (1975). This particular ability may be of no significance in cocoa. What it does mean, however, is that, if such compounds as chlorogenic acid, quinic acid and shikimic acid occur in cocoa, then there are organisms present able to alter their chemical constitution and, at the same time, perhaps modify flavour. Lactic-acid bacteria have a large number of dissimilatory and synthetic pathways, and it is possible that some of these are helping to modify cocoa in ways as yet unknown.

C. Acetic-Acid Bacteria

Without doubt the predominant aroma in any cocoa fermentary is that of acetic acid which rapidly follows the yeast fermentation. It is the substrate ethanol that enables these bacteria to bring about the following change:

$$C_2H_5OH + O_2 \rightarrow \quad CH_3COOH + H_2O + 496 \text{ kJ}$$

Ethanol $\qquad\qquad$ Acetic acid

It is an exothermic reaction producing a considerable amount of energy and is the activity mainly responsible for elevating the temperature, sometimes to $50°C$ and more. It is probably the presence of acetic acid combined with the elevated temperature that finally kills the fermenting yeasts and would explain why they can only be found in the early stages of bean fermentation. Whilst lactic acid concentration rarely exceeded 0.5% in our experiments in Africa and Malaysia, acetic acid almost invariably reached and was often higher than 2% in concentration.

The species of acetic-acid bacteria isolated from African beans were *Acetobacter ranæns*, *A. ascendens*, *A. xylinum*, *A. lavaniensis* and the closely related *Gluconobacter oxydans*. The first three were found throughout the fermentation, whereas the last-named species never survived more than the first 24 hours, probably because, with its inefficient metabolism (lack of a Krebs cycle), it could not compete with other oxidative bacteria. Surprisingly, Ostovar and Keeney (1973) isolated only one species of true acetobacter, namely *Acetobacter aceti*. In addition, they isolated what they called *A. suboxydans*, which is now known as *Gluconobacter oxydans*. Knapp (1937) reported that the known species of acetic-acid bacteria in cocoa were *Acetobacter xylinum*, *A. xylinoides*, *A. ascendens* and *A. orleanense*. It is interesting to note that two of the species correspond to those from Africa and Malaysia, whereas the other two names are no longer extant but have been absorbed into other species of acetic-acid bacteria. Roelofsen (1958) reported the presence of *Acetobacter rancens* and another strain, *A. melanogenum* which, like *A. suboxydans*, has now been grouped in the related genus *Gluconobacter* and is included in the species *G. oxydans*. In addition, he noted the presence of *A. xylinum* producing its cellulose membranes in the sweatings below the boxes.

The major end product of these bacteria is acetic acid which diffuses into the cotyledons and whose flavour and aroma are still apparent even after drying. This is said to be an essential part of the curing process. Acetification is the main activity of these bacteria, but is not the only energy-gaining mechanism. They can, for example, metabolize acetic acid whereas others metabolize malic, citric, succinic and quinic acids and nearly all strains will metabolize lactic acid. In addition, all will produce gluconic acid from glucose and some produce various oxo-forms of this acid in addition. Thus, the biochemical activities of these bacteria are varied and it is not known to what extent this secondary metabolism of available substrates affects the flavour of the cocoa.

D. Other Bacteria

The only other bacteria occurring in significant numbers in the African and Malaysian cocoas were members of the genus *Bacillus*. These aerobic spore-forming rods were more prevalent in the African cocoas than in the Malaysian ones and were identified as *B. licheniformis* and *B. subtilis* by the methods of Gordon *et al.* (1973). Knapp (1937) mentions the occurrence of bacilli in West African cocoa and Ostovar and Keeney (1973) list *B. coagulans*, *B. pumilus* and *B. stearothermophilus* as occurring in the Trinidad sweat boxes they investigated. These organisms occur somewhat sporadically and in such small numbers as to render their biochemical activities insignificant among the much larger changes taking place throughout the period of curing.

Other organisms have been listed. Ostovar and Keeney (1973) mentioned, in addition to those already discussed, *Micrococcus*, *Pediococcus*, *Streptococcus*, *Propionibacterium* and *Zymomonas* spp. Perhaps the most unusual collection of bacteria associated with cocoa is that reported by Hoynak *et al.* (1941). These include *Flavobacterium*, *Achromobacter* and *Proteus* spp., no mention being made of acetic- or lactic-acid bacteria. These investigators question their own results, since the cocoa was transported as chilled pods for three weeks before investigation. By the use of these methods, they precluded the possibility of ever isolating the typical 'bean' flora which includes fermenting yeasts, lactic- and acetic-acid bacteria.

E. Moulds

When a pile of fermenting beans is observed, there is often a layer of greyish mould growing over the surface and sometimes it is possible to see mould growth at the corners of the sweat boxes. Knapp (1937) describes moulds at some length but does not give much of an explanation of their function. Roelofsen (1958) mentions that the most common species of moulds are *Aspergillus*, *Mucor*, *Penicillium* and *Rhizopus* spp., but suggests that they are of no importance in the normal process. Knapp (1937) does mention a musty off-flavour produced by growth of actinomycetes. It would seem that further studies are required on the effects these organisms have on cocoa.

VII. CHEMISTRY OF FERMENTATION

Like all plant materials, cocoa beans contain a multiplicity of compounds, many of which will be affected by the process of bean curing. One thing that happens to the bean during the course of fermentation is that it dies. Roelofsen (1958) concluded that the time of death would differ according to the criteria applied. If germination capacity was used as a criterion of viability, then the beans might be considered dead after two hours at 45°C, whereas if colour diffusion from the cotyledons is used, then after 24 hours at 45°C would be a truer estimate. In fact, what this means is that different parts of the bean die at different rates. Death usually occurs between 30 and 36 hours according to Roelofsen (1958). At one time it was thought that heat was the major factor killing the beans, but it is now known that acetic acid at a concentration of 1% in the bean is the main cause of death, and that heat, lactic acid and ethanol only enhance its effect.

Our own work (Carr *et al.*, 1979, 1980) shows that the pH value of the cotyledon dropped from pH 6.5 to 4.5 over a period of 110 hours and that during the same period acetic acid increased from nil to 2.00% and lactic acid from 0.01 to 0.22%. At the death of the bean, a process described as maceration takes place. This is a breakdown of cell integrity allowing enzymes and substrates to mix freely.

It is not intended to describe in detail the chemical changes that occur. The reader is, however, directed to the work of Rohan (1964), Rohan and Stewart (1964, 1966) and Quesnel (1972). Even in 1958, Roelofsen listed some 17 enzymes known to occur in cocoa beans and that number has increased since then. The possible substrates are carbohydrates, lipids, phenolics and amino acids (Lopez, 1972) and these produce the potential chocolate flavour and aroma compounds that only develop fully after the cured and dried beans have been roasted. Unlike some flavours and aromas, that of chocolate is not attributable to a single compound but to a combination of many in the right proportions.

VIII. ACIDITY IN COCOA

The ideal beans should end up, after fermentation and drying, with a nib pH value of 5.4 and this is the condition of most of the Ghanaian beans when ready for export. There are, however, certain cocoa-producing areas that do not achieve this state of affairs, notably certain parts of South America and Malaysia. Generally, the beans are at about pH 4.5 and the flavour of the nib or any chocolate made from it is somewhat acidic, but this is often accompanied by a fruity flavour. The reasons for this are not well understood, and it was thought that the method of fermentation might be a contributory factor in modifying the microflora. Personal investigations showed that there were only small differences between the microflora of Africa and Malaysia cocoas. It is possible that the main difference between Ghana and Malaysia may be in the cultivars customarily grown. In West Africa, the beans of the usual cultivar, Amelonado, carry a small amount of pulp, whereas the Upper Amazon hybrids used in Malaysia carry larger amounts of pulp in each bean. It may be this difference that is a contributory cause of acidity in some fermented cocoas. However, Chong *et al.* (1978) have modified the Malaysian method in two ways and both methods have had some success in ameliorating the acidity problem. The first consisted of using a warm-air blast for two to four hours through a shallow layer of beans that had already been fermented for 20–40 hours. The second method was to place the beans under pressure to squeeze out pulp juice before it was fermented. This work on modifying fermentations is continuing.

IX. CONCLUSIONS

This account goes no further than the stage at which dried beans lie in warehouses ready to be shipped to the chocolate manufacturers of the developed World. It is there that the beans are roasted and the true chocolate flavour develops. Further processing is then applied to produce the attractive and nutritious confection which is accepted almost universally. It is a long time ago since the first Europeans discovered this relatively unattractive vegetable product as consumed by the Central American Indians, and there are several landmarks in its evolution to the present day product. Although science is being applied with increasing intensity to cocoa and its products, there are still gaps in our knowledge. In particular, the control and understanding of bean fermentation would benefit from further scientific investigation and bring about an improvement of the raw material in those parts of the world where it is not of premium quality.

X. ACKNOWLEDGEMENTS

I am indebted to Professor A.J. Willis of Sheffield University for advice on several botanical aspects of cocoa, and also to the Longman Group Limited for granting permission for the reproduction of the drawing of a cocoa flower from Wood (1975). I am also grateful to Dr. R.R.

Davenport for the work on yeast, the results of which are shown in Table 1.

REFERENCES

Carr, J.G. and Davies, P.A. (1970). *Journal of Applied Bacteriology* **33,** 768.

Carr, J.G. and Davies, P.A. (1972). *Journal of Applied Bacteriology* **35,** 463.

Carr, J.G. and Davies, P.A. (1974). *Journal of Applied Bacteriology* **37,** 471.

Carr, J.G., Davies, P.A. and Dougan, J. (1979). 'Cocoa Fermentation in Ghana and Malaysia'. A report available at Long Ashton Research Station, Long Ashton, Bristol, England.

Carr, J.G., Davies, P.A. and Dougan J. (1980). 'Cocoa Fermentation in Ghana and Malaysia' Part 2. A report available at Long Ashton Research Station, Long Ashton, Bristol, England.

Chittenden, F.J. (ed.) (1951). 'Royal Horticultural Society Dictionary of Gardening', p. 2098. Clarendon Press, Oxford.

Chong, C.F., Shepherd, R. and Poon, Y.C. (1978). *International Conference on Cocoa and Coconuts*, 22–28.

Ciferri, C. (1931). *Journal of the Department of Agriculture of Puerto Rico* **15,** 223.

Dechau, P. and Emeis, C.C. (1975). *Monatsschrift für Brauerei* **28,** 125.

Gordon, R.E., Haynes, W.C. and Hor-Nay Pang (1973). 'The genus *Bacillus*'. Agriculture Handbook No. 427, United States Department of Agriculture, Washington D.C.

Hoynak, S., Polansky, T.S. and Stone, R.W. (1941). *Food Research* **6,** 471.

Knapp, A.W. (1937). 'Cacao Fermentation'. John Bale, Sons and Curnow Ltd., London.

Lopez, A. (1972). *In* 'Fourth International Cocoa Research Conference, St. Augustine, Trinidad', p. 640. Government of Trinidad and Tobago, West Indies.

Orla-Jensen, S. (1919). 'The Lactic Acid Bacteria'. Høst, Copenhagen.

Ostovar, K. and Keeney, P.G. (1973). *Journal of Food Science* **38,** 611.

Quesnel, V.C. (1972). *In* 'Fourth International Cocoa Research Conference, St. Augustine, Trinidad', p. 602. Government of Trinidad and Tobago, West Indies.

Roelofsen, P.A. (1958). *Advances in Food Research* **8,** 225.

Rohan, T.A. (1964). *Journal of Food Science* **29,** 451.

Rohan, T.A. and Stewart, T. (1964). *International Chocolate Review* **19,** 502 + 4 + 6.

Rohan, T.A. and Stewart, T. (1966). *Journal of Food Science* **31,** 202.

Rombouts, J.E. (1952). *Proceedings of the Society of Applied Bacteriology* **15,** 103.

Urquhart, D.H. (1955). 'Cocoa'. Longmans, Green and Co., London.

Whiting, G.C. (1975). *In* 'Lactic Acid Bacteria in Beverages and Food' (J.G. Carr, C.V. Cutting and G.C. Whiting, eds.), pp. 69–85. Academic Press, London.

Whiting, G.C. and Coggins, R.A. (1964). *Report from Long Ashton Research Station 1963*, p. 151.

Wood, G.A.R. (1975). 'Cocoa'. Tropical Agriculture Series, 3rd Edition. Longmans Group Ltd., London.

10. Yeast Extracts

H. J. PEPPLER

Whitefish Bay, Wisconsin, U.S.A.

I. INTRODUCTION

A. Identity of Products

Yeast cells may be solubilized, either partially or completely, by autolysis and several other techniques. On further processing, the slurries can be converted into a variety of preparations and products which are useful in the laboratory and as ingredients in foods, feeds and fermentation media. Among the principal products are concentrates of yeast invertase and β-galactosidase, soluble yeast components in liquid, paste, powder or granular form, and isolated fractions of yeast cell constituents, such as protein and cell walls (glycan), liberated by cells fractured mechanically.

The major commercial products are clear, water-soluble extracts known generally as yeast extract, autolysed yeast extract and yeast hydrolysate. Some manufacturers fortify extracts with monosodium glutamate and 5′-nucleotides, or they may blend extracts with hydrolysed vegetable protein (Rosenthal and Pinkalla, 1960) and modified whey solids (Corbett, 1978) to produce a variety of flavour mixes. Hydrolysed vegetable proteins are known in the trade as HVP and HPP (hydrolysed plant protein). They are derived from the acid hydrolysis of soybean grits, wheat gluten, corn gluten, rice gluten, yeast and casein.

As natural flavourings approved by the Food and Drug Administration (Anon., 1973a,b), extracts are used as condiments in the preparation of meat products, sauces, soups, gravies, cheese spreads, bakery products, seasonings, vegetable products and seafoods. As reliable economical sources of peptides, amino acids, trace minerals and vitamins of the B-complex group, yeast extracts are nutritional additives in health-food formulations, baby foods, feed supplements and for enrichment of growth and production media for micro-organisms and other biological culture systems. When coupled with phenolic anti-oxidants, autolysates and hydrolysates act as synergists (Bishov and Henick, 1975).

Biomass for the manufacture of yeast extracts is obtained primarily from breweries as surplus brewer's yeasts, and secondarily from sources of primary-grown yeasts including molasses-grown baker's yeast (*Saccharomyces cerevisiae*), whey-grown *Kluyveromyces fragilis* and *Sacch. lactis*, and wood sugar- and ethanol-grown *Candida utilis*.

The release and digestion of yeast cell components can be accomplished by methods involving mechanical breakage of cells, chemical and enzymic treatments, and pasteurization, autolysis, plasmolysis and hydrolysis. Of these procedures, combinations of the last four methods are in general use for preparation of commercial yeast extracts. In the processes of mild heat treatment, autolysis and plasmolysis, the cell's endoenzymes, mainly carbohydrases and proteases, digest the cytoplasmic constituents. Such digestion procedures, in practice, are relatively slow (1 to 20 hours), require aseptic conditions, and result in low yields (about 48% of initial yeast solids). In the production of yeast hydrolysates, concentrated yeast slurries are heated (100°C) with strong acids until 50 to 60% of the protein has been converted into peptides and amino acids. After neutralization and concentration, the dried finished product may contain up to 50% sodium chloride.

In the absence of industry-wide standards (Select Committee on Generally Regarded as Safe Substances, 1976), the specifications and compositions of protein hydrolysate products vary among manufacturing plants and in products finished in the same plant. Autolysed yeast extracts, however, which are the main topic of this chapter, have been defined by the International Hydrolyzed Protein Council (1977) as follows: 'Autolyzed yeast extracts are food ingredients used as natural food flavors throughout the world. They are composed primarily of (a) amino acids, peptides and polypeptides resulting from the enzymatic splitting of peptide bonds due to naturally occurring enzymes present in the edible yeast and (b) the water soluble components of the yeast cell. Food grade salt may be added during processing. Autolyzed Yeast Extract may be in a liquid, paste, powder or granular form'. Chemical specifications proposed by the Council are listed in Table 1.

B. Production and Marketing

Compilation of production data on yeast extracts began in 1966 when the Fermentation Section of the International Union of Pure and Applied Chemistry conducted a world-wide survey of fermentation

Table 1

Chemical specifications of autolysed yeast extract. From International Hydrolyzed Protein Council (1977).

Total nitrogen, dry and sodium chloride-free basis (%)	9.0 min.
α-Amino nitrogen, dry and sodium chloride-free basis (%)	3.5 min.
Glutamic acid, dry basis (%)	12.0 max.
Total sodium, dry basis (%)	20.0 max.
Total heavy metals, dry basis (p.p.m.)	20.0 max.
Lead, dry basis (p.p.m.)	10.0 max.
Arsenic, dry basis (p.p.m.)	3.0 max.
Solubility[a] (%)	99.0 min.
pH Value[b]	4.5–6.0

[a] 5g per 100 ml of water at 20°C.
[b] 2% solution in distilled water.

industries (International Union of Pure and Applied Chemistry, 1966, 1971). With only five producing countries reporting a total of 1700 metric tonnes of yeast extract, the accounting was quite incomplete. It lacked input from the large extract manufacturers based in England, France and Switzerland. Currently, production statistics on extracts are gathered annually by the International Hydrolyzed Protein Council located in Washington, D.C., U.S.A. In 1976, 11 member producers reported a total output of 17 000 metric tonnes (dry basis) of hydrolysed vegetable protein in the classification of 38 to 50% sodium chloride (International Hydrolyzed Protein Council, 1977). The same reporting group produced about 3500 metric tonnes of autolysed yeast extract classed in the product group containing up to 38% sodium chloride. This level of output of autolysed yeast extract is essentially unchanged from earlier estimates (Peppler, 1970), but is likely to increase by about 800 metric tonnes when a new plant in St. Louis, Missouri, U.S.A. attains full production (Anon., 1978). Additionally, about 1300 metric tonnes of blended products (hydrolysed vegetable protein and autolysed yeast extract) are marketed annually (International Hydrolysed Protein Council, 1977).

Major producers of yeast extracts, autolysed yeast extracts and yeast hydrolysates include The English Grains Co., Ltd. (England), Fould-Springer (France), Gist-Brocades (The Netherlands), Nestlé (Switzerland); and in the United States, Amber Laboratories, Anheuser-Busch, Campbell Soup Co., Miles Laboratories, Pure Culture Products, Inc., Staley Co., Stauffer Chemical Co., Universal Foods Corp. and Yeast Products, Inc. Some of the familiar trade names, many recognized in the international trade, include Amberex, Barmene, Gistex, Maggi, Max-arome, Tureen, Yeastor, Yeatex and Zyest.

II. EXTRACTS DERIVED FROM LIVE YEASTS

Preparation of autolysed yeast extract begins with living yeast cells in aqueous suspension (cream), yeast press cake (70% moisture) or active dry yeast granules (8% moisture). Brewer's yeast slurries and primary yeast creams are the principal starting materials. Press cake is usually subjected to a preliminary plasmolytic action of sodium chloride or sucrose before autolysis is accelerated (Nolf, 1911; Cregor et al., 1941; Soderstrom, 1948). Active dry yeast can be extracted with water (Herrera et al., 1956; Cooper and Peppler, 1959).

Autolysis, or self-digestion, is a complex process. Details of its biochemical aspects have been studied and reviewed extensively by Drews (1937), Vosti and Joslyn (1954), Joslyn and Vosti (1955), Maddox and Hough (1969) and Hough and Maddox (1970). With yeast, the reactions develop as the cream is heated to a temperature at which autofermentation of reserve carbohydrates (mainly glycogen) begins (usually 40°C to 60°C). As other endogenous enzymes (proteases and nucleases) increase in activity, and the cell membrane adjacent to the cell wall gradually loses its selectivity, leakage of soluble compounds ensues. With increased permeability of the cell membrane and the cell wall, proteolysis also occurs outside the cell.

Autolysis can be accelerated by plasmolysing agents such as sodium chloride, sucrose, ethanol, ethyl acetate, amyl acetate, chloroform, toluene and combinations of salt and ethanol (Sugimoto, 1974; Sugimoto *et al.*, 1976) and potassium chloride (Moore, 1980). The organic solvents referred to alter cell permeability and also suppress bacterial putrefaction.

The degree of extraction of yeast cells depends on temperature, pH value, time and type of yeast. Complete extraction is never achieved because cell walls, which account for about 20% of yeast solids (Robbins and Seeley, 1978), remain largely intact and variable amounts of carbohydrate (glycogen, mannan) and nitrogenous material remain insoluble (International Hydrolyzed Protein Council, 1977).

Efforts to increase extraction yields have resulted in several innovative treatments. These include (a) digestion of intact yeast cells with lysozyme and microbial glucanase (Funatsu *et al.*, 1978; Knorr *et al.*, 1979), (b) addition of proteases which act on protein leaked out of the cell (Bavisotta, 1965; Knorr *et al.*, 1979b), and (c) mechanical rupture and comminution of cells (Wiseman, 1969; Lindblom, 1974, 1977; Mogren *et al.*, 1974; Newell *et al.*, 1975; Robbins and Seeley, 1978). In the last treatment, activity of endoenzymes is retained (Follows *et al.*, 1971; Lindblom and Mogren, 1974). Large-scale usage of these processes is limited, owing to low enzyme-reaction rates and the need for aseptic operations. The merits of such methods for increasing the availability of cell proteins and enzymes have been discussed by Tannenbaum (1968).

A. Autolysates

The attributes of yeast of one form or another have been asserted for more than two centuries. Initially, yeast froth (barm) from fermenting mashes was used in the treatment of gastric distress and fever (Anon., 1799). By 1874, a debittered brewer's yeast extract, made by a patented process, was used as a tonic (McClary, 1948). Since 1892, brewer's yeast has been added to animal feeds (Lyall, 1965). Following the discovery of vitamins in 1901, and learning of their abundance in yeast, many new modifications of yeast were developed in the following 40-year period for medical, pharmaceutical and nutritional uses. For example, Willstätter and Sobotka (1923) patented a process for preparing a sugar plasmolysate of brewer's yeast. The honey-flavoured mildly-heated enzyme-rich and vitamin-laden liquid was recommended for therapeutic uses and as an ingredient for bakery products and malt extracts. A similar flavourful product, which added autolysis to the process, was patented by Lendvai (1962), nearly 40 years later. In a similar manner, Cregor et al. (1941) plasmolysed a mixture of baker's and brewer's yeast with sucrose, added papain, and obtained an extract with high vitamin B-complex potency.

In 1902, a yeast extract factory was established (Acraman, 1966). One of the earlier improvements in the flavour and nutritional values of yeast was patented by Nolf (1911). His process embraced most of the essential steps in industrial practice today, namely, heat yeast slurries to 45–60°C, agitate during autolysis for three days, separate extract from insoluble components, subject extract to a fining process (to decrease bitterness) and filter the extract. Additional historical notes on the expanded interest and growth of yeast extract technology also appear in the reviews by Vosti and Joslyn (1954), Pyke (1958), Lyall (1964, 1965) and Meister (1965).

1. Flavour Enhancers

Yeast extracts are versatile, multifunctional food ingredients. As additives in meat and poultry products, extracts accentuate the protein flavour initially present according to their content of glutamic acid and nucleotides. They also contribute flavour and aroma, depending on the strength of the extract in flavour-imparting peptides and amino acids, and the presence of other components in a given producer's formula (Albrecht and Deindoerfer, 1966). Clarity of the extract, and its lightness of colour, are also important marks of desired quality.

In general, the process parameters favouring uniform and flavourful quality are: (a) selected yeast strains of high protein content; (b) mild autolysing conditions (about 50°C); (c) efficient stainless-steel equipment; and (d) appropriate and dependable process controls. The principal steps in typical autolysate production are detailed in Figure 1. Fermentation and autolysis begin rather quickly in yeast cream (15 to 18% solids) as it is agitated and slowly brought to 50–55°C. Autolysis is terminated by pasteurization after 8 to 24 hours, or when the desired level of α-amino nitrogen is reached. Figure 2 illustrates the changes in yeast-cell weight as leakage of cell constituents and their derivatives occurs during autolysis (Hough and Maddox, 1970). The course of changes in yeast protein and amino acid content during autolysis are shown in Figure 3.

At the option of the manufacturer, small amounts of sodium chloride (3–5%, yeast solids basis) may be added to accelerate autolysis. Recently, chloroform, a commonly used inducer of autolysis, was banned in food processing. However, in non-food products, chloroform and other organic solvents, as well as high concentrations of salt, may be used.

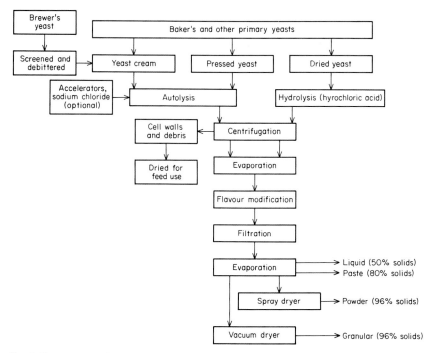

Fig. 1. Process routes in the manufacture of yeast extracts. In part from Acraman (1966) and International Hydrolyzed Protein Council (1977).

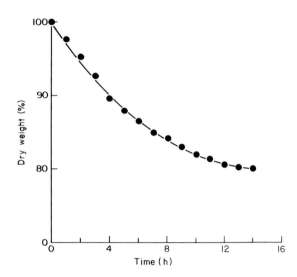

Fig. 2. Solubilization of yeast solids during autolysis in water at pH 6.5 and 45°C. From Hough and Maddox (1970).

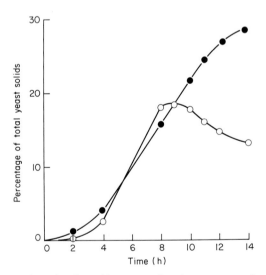

Fig. 3. Changes in protein and amino acid concentrations (as percentage of total solids) in yeast extract during autolysis. O indicates changes in protein concentration, ● changes in amino acid concentration. After Hough and Maddox (1970).

In the usual practice, the clear extract obtained is concentrated in a wiped-film evaporator to a paste (70–80% solids), or is spray-dried to hygroscopic, water-soluble powders. A new, improved drying technique, centrifugal atomization with steam injection, has been described by Stinson (1980). It produces autolysed yeast-extract powder with high bulk density and extended shelf life. Extract yields of about 48% of the initial yeast solids are attainable. When the whole unseparated autolysate is dried (usually on drum dryers), as much as 70% of the original yeast solids may be recovered. Details of extract manufacture have been reported by Acraman (1966), Reed and Peppler (1973) and International Hydrolyzed Protein Council (1977).

Recent chemical analyses pertaining to six extracts produced by five different manufacturers are compared in Tables 2 to 6. The gross composition of selected commercial autolysed yeast powders (Table 2) varies markedly in protein and ash content. This indicates the degree of variability in initial yeast composition and differences in process conditions. Different strains of yeast are known to exhibit specific pH optima and a wide range in proteolytic capability (Drews, 1937; Vosti and Joslyn, 1954; Hough and Maddox, 1970).

A more useful index of flavour potential is indicated by the free amino acid contents in the extract, as shown for the same series of products in Table 3. On a protein basis, the amino acid profiles are relatively uniform, and the glutamic acid analyses are within prescribed specifications (Table 1). Glutamic acid, known for its beefy taste, is a key indicator of flavour potential. No other amino acid tastes like it.

Table 2

Proximate analysis of commercial autolysed yeast extracts from five manufacturers. From Select Committee on Generally Regarded as Safe Substances (1977).

Component (%, w/w)	A	B	C	D (standard extract)	D (low sodium extract)	E
Moisture	3.4	3.1	3.4	3.3	4.4	5.8
Carbohydrates	27.3	19.3	20.5	12.0	15.4	6.2
Protein (total nitrogen times 6.25)	55.5	50.8	65.1	57.0	71.7	67.2
Ash	9.3	22.7	9.7	25.8	12.2	10.9
Lipids	0.2	0.3	0.1	0.1	0.2	0.5
Organic acids	1.6	3.9	2.6	2.0	2.5	4.8
Ammonium chloride	0.7	0.9	1.0	0.7	0.5	2.9

Table 3

Proximate amino acid composition of commercial autolysed yeast extracts[a]

Amino Acids	Composition (g per 100 g of extract)					
	A	B	C	D (standard extract)	D (low sodium extract)	E
Arginine	2.9	1.5	3.7	2.8	3.4	0.9
Cystine	NR[b]	NR	NR	NR	NR	0.6
Histidine	1.3	0.8	1.4	1.1	1.4	1.6
Isoleucine	2.4	2.2	2.6	2.7	3.5	3.4
Leucine	3.7	3.4	4.0	3.9	5.1	4.6
Lysine	4.2	3.8	4.3	4.0	5.2	5.0
Methionine	0.8	0.7	0.9	0.8	1.1	1.1
Phenylalanine	2.1	1.9	2.4	2.2	2.8	2.7
Threonine	2.4	2.2	2.6	2.5	3.2	2.5
Trytophan	0.5	0.4	0.5	0.6	0.7	1.1
Tyrosine	1.8	1.5	2.0	1.3	0.9	1.8
Valine	2.8	2.6	3.2	3.2	4.0	3.8
Alanine	3.5	3.3	4.1	4.1	5.2	5.8
Aspartic acid	5.5	4.7	5.9	5.2	6.8	7.1
Glutamic acid	7.0	5.6	8.0	9.9	11.6	9.1
Glycine	2.7	2.3	2.8	2.3	3.0	3.3
Proline	2.4	2.4	2.8	2.1	2.5	2.5
Serine	2.6	2.3	2.8	2.6	3.4	3.0
Total amino acids	49.5	42.5	55.0	52.3	64.8	61.0
Free amino acids (percentage of total)	53.4	44.9	46.7	62.5	77.6	NR

[a] In part from Select Committee on Generally Regarded as Safe Substances (1977); same products as listed in Table 2.
[b] NR indicates that the analysis was not reported.

According to Solms (1969), many amino acids are tasteless, or nearly so; these are the D- and L-isomers of arginine, aspartic acid, isoleucine, lysine, proline, serine, threonine and valine. L-Alanine is sweet, whereas L-trytophan, L-phenylalanine, L-tryosine and L-leucine taste bitter.

The nucleic acid components found in two commercial autolysates (Table 4) include GMP (5′-guanylic acid), a highly desired flavour booster derived from yeast RNA. Deoxyribonucleic acid is not degraded during autolysis of bakers' yeast (Trevelyan, 1978). One manufacturer markets a premium grade of autolysed yeast extract which is sevenfold greater in GMP content than Product A in Table 4. The product is said to be 'meatier' in flavour and devoid of 'yeasty' flavour (Moore, 1977;

Table 4

Nucleic acid components found in two commercial autolysed yeast extracts.[a]

Component	Content (percentage of product)	
	A	E
Adenosine monophosphate	ND[b]	0.93
Adenosine	0.48	0.18
Adenine	0.50	ND
Cytidine monophosphate	0.40	0.65
Cytidine	0.57	0.18
Cytosine	0.03	ND
Guanosine monophosphate	0.28	1.32
Guanosine	0.85	0.15
Guanine	0.15	ND
Uridine monophosphate	0.39	1.14
Uridine	0.33	0.24
Uracil	0.65	0.30
Xanthine	ND	0.19
Total	4.63	5.28

[a] In part from Select Committee on Generally Regarded as Safe Substances (1977); analyses made on same products for which data are listed in Table 2.
[b] ND indicates that the component was not detected.

Anon., 1977). Hoehn and Solms (1975) fractionated baker's yeast homogenates and attributed the disagreeable 'yeasty' taste to thiamin and thiamin diphosphate (cocarboxylase). Oota and Kitamura (1973) deodorized yeast before autolysis by steeping the cells in saturated potassium and sodium chloride solutions. A novel product, yeast-malt sprout extract, containing up to 6% 5′-nucleotides, is an approved food additive (Anon., 1973a,b). It is prepared from partially autolysed *Candida utilis* or *Kluyveromyces fragilis* treated with enzymes found in barley-malt sprouts.

Manufacturers'-suggested levels of usage of autolysed yeast extract products range from 0.5 to 1.5%, by weight, for augmentation of flavours in soups and meat products. For snack foods, vegetable products, sauces and seafoods, usage is recommended at 0.1 to 0.5%. Highest supplements (1 to 2.5%) are suggested for breadings and batters. A newly developed product coming on market is a low-sodium and high-potassium (32% potassium chloride) autolysate of *Candida utilis* grown on ethanol. The novel product is recommended for

Table 5

Ash composition of commercial autolysed yeast extracts.[a]

Element	A	B	C	D (standard extract)	D (low sodium extract)
Percentage (w/w)					
Calcium	0.1	.03	0.2	.02	0.04
Chloride	2.2	9.3	0.8	11.1	0.9
Magnesium	0.3	.1	0.3	0.1	0.1
Phosphorous	1.4	1.2	2.0	0.7	1.0
Potassium	2.4	3.0	3.0	5.6	5.7
Sodium	1.0	6.1	0.1	7.4	0.2
Parts per million					
Aluminium	< 5.0	< 10.0	< 10.0	< 5.0	< 5.0
Arsenic	< 0.5	< 0.5	< 0.5	< 0.5	< 0.5
Barium	< 1.0	< 2.0	< 2.0	< 1.0	< 1.0
Boron	4.4	1.6	5.2	1.7	2.4
Chromium	2.3	< 3.0	< 3.0	1.6	< 1.5
Copper	2.3	1.6	NR[b]	1.4	< 0.5
Iron	25.3	7.5	48.3	5.4	34.3
Lead	0.2	< 0.1	0.2	< 0.1	< 0.1
Manganese	7.7	< 2.0	10.3	< 1.0	1.0
Strontium	6.1	1.3	1.9	8.2	5.2
Zinc	53.3	4.0	NR[b]	21.5	53.4

[a] Adapted from Select Committee on Generally Regarded as Safe Substances (1977); same as products listed in Table 2.
[b] NR indicates that the analysis was not reported.

Table 6

Range of vitamin content of commercial autolysed yeast extracts. Taken from Technical Bulletins of Manufacturers in the U.S.A.

Vitamin	Range of content (μg per g of extract)
Thiamin	10–50
Riboflavin	15–75
Nicotinamide	125–550
Calcium pantothenate	30–120
Pyridoxine	1–25
Biotin	0.05–2

improving the flavour of pork and poultry products (Anon., 1980). Usage levels range from 1 to 2.5% in meats and 0.05 to 0.5% in soups and sauces.

2. *Nutritional Supplements*

The nutritional importance of yeast extract centres about its high content of soluble nitrogen (Table 3) and significant contents of the B-vitamins, phosphorous and trace mineral elements. Most of the autolysate production is consumed as nutritional supplements in feeds and in industrial fermentations, chiefly for production of wine, anti-biotics, pharmaceuticals and bacterial biomass. When prepared for human use, autolysed yeast extracts may be used as general dietary supplements and to assure amino acid balance in cereal products. Numerous special products are derived from yeasts that are propagated to biosynthesize high concentrations of vitamins and chelated trace elements. These dietary aids are popular among health food consumers. Some organic complexes of chromium, zinc and selenium may well have major medical significance, according to Mertz (1975, 1977), Schwarz (1976) and Tuman *et al.* (1978).

3. *Enzyme Concentrates*

Mild yeast autolysis is a primary step in the release of invertase and β-galactosidase. These enzymes are produced in large volume and are used mainly in the food industry.

a. *Invertase.* Invertase (saccharase, sucrase, β-D-fructofuranoside fructo-hydrolase, EC 3.2.1.26) catalyses hydrolysis of sucrose to fructose and glucose, and converts raffinose into fructose and melibiose. It occurs in many yeast species as part of the cell-wall polysaccharides. Baker's yeast is the preferred commercial source (Meister, 1965; Harrison, 1968; Reed and Peppler, 1973).

Invertase products derived from yeast are available in two forms. One is as clear, stabilized, liquid concentrates used in the preparation of fondant-centred candies for softening or liquefying the sucrose-cream centres after they are enrobed with chocolate. The other is as an invertase-rich, active dry baker's-type yeast of interest to the sugar industry for inversion of concentrated cane and beet sugar solutions and molasses to produce stable soluble concentrates known as 'high test' molasses.

The liquid products are essentially autolysed yeast extracts obtained from baker's yeast processed under mild conditions (40°C). The extract is concentrated under vacuum in a wiped-film evaporator, filtered and stabilized at pH 4.5, usually with glycerol (about 50% by weight). Solutions with invertase activities of 0.3 and 0.6 k values are marketed. Activity expressed as 1.0 k value (Association of Official Agricultural Chemists, 1980) is equivalent to 570 inverton units and 15,500 Sumner units. An inverton unit equals the quantity of invertase that will hydrolyse 5 mg of sucrose per minute at 25°C (Johnston et al., 1935). A Sumner unit equals the quantity of invertase that forms 1 mg of invert sugar from 162.5 mg of sucrose in 6 ml of acetate buffer (pH 4.7) at 20°C (Swiss Ferment Co. Ltd., 1977).

In the manufacture of the dry, high-invertase product, a baker's yeast cream is subjected to a short period of incipient autolysis, and then dewatered rapidly to 28% solids. The yeast cake is extruded, as in the preparation of active dry yeast, and dried to 92% solids (Peppler and Thorn, 1960).

Successful techniques for releasing invertase by mechanical disruption and treatment of live yeast with cell-wall lytic enzymes (β-(1→3)-glucanase; Mann et al., 1978) have been described and evaluated for invertase activity, but none appears in commercial development (Wiseman and Jones, 1971; Williams and Wiseman, 1974). The chemical properties of purified invertase have been studied extensively by Neumann and Lampen (1967).

b. *Lactase.* Lactase (β-galactosidase, β-D-galactoside galactohydrolase, EC 3.2.1.23) catalyses hydrolysis of lactose (milk sugar) to glucose and galactose. *Saccharomyces lactis, K. fragilis* and *C. pseudotropicalis* are the most common lactase-producing yeasts. The principal commercial lactase-containing product is prepared from the extract of partially autolysed cells of *Sacch. lactis* (Bouvy, 1975). The active material is precipitated from the filtered extract and dried. The odourless, tasteless, high-potency powder has a lactase activity of 40,000 o-nitrophenyl-β-D-galactoside (oNPG) units per gram. One oNPG unit equals the quantity of enzyme that will catalyse formation of one μmol of *ortho*nitrophenol (oNP) per minute from the artificial substrate *ortho*nitrophenyl-β-D-galactoside (oNPG) in phosphate buffer (pH 6.5) at 30°C (Dooley, 1980); four oNPG units equal one neutral lactase unit (Nijpels, 1977a,b). The enzyme is used mainly in milk, whey and milk products to increase sweetness, prevent lactose crystallization, and alleviate lactose intolerance in humans, poultry and cattle (Pomeranz, 1964a,b; Bouvy, 1975). Although lactase usage is comparatively low, interest in new applications and better strains of yeast as source material for practical development are continuing (Mahoney *et al.*, 1975; Guy and Bingham, 1978; Holsinger, 1978; Baer and Lowenstein, 1979; De Bales and Castillo, 1979).

B. Homogenates

Spurred by the global concern for increased supplies of protein, many investigators have developed and tested a variety of techniques for instantaneous disintegration of microbial cells to release protein and other cytoplasmic components. Yeast, an efficient converter of ammonia into protein, has been in the forefront of those efforts directed toward low-cost sources of single-cell protein. Some of the devices and methods investigated include stone and colloid milling, high-pressure homogenization (Follows *et al.*, 1971; Bauer-Staeb and Bouvard, 1973), grinding, rapid gas decompression, freeze–thawing, sonic oscillation (Cunningham *et al.*, 1975), bead homogenization (Lindblom, 1974; Lindblom and Mogren, 1974) and rasping frozen biomass (Bleeg and Christensen, 1979). Among the numerous processes in development, one system of total yeast solids recovery is outstanding (McCormick, 1973; Seeley, 1975; Robbins and Seeley, 1978). Since yeast extract is part of this proven scheme, a synopsis of the process will be described.

Figure 4 depicts the main steps of this process. It begins with baker's yeast cream (6–20% solids) which undergoes autolysis (indicated as post treatment in Fig. 4), then homogenization and separation into three fractions, namely yeast extract, protein and cell walls (mainly mannan,

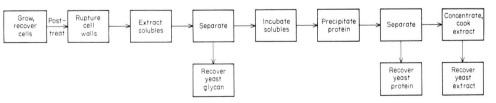

Fig. 4. Process flow for isolation of products from comminuted baker's yeast. From Seeley (1975).

but termed glycan). Following autolysis at 20–60°C for 10–72 hours (Robbins and Seeley, 1978), the slurry is comminuted by repeated passage through a Manton-Gaulin homogenizer. After the homogenate is given a mild alkaline treatment to solubilize protein (Seeley, 1975, 1977), the insoluble cell walls (glycan) are separated from the soluble fraction. This extract is incubated to allow hydrolysis of nucleic acids and precipitation of protein. A final centrifugal separation recovers the protein and the soluble extract. The latter is evaporated to form a meaty-flavoured paste which contains approximately 48% protein, 27% carbohydrate and 24% ash (Robbins *et al.*, 1975). Functional properties of the protein and cell-wall products were reviewed by McCormick (1973) and Seeley (1975). Except for the yeast extract product, the novel fractions appear to pose marketing problems. These include the high cost of raw material (baker's yeast), a long process time, and expensive market development.

III. EXTRACTS DERIVED FROM DRIED YEASTS

A. Acid Hydrolysates

Commercial yeast hydrolysates are grouped with hydrolysed vegetable proteins. In their manufacture, yeast alone is the protein source, or it may be mixed with other plant proteins (soybean and the glutens of corn, rice and wheat) (Rosenthal and Pinkalla, 1960; Select Committee on Generally Regarded as Safe Substances, 1976). In terms of yield of extract, acid hydrolysis is the more efficient method of processing than the preparation of autolysates. In contrast to autolysed yeast extracts, the proteins in hydrolysates are almost completely degraded to amino acids, and yeast cell walls are solubilized. However, during the strong acid hydrolysis, tryptophan is lost and drastic decreases occur in the contents of tyrosine, cysteine and methionine (Select Committee on Generally Regarded as Safe Substances, 1976). The high salt content (38% or more of sodium chloride), loss of some vitamins and poorer flavour of extracts are other disadvantages which limit their use as food ingredients. Recently one manufacturer developed a premium yeast hydrolysate aptly named SMAK.

A common process of hydrolysis begins with dried yeast, primarily brewer's, baker's or *Candida utilis* strains, which is reslurried in water

(65–85% solids). After acidification with hydrochloric acid, usually to 6M concentration, hydrolysis proceeds in a wiped-film evaporator at 100–120°C. Reaction time varies from 4 to 12 hours depending on the temperature control and level of α-amino nitrogen desired (Ziemba, 1967). After adjustment of pH value with sodium hydroxide (to pH 5–6), the mixture is filtered, decolourized and evaporated to a paste (about 85% solids) or spray dried (95 to 97% solids). Alternatively, hydrolysates may be diluted with two parts of yeast cream (18 to 20% solids) and dried on drum dryers. The finished yeast hydrolysate generally contains, on a solids basis, 13% total nitrogen, 8% α-amino nitrogen, 5–8% glutamic acid and 40% (or more) sodium chloride.

B. Water Extracts

Inactive (non-fermenting) dried yeasts, as well as active dry baker's yeast, can be extracted with water. About 20 to 25% of the cell contents may be recovered (Herrera *et al.*, 1956; Cooper and Peppler, 1959). The highest yield of solids is released by active dry yeast stirred in cold (4°C) water, buffer solutions (0.1M phosphate) or dilute saline (0.85% sodium chloride). Extract yields on inactive dry yeasts are not temperature dependent.

The inefficiency of water extraction limits its application to special pilot processes for recovery of coenzymes, certain vitamins and some active cell constituents (Cooper and Peppler, 1959; Harrison, 1968; Reed and Peppler, 1973).

IV. OUTLOOK

Production levels of autolysed yeast extracts and hydrolysed vegetable proteins have shown moderate increases. Since 1976, autolysed yeast extracts production increased only about 0.5%, hydrolysed vegetable proteins gained nearly 11% and blends of the two products showed minor fluctuations. The static condition in autolysed yeast extracts production may show a downturn as supplies of brewer's yeast become less available, and production costs of primary yeasts continue to escalate. As fewer small breweries survive economic recessions, and the larger brewers are finding a better market for their spent yeast in the feeder lots and feed-formulation industry, less brewer's yeast is likely to be available for extract manufacture. Menegazzi and Ingledew (1980) have addressed their studies to meet this trend.

Among primary yeast producers around the World, developments in single-cell protein processes have fallen short of expectations (Chen and Peppler, 1978). The yeast extract industry may gain little raw material from single-cell protein production. Thus, at best, autolysed yeast extracts production will probably remain static, and hydrolysed vegetable proteins manufacture is likely to increase steadily.

REFERENCES

Anon. (1799). *The Spectator*, Sept. 18, p. 1, New York.
Anon. (1973a). *Federal Register* **38**(9), 12397.
Anon. (1973b). *Federal Register* **38**(231), 33284.
Anon. (1977). *Food Product Development* **11**(9), 60.
Anon. (1978). *Food Engineering* **50**(9), 92.
Anon. (1980). *Food Engineering* **52**(3), 70.
Acraman, A.R. (1966). *Process Biochemistry* **1**(6), 313.
Albrecht, J.J. and Diendoerfer, F.H. (1966). *Food Engineering* **38**(10), 92.
Association of Official Agricultural Chemists (1980). 'Methods of Analysis', 13th edn., Section 31.024. Association of Official Analytical Chemists, Washington, D.C.
Baer, R.J. and Lowenstein, M. (1979). *Journal of Dairy Science* **62**, 1041.
Bauer-Staeb, G. and Bouvard, F. (1973). *Lebensmittel-Wissenschaft Technologie* **6**, 219.
Bavisotta, V.S. (1965). United States Patent 3 212 902.
Bishov, S.J. and Henick, A.S. (1975). *Journal of Food Science* **40**, 345.
Bleeg, H.S. and Christensen, F. (1979). *In* 'Yeast—a Newsletter' (H.J. Phaff, ed.), No. 28(1), pp. 23–33; University of California, Davis, California.
Bouvy, F.A.M. (1975). *Food Product Development* **9**(2), 10.
Chen, S.L. and Peppler, H.J. (1978). *Developments in Industrial Microbiology* **19**, 79.
Cooper, E.J. and Peppler, H.J. (1959). United States Patent 2,904,439.
Corbett, C.R. (1978). United States Patent 4,165,391.
Cregor, N.M., Timmer, F.E. and Allen, R.M. (1941). United States Patent 2,235,827.
Cunningham, S.D., Cater, C.M., Mattil, K.F. and Vanderzant, C. (1975). *Journal of Food Science* **40**, 732.
De Bales, S.A. and Castillo, F.J. (1979). *Applied and Environmental Microbiology* **37**, 1201.
Dooley, J.G. (1980). Technical Bulletin, GB Fermentation Industries, Des Plaines, Illinois.
Drews, B. (1937). *Biochemische Zeitschrift* **236**, 207.
Follows, M., Hetherington, P.J., Dunnill, P. and Lilly, M.D. (1971). *Biotechnology and Bioengineering* **13**, 549.
Funatsu, M., Oh, H., Aizono, Y. and Shimoda, T. (1978). *Agricultural Biological Chemistry* **42**, 1975.
Guy, E.J. and Bingham, E.W. (1978). *Journal of Dairy Science* **61**, 147.
Harrison, J.S. (1968). *Process Biochemistry* **3**, 89.
Herrera, T., Peterson, W.H., Cooper, E.J. and Peppler, H.J. (1956). *Archives of Biochemistry and Biophysics* **63**, 131.

Hoehn, J.E. and Solms, J. (1975). *Lebensmittel-Wissenschaft Technologie* **8**, 206.

Holsinger, V.H. (1978). *Food Technology* **32**(3), 35.

Hough, J.S. and Maddox, I.S. (1970). *Process Biochemistry* **5**, 50.

International Hydrolyzed Protein Council (1977). 'Comments with Additional Data on SCOGS Tentative Evaluation of the Health Aspects of Protein Hydrolyzates as Food Ingredients—Report 37b.' Washington, D.C.

International Union of Pure and Applied Chemistry (1966). *Pure and Applied Chemistry* **13**, 405.

International Union of Pure and Applied Chemistry (1971). 'Survey of Fermentation Industries, 1967' Technical Reports, No. 3.

Joslyn, M.A. and Vosti, D.C. (1955). *Wallerstein Laboratory Communications* **18**, 191.

Johnston, W.R., Redfern, S. and Miller, G.E. (1935). *Industrial and Engineering Chemistry, Analytical Edition* **7**, 82.

Knorr, D., Shetty, K.J. and Kinsella, J.E. (1979a). *Biotechnology and Bioengineering* **21**, 2011.

Knorr, D., Shetty, K.J., Hood, L.F. and Kinsella, J.E. (1979b). *Journal of Food Science* **44**, 1362.

Lendvai, A. (1962). United States Patent 3,051,576.

Lindblom, M. (1974). *Biotechnology and Bioengineering* **16**, 1495.

Lindblom, M. (1977). *Biotechnology and Bioengineering* **19**, 199.

Lindblom, M. and Mogren, H. (1974). *Biotechnology and Bioengineering* **16**, 1123.

Lyall, N. (1964). *Food Engineering* **36**(5), 98.

Lyall, N. (1965). *Food Trade Review* **34**(4), 1, 6.

Maddox, I.S. and Hough, J.S. (1969). *Proceedings European Brewers Convention, Interlaken* **12**, 315.

Mahoney, R.R., Nickerson, R.A. and Whitaker, J.R. (1975). *Journal of Dairy Science* **58**, 1620.

Mann, J.W., Jeffries, T.W. and MacMillan, J.D. (1978). *Applied and Environmental Microbiology* **36**, 594.

McClary, J.E. (1948). *In* 'Yeasts in Feeding' (S. Brenner, ed.), pp. 80–82. Quartermaster Food and Container Institute, Chicago.

McCormick, R.D. (1973). *Food Product Development* **7**(6), 17, 20.

Meister, H. (1965). *Wallerstein Laboratory Communications* **28**, 7.

Menegazzi, G.S. and Ingledew, W.M. (1980). *Journal of Food Science* **45**, 182.

Mertz, W. (1975). *Nutrition Review* **33**, 129.

Mertz, W. (1977). *Food Product Development* **11**(6), 62.

Mogren, H., Lindblom, M. and Hedenskog, G. (1974). *Biotechnology and Bioengineering* **16**, 261.

Moore, K. (1977). *Food Product Development* **11**(9), 60.

Moore, K. (1980) *Food Product Development* **14**(7), 16.

Neumann, N.P. and Lampen, J.O. (1967). *Biochemistry* **6**, 468.

Newell, J.A., Seeley, R.D. and Robbins, E.A. (1975). United States Patents 3 867 555 and 3 888 839.

Nijpels, H.H. (1977a). *North European Dairy Journal* **10**, 358.

Nijpels, H.H. (1977b). *North European Dairy Journal* **11**, 382.

Nolf, P. (1911). United States Patent 1,012,147.

Oota, S. and Kitamura, S. (1973). Japan Patent 48/26,237.

Peppler, H.J. (1970). *In* 'The Yeasts' (A.H. Rose and J.S. Harrison, eds.), vol. 3, pp. 421–462. Academic Press, London.

Peppler, H.J. and Thorn, J.A. (1960). United States Patent 2,922,748.

Pomeranz, Y. (1964a). *Food Technology* **18**, 682.
Pomeranz, Y. (1964b). *Food Technology* **18**, 690.
Pyke, M. (1958). *In* 'The Chemistry and Biology of Yeasts' (A.H. Cook, ed.), pp. 535–586. Academic Press, New York.
Reed, G. and Peppler, H.J. (1973). 'Yeast Technology'. AVI Publishing Co., Westport, Connecticut.
Robbins, E.A. and Seeley, R.D. (1978). United States Patent 4,122,196.
Robbins, E.A., Sucher, R.W., Seeley, R.D., Schuldt, E.H., Newell, J.A. and Sidoti, D.R. (1975). United States Patent 3,914,450.
Rosenthal, W.A. and Pinkalla, H.A. (1960). United States Patent 2,928,740.
Schwarz, K. (1976). *Medical Clinics of North America* **60**(4), 745.
Seeley, R.D. (1975). *Food Product Development* **9**(7), 46.
Seeley, R.D. (1977). *MBAA Technical Quarterly* **14**(1), 35.
Select Committee on Generally Regarded as Safe Substances (SCOGS) (1976). 'Tentative Evaluation of the Health Aspects of Protein Hydrolysates As Food Ingredients', Report 37b, pp. 1–28. Life Sciences Research Office, FASEB, Bethesda, Maryland.
Soderstrom, N.M.G. (1948). British Patent 596,847.
Solms, J. (1969). *Journal of Agricultural and Food Chemistry* **17**, 686.
Stinson, W.S. (1980). *Food Processing* **41**(1), 90.
Sugimoto, H. (1974). *Journal of Food Science* **39**, 939.
Sugimoto, H., Takeuchi, H. and Yokotsuka, T. (1976). United States Patent 3,961,080.
Swiss Ferment Co. Ltd., Basel (1977). Sumner Units Bulletin From Sucrest Corp, Woodbridge, New Jersey.
Tannenbaum, S.R. (1968). *In* 'Single-Cell Protein' (R.I. Mateles and S.R. Tannenbaum, eds.), pp. 343–352. The M.I.T. Press, Cambridge, Massachusetts.
Trevelyan, W.E. (1978). *Journal of Science and Food Agriculture* **29**, 903.
Tuman, R.W., Bilbo, J.T. and Doisy, R.J. (1978). *Diabetes* **27**, 49.
Vosti, D.C. and Joslyn, M.A. (1954). *Applied Microbiology* **2**, 70: and **2**, 79.
Williams, N.J. and Wiseman, A. (1974). *Biochemical Society Transactions* **1**, 1299.
Wiseman, A. (1969). *Process Biochemistry* **4**(5), 63.
Wiseman, A. and Jones, P.R. (1971). *Journal of Applied Chemistry and Biotechnology* **21**, 26.
Willstätter, R. and Sobotka, H. (1923). United States Patent 1,538,360.
Ziemba, J.V. (1967). *Food Engineering* **39** (1), 82.

Note Added in Proof
Exogenous sulphhydryl proteases, thiamin, fatty acids and their glycerol esters added to yeast undergoing autolysis increase the rate of digestion and the yield of extract. Papain additions (0.01–0.3% of yeast solids) to autolysing *Candida utilis* double the yield of extract obtained within 10 hours at 55°C (Chao *et al.*, 1980). With added thiamin (0.01–0.4% of yeast solids), baker's yeast autolysates produce 34% more extract within three hours at 55°C (Akin and Murphy, 1981). Capric and caprylic acids (0.1–6.0% of yeast solids) also accelerate autolysis, nearly doubling the extract yield (Hill, 1981).

References
Akin, C. and Murphy, R.M. (1981). United States Patent 4,285,976.
Chao, K.C., McCarthy, E.F. and McConaghy, G.A. (1980). United States Patent 4,218,481.
Hill, F.F. (1981). United States Patent 4,264,628.

AUTHOR INDEX

Numbers in italics are those on which References are listed

SUBJECT INDEX

A

Acetaldehyde
  fermented milk flavour component,
    206, 207, 212
  yogurt flavour component, 214–216
Acetic-acid bacteria, in cocoa fermentation, 287–291
Acidophilus milk
  production method, 220
  therapeutic value of, 219
Aflatoxin, protective effect of *Rhizopus* spp.
  during fermentation processes, 19
Agidi, African fermented maize product,
  24
Akpler, African fermented maize product,
  24
Alcohols, aroma of foods improved by, 35
Amino acids
  cheese starter requirements, 165–168
  content in grain and legumes, 44
  content of individual acids in commer-
    cial autolysed yeast extracts, 302
  production of free acids during yeast
    autolysis, 300
Angkak, fermented rice product, 18, 34
Animal feed
  soy sauce production residues for, 83
Antibiotics, food fermentations affected
  by milk containing, 159, 164, 165,
  200
Appam, Indian fermented rice product, 25
Arroz fermentado (amarillo, requemado), fermented rice product, 2

Aspergillus spp.
  in cocoa bean fermentations, 289
  depulped coffee as source of, 269
  miso production by, 57, 61
  protease activity of, 48
  soy bean fermentation by, 31, 32, 34,
    41, 58
Autolysed yeast extract, *see also* Yeast
  extract
  amino-acid composition of commercial
    products, 302
  chemical composition of commercial
    products, 301–304
  chemical specifications for, 295
  dried products flow sheet, 299
  drying processes for, 299, 301
  flavour and aroma enhancement of
    foods by, 298–305
  β-galactoside production from, 306
  historical survey of uses, 297, 298
  invertase production from, 305,
    306
  metal content of commercial products,
    304
  nucleic acid composition of commercial
    products, 302, 303
  nutritional importance, 305
  official definition of, 295
  process parameters for quality produc-
    tion of, 298, 299
  protein and amino-acid concentration
    changes during autolysis, 300